David Abraham • Clive Handler
Michael Dashwood • Gerry Coghlan
Editors

Translational Vascular Medicine

Pathogenesis, Diagnosis, and Treatment

Editors
David Abraham, PhD
Professor of Cellular and Molecular Biology
UCL Medical School
London
UK

Michael Dashwood, PhD
Principal Research Fellow
Clinical Biochemistry
UCL Medical School
London
UK

Clive Handler, BSc, MD, MRCP
FACC, FESC
Consultant in Pulmonary Hypertension
Royal Free Hospital
London
UK

Honorary Senior Lecturer
Division of Medicine
UCL Medical School
London
UK

Gerry Coghlan, MD, FRCP
Consultant Cardiologist
Royal Free Hospital
London
UK

ISBN 978-0-85729-919-2 e-ISBN 978-0-85729-920-8
DOI 10.1007/978-0-85729-920-8
Springer London Dordrecht Heidelberg New York

British Library Cataloguing in Publication Data
A catalogue record for this book is available from the British Library

Library of Congress Control Number: 2011942266

Printed on acid-free paper

Springer is part of Springer Science+Business Media (www.springer.com)

Preface

This is the third volume in the series of books on translational medicine gleaned from the annual vascular biology and clinical medicine workshop held at the Royal College of Physicians. The chapters are invited papers presented by internationally recognized basic science and clinical experts. The aim of the workshop is to bring basic scientists and clinicians together to discuss their work and perspectives in areas of cardiovascular medicine and biology. We ask them to address the areas which are likely to be important in the future and the associated challenges.

Our previous books, *Vascular Complications in Human Disease* (2008) and *Advances in Vascular Medicine* (2010), both also published by Springer, have dealt with other key and developing areas of basic science and its clinical applications. This volume covers new and exciting advances in cardiovascular medicine. As before, we have tried to explore the bi-directional and integrated approaches of translational cardiovascular medicine, linking basic science to patient care.

The chapters in this book span a number of translational themes in cardiovascular medicine. There is a section on surgery and non-pharmacological treatments for atherosclerotic disease of the aorta. Pulmonary arterial hypertension is a rapidly evolving area following recent discoveries of some of the molecular pathways implicated in its pathogenesis which have led to some promising drug development and clinical optimism. Some of the trials underpinning clinical guidelines are described. Other chapters include "Cytoprotective Mechanisms in the Vasculature," "Potassium Channels Regulating the Electrical Activity of the Heart," and "Novel Molecular Mediators Regulating the Cardiovascular System." We are particularly pleased to include a chapter on "The Broken Heart Syndrome" by our friend and colleague, Professor Larry Cohen, from Yale University School of Medicine, with which UCL has recently established a collegiate and collaborative relationship.

We hope that this book, a formal record and reference of our annual workshop, is a useful way to transmit the information from the excellent papers presented at the meeting to a wider readership. Our authors provide their expert insight into important areas of translational cardiovascular medicine and key bibliographies for the reader.

We hope that this book, like its predecessors, is a useful contribution to the literature in this fascinating field.

David Abraham
Clive Handler
Michael Dashwood
Gerry Coghlan

Contents

Contributors

B. Ahmetaj, MD, Phd, FRCP School of Pharmacy and Chemistry, Kingston University, Kingston-Upon-Thames, Surrey, UK

F. Arrigoni, School of Pharmacy and Chemistry, Kingston University, Kingston-Upon-Thames, Surrey, UK

Matthias Barton, MD Molecular Internal Medicine, University of Zurich, Zürich, Switzerland

Tammie Bishop, Wellcome Trust Centre for Human Genetics, University of Oxford, Oxford, UK

K. Richard Bruckdorfer, Structural and Molecular Biology, Faculty of Life Sciences, University College London, London, UK

Ann E. Canfield, Wellcome Trust Centre for Cell-Matrix Research & Cardiovascular Research Group, The Michael Smith Building, School of Biomedicine, Faculty of Medical & Human Sciences, University of Manchester, Manchester, UK

Marie-Camille Chaumais, Faculté de médecine, Univ. Paris-Sud, Kremlin-Bicêtre, France

INSERM U999, Hypertension Artérielle Pulmonaire: Physiopathologie et Innovation Thérapeutique, Le Plessis-Robinson, France

Centre Chirurgical Marie Lannelongue, Le Plessis-Robinson, France

Service de pharmacie, Hôpital Antoine Béclère, Assistance Publique des Hôpitaux de Paris, Clamart, France

Lawrence S. Cohen, MD Department of Cardiology, Yale University School of Medicine, New Haven, CT, USA

Sylvia Cohen-Kaminsky, Faculté de médecine, Univ. Paris-Sud, Kremlin-Bicêtre, France

Service de Pneumologie et Réanimation Respiratoire, AP-HP, Centre National de Référence de l'Hypertension Pulmonaire Sévère, Hôpital Antoine Béclère, Clamart, France

INSERM U999, Hypertension Artérielle Pulmonaire: Physiopathologie et Innovation Thérapeutique, Le Plessis-Robinson, France

Centre Chirurgical Marie Lannelongue, Le Plessis-Robinson, France

Anthony P. Davenport, Clinical Pharmacology Unit, University of Cambridge, Centre for Clinical Investigation, Addenbrooke's Hospital, Cambridge, UK

Peter Dorfmüller, Faculté de médecine, Univ. Paris-Sud, Kremlin-Bicêtre, France

Service de Pneumologie et Réanimation Respiratoire, AP-HP, Centre National de Référence de l'Hypertension Pulmonaire Sévère, Hôpital Antoine Béclère, Clamart, France

INSERM U999, Hypertension Artérielle Pulmonaire: Physiopathologie et Innovation Thérapeutique, Le Plessis-Robinson, France

Centre Chirurgical Marie Lannelongue, Le Plessis-Robinson, France

Marcus Fruttiger, Cell Biology, UCL Institute of Ophthalmology, London, UK

Erica Gabbrielli, MS Department of Neuroscience, Molecular Medicine Section, University of Siena, Siena, Italy

George Hamilton, MD, FRCS Royal Free Vascular Unit, Royal Free Hampstead NHS Trust, London, UK

Stephen C. Harmer, Department of Medicine, University College London, London, UK

Kristin B. Highland, MD, MSCR Pulmonary Hypertension Program, Medical University of South Carolina, Charleston, SC, USA

Rob Hinchliffe, MD, FRCS Vascular Surgery, St George's Vascular Institute, St George's Hospital, London, UK

Peter Holt, PhD, FRCS Vascular Surgery, St George's Vascular Institute, St George's Hospital, London, UK

Alice Huertas, Faculté de médecine, Univ. Paris-Sud, Kremlin-Bicêtre, France

Service de Pneumologie et Réanimation Respiratoire, AP-HP, Centre National de Référence de l'Hypertension Pulmonaire Sévère, Hôpital Antoine Béclère, Clamart, France

INSERM U999, Hypertension Artérielle Pulmonaire: Physiopathologie et Innovation Thérapeutique, Le Plessis-Robinson, France

Centre Chirurgical Marie Lannelongue, Le Plessis-Robinson, France

Marc Humbert, Faculté de médecine, Univ. Paris-Sud, Kremlin-Bicêtre, France

Service de Pneumologie et Réanimation Respiratoire, AP-HP,
Centre National de Référence de l'Hypertension Pulmonaire Sévère,
Hôpital Antoine Béclère, Clamart, France

INSERM U999, Hypertension Artérielle Pulmonaire: Physiopathologie et Innovation Thérapeutique, Le Plessis-Robinson, France

Centre Chirurgical Marie Lannelongue, Le Plessis-Robinson, France

Gareth D. Hyde, Wellcome Trust Centre for Cell-Matrix Research & Cardiovascular Research Group, The Michael Smith Building, School of Biomedicine, Faculty of Medical & Human Sciences, University of Manchester, Manchester, UK

Shalini Jadeja, Medical and Developmental Genetics, MRC Human Genetics Unit, Western General Hospital, Edinburgh, UK

James Leiper, MRC Clinical Sciences Centre, Hammersmith Hospital, Imperial College London, London, UK

Ian Loftus, MD, FRCS Vascular Surgery, St George's Vascular Institute, St George's Hospital, London, UK

Janet J. Maguire, University of Cambridge, Centre for Clinical Investigation, Addenbrooke's Hospital, Cambridge, UK

Justin C. Mason, Bywaters Centre for Vascular Inflammation, National Heart and Lung Institute, Imperial College London, Hammersmith Hospital, London, UK

David Montani, Faculté de médecine, Univ. Paris-Sud, Kremlin-Bicêtre, France

Service de Pneumologie et Réanimation Respiratoire, AP-HP,
Centre National de Référence de l'Hypertension Pulmonaire Sévère,
Hôpital Antoine Béclère, Clamart, France

INSERM U999, Hypertension Artérielle Pulmonaire: Physiopathologie et Innovation Thérapeutique, Le Plessis-Robinson, France

Centre Chirurgical Marie Lannelongue, Le Plessis-Robinson, France

Frédéric Perros, Faculté de médecine, Univ. Paris-Sud, Kremlin-Bicêtre, France

Service de Pneumologie et Réanimation Respiratoire, AP-HP,
Centre National de Référence de l'Hypertension Pulmonaire Sévère,
Hôpital Antoine Béclère, Clamart, France

INSERM U999, Hypertension Artérielle Pulmonaire:
Physiopathologie et Innovation Thérapeutique, Le Plessis-Robinson, France

Centre Chirurgical Marie Lannelongue, Le Plessis-Robinson, France

INSERM U999, Centre Chirurgical Marie Lannelongue,
Le Plessis-Robinson, France

Peter J. Ratcliffe, Wellcome Trust Centre for Human Genetics,
University of Oxford, Oxford, UK

Antonella Rossi, PhD Department of Neuroscience,
Molecular Medicine Section, University of Siena, Siena, Italy

Francesca Sozio, MS Department of Neuroscience,
Molecular Medicine Section, University of Siena, Siena, Italy

Matt Thompson, MD, FRCS Vascular Surgery, St George's Vascular Institute,
St George's Hospital, London, UK

Andrew Tinker, Department of Medicine, University College London,
London, UK

Janice Tsui, MD, FRCS Royal Free Vascular Unit,
Royal Free Hampstead NHS Trust, London, UK

Elisabetta Weber, MD Department of Neuroscience,
Molecular Medicine Section, University of Siena, Siena, Italy

Abbreviations

5-HT	Serotonin
5-HTT	Serotonin transporter
AGM	Aorta-gonad mesonephros
ALK-1	Active-like kinase type-1
AMP	Adenosine monophosphate
bHLH	Basic helix-loop-helix
BL-CFC	Blast colony–forming cells
BMP	Bone morphogenetic protein
BMPR2	Bone morphogenetic protein receptor II
CADASIL	Cerebral arteriopathy with subcortical infarcts and leukoencephalopathy
cAMP	Cyclic adenosine monophosphate
cGMP	Cyclic guanosine monophosphate
CSL	CBF1 Suppressor of Hairless Lag-1
DAPT	*N*-[(3,5-Difluoro phenyl)acetyl]-L-alanyl-2-phenyl]glycine-1, 1-dimethylethyl ester
Dll	Delta-like ligand
DMOG	Dimethyl-oxalylglycine
E	Embryonic day
EC	Endothelial cell
ECE-1	Endothelin converting enzyme-1
ECGS	Endothelial cell growth supplement
eNOS	Endothelial nitric oxide synthase
ET	Endothelin
ET-1	Endothelin-1
ET-2	Endothelin-2
ET-3	Endothelin-3
ETA	Endothelin receptor type A
ETB	Endothelin receptor type B
FAK	Focal adhesion kinase
FGF	Fibroblast growth factor
FIH	Factor inhibiting HIF

GLUT4	Glucose transporter type 4
HGF	Hepatocyte growth factor
HIF	Hypoxia-inducible factor
HSC	Hematopoietic stem cell
IPAH	Idiopathic pulmonary arterial hypertension
IPC	Ischemic preconditioning
IR	Ischemia-reperfusion
Jag	Jagged
KO	Knockout
Kv1.5	Voltage-gated potassium channels subunit 1.5
MAGP-1	Microfibril-associated glycoprotein-1
MAPK	Mitogen-activated protein kinases
mmHg	Millimeters of mercury
mPAP	Mean pulmonary artery pressure
MSC	Mesenchymal stem/stromal cell
NEP	Neutral endopeptidase
NG2	Neuron glial 2
NICD	Notch intracellular domain
NO	Nitric oxide
PAH	Pulmonary arterial hypertension
PDE-5	Phosphodiesterase type
PDGF	Platelet-derived growth factor
PGI_2	Prostacyclin
PH	Pulmonary hypertension
PHD	HIF prolyl hydroxylases
PVR	Pulmonary vascular resistance
Rbpj	Recombination signal binding protein for immunoglobulin kappa J region
RGD	arginine-glycine-aspartic acid
RGS-5	Regulator of G protein signaling 5
RV	Right ventricle
RVH	Right ventricular hypertrophy
SERT	Serotonin transporter
SSc	Systemic sclerosis
SSRI	Selective serotonin reuptake inhibitor
TGF-β	Transforming growth factor
TRPC6	Transient receptor potential cation channel subfamily C, member 6
VEGF	Vascular endothelial growth factor
VEGFR	Vascular endothelial growth factor receptor
VegfR2	Vascular endothelial growth factor receptor 2
VHL	von Hippel–Lindau tumor suppressor
VIP	Vasoactive intestinal peptide
VSMC	Vascular smooth muscle cells
vWF	von Willebrand factor

Section I

Hot Topics in Vascular Biology

1 Pericytes: Adaptable Vascular Progenitors

Gareth D. Hyde and Ann E. Canfield

1.1 General Introduction

The existence of perivascular cells associated with capillaries was first reported by Eberth and Rouget in the late nineteenth century. Since then, these cells have been given a variety of names, including Rouget cells, mural cells, deep cells, adventitial cells, perivascular cells, and periendothelial cells. Zimmermann introduced the name "pericyte" (*peri*=around; *cyte*=cell) in 1923, and it is this term which is still used most frequently.

In this chapter, we will discuss the morphological characteristics of pericytes, their frequency and distribution within the vasculature, the markers that can be used to identify pericytes, and the theories about the origin of these cells. In addition, we shall discuss pericyte function and review the evidence that pericytes are adaptable vascular progenitor cells with potential therapeutic use. Readers are referred to other excellent recent reviews for information on additional pericyte functions, including regulating microvascular blood flow, and pericyte involvement in diseases such as cancer, hypertension, and diabetic retinopathy [1–3].

1.2 Pericyte Morphology, Frequency, and Distribution

Although pericytes are an extremely heterogeneous population of cells, they can be characterized by several morphological properties. For example, pericytes are typically elongated, stellate-shaped cells with multiple processes that extend along the length and, sometimes, the circumference of the vessel. In addition, pericytes often

G.D. Hyde • A.E. Canfield (✉)
Wellcome Trust Centre for Cell-Matrix Research & Cardiovascular Research Group, The Michael Smith Building, School of Biomedicine, Faculty of Medical & Human Sciences, University of Manchester, Oxford Road, Manchester, M13 9PT, UK
e-mail: ann.canfield@manchester.ac.uk

D. Abraham et al. (eds.), *Translational Vascular Medicine*,
DOI 10.1007/978-0-85729-920-8_1, © Springer-Verlag London Limited 2012

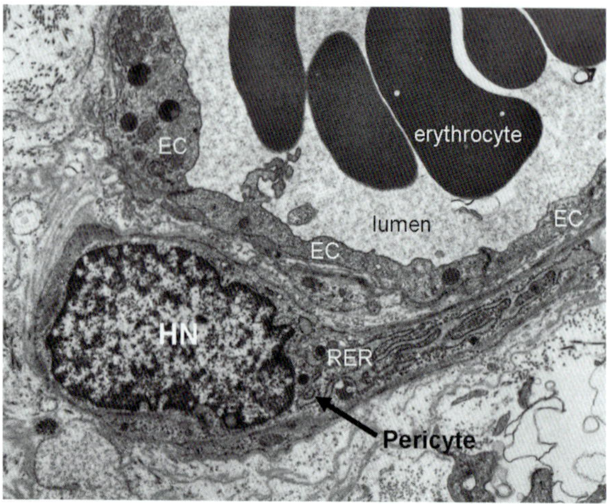

Fig. 1.1 Transmission electron micrograph of a capillary. A ring of endothelial cells (*EC*) forms the lumen of the capillary which contains several erythrocytes. On the abluminal surface of the capillary, a pericyte can be seen. The pericyte has several characteristic features including a large heterochromatic nucleus (*HN*), and an elongated cellular process containing large amounts of rough endoplasmic reticulum (*RER*) (Image kindly provided by Dr. C. Jones, University of Manchester)

possess a heterochromatic nucleus, large numbers of plasmalemmal vesicles, and contractile microfilament bundles (see Fig. 1.1).

Interestingly, the actual shape and size of pericytes can vary markedly depending on their anatomical location. The relative frequency of pericytes also varies between vessel type, developmental stage, and species. For example, the human retina has been shown to have a higher pericyte to endothelial cell ratio than rats (1:1 and 1:3 respectively) [4] and retinal microvessels have been reported to contain a higher ratio (1:1) compared to those in striated muscle (1:100) [5]. It is also noteworthy that alterations in pericyte frequency and distribution can contribute to the development and progression of several pathologies, including diabetic retinopathy (loss), myopathy (gain), fibrosis, and cancer (distribution) [1].

In arterioles, capillaries, and venules, pericytes are closely associated with endothelial cells and are embedded within a shared basement membrane. Via their long processes, pericytes can make contact with multiple endothelial cells, resulting in the partial coverage of the abluminal surface, and can also connect vessels within the microcirculation. Pericytes are frequently found adjacent to endothelial cell junctions and themselves form multiple connections with endothelial cells via peg and socket arrangements, adherens junctions, gap junctions, and tight junctions.

Pericyte or pericyte-like cells have also been identified in larger vessels by immunohistochemistry using the 3G5 antibody [6–8] which recognizes a cell surface ganglioside present on pericytes but not on endothelial cells, smooth muscle cells, or fibroblasts [9]. Using this antibody, pericytes have been shown to be present in the subendothelial layer of the intima; in the media and in the vaso vasora of the adventitia; in large, medium, and small arteries and veins. Furthermore, the pericyte-like cells identified in these locations were shown to contact each other via their processes forming a subendothelial network in the vascular bed [6].

Table 1.1 Most commonly used pericyte markers

Marker	Description	Example References
Alpha-smooth muscle actin	Cytoskeletal contractile protein	[10–13]
Aminopeptidase A+N	Zinc-dependent peptidase	[14–16]
Desmin	Intermediate filament protein predominantly expressed in muscle cells	[17, 18]
Nestin	Intermediate filament protein predominantly expressed in nerve cells	[14]
Neuron glial 2 (NG2) (HMWMAA)	Chondroitin sulfate proteoglycan	[19–21]
Platelet-derived growth factor (PDGF) receptor-beta	Transmembrane receptor tyrosine kinase	[22]
Regulator of G protein signaling 5 (RGS-5)	GTPase-activating protein	[23, 24]
3G5	Cell surface ganglioside	[6, 9]

1.3 Pericyte Markers

The heterogeneous nature of pericytes has made the identification of cell markers difficult. Indeed, the identification of a marker exclusively expressed by pericytes and expressed by all pericytes at all times remains elusive. In the absence of such a maker, many other antigens have been used (see Table 1.1).

It should be stressed that the expression of these markers by pericytes is species, tissue, developmental stage, and disease dependent. For example, NG2 is present on the surface of arteriolar and capillary pericytes but is absent in venular pericytes [19]. Alpha-smooth muscle actin is absent in pericytes in many tissues but is present in pericytes isolated from chick embryonic brains [25] and appears to be upregulated in pericytes within tumors [21, 26].

1.4 Pericyte Origin

One reason for the heterogeneity in pericyte marker expression may be their differing origins. As with vascular smooth muscle cells (VSMCs) [27], pericytes have been proposed to arise from multiple embryonic and cellular progenitors. Pericytes are often thought of as having a mesenchymal origin. However, studies using avian embryos have shown that the pericytes of the face and forebrain develop from the neural crest, whereas the endothelial cells are mesoderm-derived [28]. It has also been reported that perivascular mural cells (pericytes and VSMC) and endothelial cells can both develop from Flk1-positive embryonic stem cells [29]. As Flk1 is a marker of the embryonic lateral plate mesoderm, this work suggests that both endothelial and perivascular cells have a common mesodermal origin. These two theories are not mutually exclusive, and it is therefore possible that in the face and forebrain, pericytes arise from the neural crest, while in other parts of the body they develop from a more mesodermal progenitor that can also give rise to endothelial cells.

The study that proposed a common ontogeny for both perivascular mural cells and endothelial cells went on to show that Flk1-positive embryonic stem cell differentiation into these cell types is dependent on PDGF-BB and vascular endothelial growth factor (VEGF) respectively. As a result of this, and many other studies, it is now well known that PDGF-BB and its receptors are critical for pericyte differentiation, recruitment to endothelial tubes, and normal vessel morphogenesis and function [30–33].

In addition to having multiple embryonic origins, it has also been suggested that pericytes can be derived from several adult cell types. These include VSMC [34], endothelial cells [35], and bone marrow–derived cells [36–39]. Pericyte progenitor cells have also been isolated from the rat aorta using suspension culture. This method led to the isolation of an anchorage-independent population of cells that formed spheroidal colonies in suspension and that expressed several pericyte markers [40].

1.5 Pericyte Function

Pericytes have multiple functions within the vasculature. These include:

- Giving structural rigidity to the vessel wall
- Regulating the contractile ability, blood flow, and permeability of the vessel
- Regulating endothelial cell proliferation and differentiation
- Maintaining the functional integrity of the blood–brain barrier
- Phagocytic and antigen-presenting functions

For further details on these functions, readers are referred to other recent excellent reviews [1, 3]. This chapter will focus on pericyte function as vascular progenitor cells.

1.5.1 Pericytes as Progenitor Cells: An Historical Perspective

In 1927, Maximov described pericytes as "resting wandering cells" and "primitive mesenchymal cells" [41]. After Maximov's ideas of the 1920s, the concept that pericytes could act as progenitor cells failed to receive much attention until the early 1980s. At this time, it was proposed that pericytes could give rise to immature adipocytes in response to thermal lesions in the rat inguinal fat pad [42], and that pericytes were the target of bone morphogenetic protein (BMP) signaling during cranial bone regeneration, resulting in pericyte differentiation into osteoprogenitor cells [43]. These early analyses of animal injury models generated the first data indicating that pericytes had the ability to differentiate into other cell types.

In a series of elegant studies performed in the early 1990s, Diaz-Flores and colleagues labeled vascular cells with Monastral Blue and monitored their fate in vivo. Their studies investigating neochondrogenesis in grafted perichondrium [44] and periosteal osteogenesis [45] indicated that pericytes could differentiate down the chondrogenic and osteogenic lineages, respectively. Subsequent ultrastructural studies during post-injury bone formation supported these conclusions [46, 47].

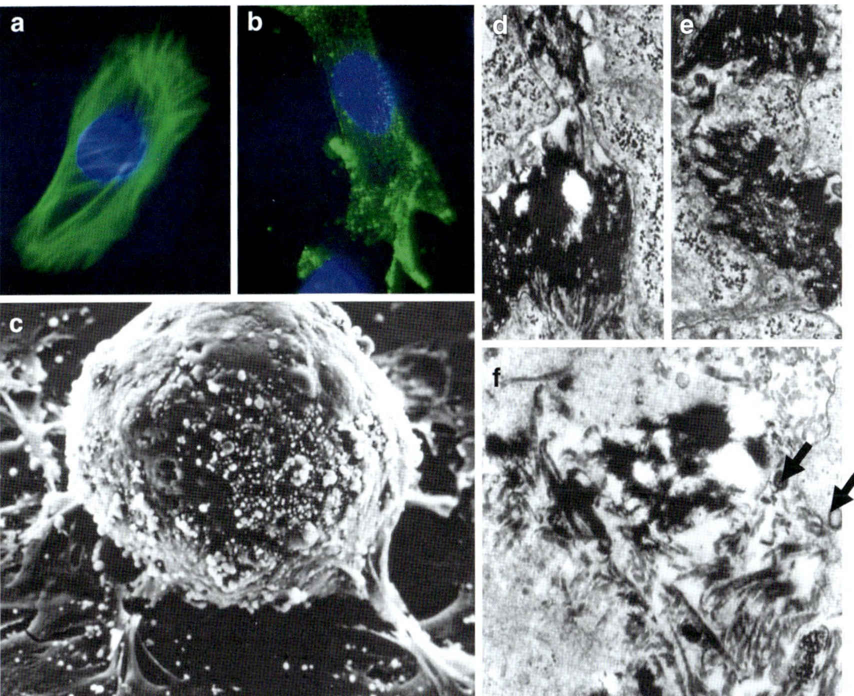

Fig. 1.2 Pericytes cultured in vitro deposit a calcified matrix. Immunocytochemical detection of alpha-smooth muscle actin (**a**) and the cell surface ganglioside recognized by the 3G5 monoclonal antibody (**b**) in pericytes isolated from bovine retinal microvessels. Scanning electron micrograph of a multicellular nodule formed by bovine retinal pericytes during in vitro culture (**c**). Transmission electron micrographs showing matrix calcification in sections cut through pericyte nodules (**d–f**). Areas of dense calcification can be seen in many sections (**d–e**). In addition, matrix vesicles (*arrowed*) and needle-like crystals of hydroxyapatite are apparent (**f**) (Figures **a** and **b** are reproduced from Farrington-Rock et al. [49]. Figures **c–f** are reproduced from Schor et al. [48])

The first direct evidence that pericytes could undergo osteogenic differentiation was published in 1990, when it was demonstrated that isolated bovine retinal pericytes could deposit a calcified matrix which resembled bone in vitro [48]. After reaching confluence, pericytes cultured on either plastic or a collagen substratum formed multilayered areas that retracted away from each other, leading to the formation of multicellular nodules containing needle-like crystals of hydroxyapatite (see Figs. 1.2 and 1.3). Furthermore, the cells within these nodules expressed markers of the osteoblastic lineage including bone sialoprotein, osteocalcin, osteonectin, and osteopontin [50].

In addition to undergoing osteogenic differentiation, cultured pericytes were shown to be able to differentiate along the chondrogenic and adipogenic lineages. When grown as pellets in chondrogenic medium, pericytes deposited an extracellular matrix rich in sulfated proteoglycans and expressed the chondrogenic markers Sox9, aggrecan, and type II collagen (see Fig. 1.3). In adipogenic medium, pericytes

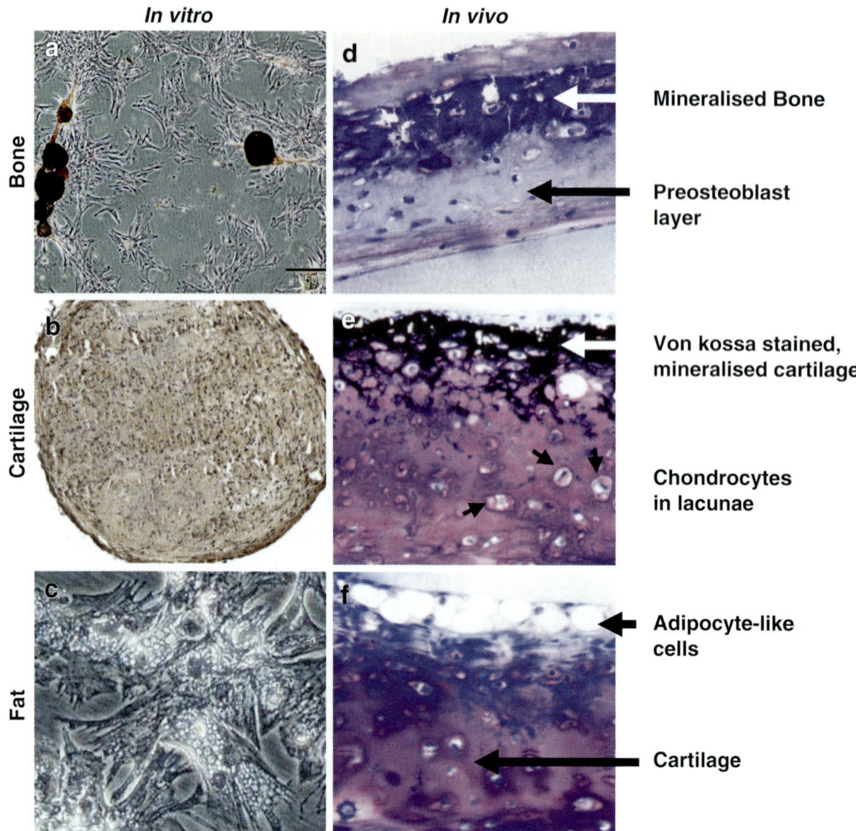

Fig. 1.3 Pericytes can undergo osteogenic, chondrogenic, and adipogenic differentiation in vitro and in vivo. In vitro differentiation of pericytes (**a–c**). Pericytes grown in monolayer in vitro form multicellular nodules that stain positive with alizarin red, indicating the presence of calcium deposits (**a**). Pericytes grown as a pellet in chondrogenic medium produce type II collagen that can be detected immunohistochemically (*brown staining*) (**b**). Pericytes cultured in adipogenic medium accumulate intracellular lipid droplets (**e**). In vivo differentiation of pericytes (**d–f**). Pericytes inoculated into diffusion chambers and implanted into athymic mice could be seen to form mineralized bone (**d**), cartilage and mineralized cartilage that stained with Von Kossa indicating the presence of mineral (**e**), and adipocyte-like cells (**f**) (Figures **b, e–f** are reproduced from Farrington-Rock et al. [49])

accumulated oil red O positive lipid droplets and expressed the adipocyte transcription factor proliferator-activated receptor-gamma [49].

Direct evidence that pericytes could undergo multi-lineage differentiation in vivo was generated when isolated pericytes were inoculated into diffusion chambers and implanted into athymic mice. When recovered, the chambers containing pericytes were found to contain tissue resembling bone, mineralized cartilage, fibrocartilage, non-mineralized cartilage with lacunae containing chondrocytes and small clusters of cells that resembled adipocytes [49, 50] (see Fig. 1.3).

There is now evidence that in addition to being able to differentiate along the "classical" osteogenic, chondrogenic, and adipogenic lineages, pericytes can also differentiate into VSMCs [51], Leydig cells [52], fibroblasts [53], myoblasts [54], myofibroblasts [55, 56], odontoblasts [57], and neuronal cell types [58], suggesting that these cells have enormous therapeutic potential.

1.5.2 Regulation of Pericyte Differentiation

Aberrant pericyte differentiation has been implicated in multiple disorders including chondro/osteoblastic differentiation in calcific vasculopathies [7] and myofibroblastic differentiation in kidney fibrosis [55], dermal scarring [53], spinal cord scarring [59], and systemic sclerosis [56]. Understanding what regulates pericyte differentiation would not only be of potential therapeutic use in these conditions but would also be of use in tissue regeneration and engineering strategies that use pericytes as a source of progenitor cells.

Despite the potential value of understanding how pericyte differentiation is regulated, little is currently known. One signaling pathway that has been implicated in pericyte differentiation is the canonical Wnt pathway [60]. In these studies, Wnt signaling was activated by the addition of either Wnt3a or LiCl, or inhibited (by adenovirus mediated overexpression of dominant negative TCF-4) during pericyte in vitro differentiation. Using this approach, it was demonstrated that Wnt signaling promoted pericyte chondrogenic differentiation, and inhibited pericyte adipogenic differentiation [60]. In support of this finding, it has been demonstrated that endothelial cells repress the adipogenic potential of adipose stromal cells (which have a functional and phenotypic overlap with pericytes [61, 62]) by the secretion of Wnt ligands [61]. Recent studies have also shown that Wnt signaling regulates the osteogenic differentiation of pericytes, although this effect is highly dependent upon the stage at which Wnt signaling is activated (Canfield and Brennan, unpublished information). BMPs and fibroblast growth factors (FGFs) have also been implicated in pericyte differentiation. BMP signaling has been suggested to promote the osteogenic differentiation of pericytes [43] whereas basic FGF has been shown to promote the neuronal differentiation of central nervous system–derived pericytes [58].

In addition to secreted signaling molecules, dexamethasone, a synthetic glucocorticoid, has been shown to downregulate pericyte expression of calcification inhibitor molecules and thereby promote pericyte osteogenic differentiation [63]. Similarly, dexamethasone has been shown to stimulate odontoblastic differentiation of pericytes isolated from human dental pulp [57]. However, it is clear that we still have much to learn about what regulates pericyte differentiation, both in disease states and in potential tissue regeneration strategies.

1.6 Progenitor Cells and the Perivascular Niche

The perivascular niche is a 3-dimensional microenvironment that includes the progenitor cells, their neighboring differentiated cells, the extracellular matrix, and soluble secreted molecules. It is proposed that residing within this specific niche allows adult progenitor cells to retain their multi-lineage potential and self-renewal capacity.

Many studies have suggested that in different tissues and organs, adult progenitor cells or mesenchymal stromal/stem cells (MSCs) reside within a perivascular niche. These include: bone marrow [64, 65], dental pulp [66], periodontal ligament [67], aorta [7, 68], umbilical cord Wharton's jelly [69], skeletal muscle [54], adipose tissue [62, 70], neural tissue [71], infrapatellar fat pads [72], chorionic villi [73], bone [74], and saphenous vein [75]. Indeed, it has now been established that MSCs reside in a perivascular niche in virtually all postnatal tissues and organs [71, 76, 77].

In many of these cases, the population of adult stem cells isolated from the tissue or organ has been found to express markers of pericytes. For example, dental pulp stem cells were found to be positive for alpha-smooth muscle actin and the cell surface ganglioside recognized by the 3G5 antibody [66]. Skeletal muscle progenitors were shown to express NG2 and alkaline phosphatase [54], adipose-derived stem cells have been shown to express the 3G5 epitope [70] and other pericyte markers [62], and stem cells in human placental chorionic villi [73] and infrapatellar fat pads [72] were shown to express the 3G5 epitope. Indeed, adult stem cells have been isolated from many human tissues on the basis of the expression of pericyte markers [66, 67, 76, 77].

In addition to adult stem cells being shown to express pericyte markers, pericytes have been shown to express markers normally associated with mesenchymal stem cells, such as STRO-1 [50, 66, 77]. Furthermore, pericytes isolated from multiple human tissues have been shown to have clonal multi-lineage potential during long-term culture [77], and such data has led Caplan to ask the question: "are all MSCs pericytes?" [78] In 2008, Covas and colleagues performed gene expression profiles and other characterizations on MSCs isolated from adult and fetal human tissues, differentiated cell types, and retinal pericytes [79]. A comparison of the gene expression profiles demonstrated that MSCs and pericytes are very similar, more similar then pericytes and smooth muscle cells or fibroblasts, for example [79]. Taken together, these data demonstrate that pericytes and adult mesenchymal stem cells have many common characteristics including their perivascular location, their distribution throughout the body, their cellular phenotype, and their differentiation potentials.

1.6.1 Therapeutic Potential of Pericytes

Several groups have started to explore the potential of using pericytes or pericyte-like cells as a source of progenitor cells for tissue regeneration and repair. Promising results have been achieved using human skeletal muscle–derived pericytes for the treatment of Duchenne muscular dystrophy [54]. In this study, human skeletal muscle–derived pericytes were inoculated into a murine model of Duchenne muscular dystrophy and their fate, effect on muscle regeneration and functional consequence, was analyzed. The implanted pericytes where shown to colonize host muscle, generate muscle fibers containing human dystrophin, and to result in partial but significant functional recovery as judged by frequency of falling and treadmill exhaustion tests.

Another group has also demonstrated that in addition to being able to repair dystrophic muscle, human pericytes derived from either muscle, placenta, or pancreas can regenerate cardiotoxin-injured muscle [77]. The same group has also reported that human skeletal muscle–derived pericytes improve cardiac function in acutely infarcted mouse hearts, and they suggested that this improvement may be due to increased angiogenesis and reduced fibrosis [80, 81]. These suggestions are consistent with recent studies demonstrating that pericyte-like progenitor cells increase neovascularization in a mouse model of muscle ischaemia [75] and improve repair of infarcted mouse hearts through pro-angiogenic and anti-fibrotic programs [82].

Beyond muscle regeneration, there is evidence for pericyte therapeutic potential in bone fracture repair. Over twenty-five years ago, pericytes were suggested to be the target of BMP signaling and the source of osteoprogenitor cells during cranial bone regeneration; much more recently, it was shown that inoculation with human umbilical cord perivascular cells (a cell population with many similarities to pericytes [69]) increases the rate of bone and cartilage regeneration in mice. A recent study has also shown that pericytes can promote epidermal tissue renewal by modifying the extracellular microenvironment of epithelial stem cells, suggesting that these cells may also be of therapeutic use in skin regeneration [83].

The potential use of pericytes for therapeutic tissue engineering is also starting to be explored [84]. He and colleagues seeded human pericytes onto bi-layered tubular, elastomeric, biodegradable scaffolds and implanted them into rats as aortic interposition grafts. Interestingly, the grafts initially seeded with pericytes had a higher patency rate than unseeded controls. There was evidence of extensive tissue remodeling, together with the deposition of collagen and elastin, and the presence of cells expressing VSMC and endothelial cell markers. Intriguingly, these cells appeared to originate from the host tissue, rather than from the pericytes themselves [84], which suggests that pericytes may improve the patency of vascular grafts by promoting the recruitment of host progenitor cells through the secretion of specific growth factors.

1.7 Conclusion

That pericytes closely resemble MSCs and are adaptable progenitor cells with great potential for tissue regeneration and repair is without question. The therapeutic potential of these cells may result from their ability to differentiate along multiple lineages, but it may also be due to their ability to evoke a host response, by releasing specific growth factors, cytokines, or matrix proteins, or by inducing angiogenesis [75, 80, 83]. However, as uncontrolled differentiation of pericytes can also contribute to calcific vasculopathies and fibrosis (for example), it is important that long-term follow-up studies are performed when the therapeutic potential of these cells is evaluated in vivo.

Several key questions remain to be resolved. For example: Do all pericytes have multi-lineage potential? What is the nature of the perivascular niche? How is the stemness of pericytes maintained and controlled in vivo? How are pericytes liberated from their niche? How is pericyte differentiation regulated? Do pericytes really contribute to repair and regeneration in vivo and, perhaps most importantly, do these cells have therapeutic potential in humans? The answer to all of these questions is eagerly awaited.

Acknowledgments The financial support of the British Heart Foundation is gratefully acknowledged. We would also like to thank Dr. Carolyn Jones (University of Manchester) for providing the electron micrograph.

References

1. Diaz-Flores L, Gutierrez R, Madrid JF, et al. Pericytes. Morphofunction, interactions and pathology in a quiescent and activated mesenchymal cell niche. Histol Histopathol. 2009;24:909–69.
2. Hall AP. Review of the pericyte during angiogenesis and its role in cancer and diabetic retinopathy. Toxicol Pathol. 2006;34:763–75.
3. Kutcher ME, Herman IM. The pericyte: cellular regulator of microvascular blood flow. Microvasc Res. 2009;77:235–46.
4. Tilton RG, Miller EJ, Kilo C, Williamson JR. Pericyte form and distribution in rat retinal and uveal capillaries. Invest Ophthalmol Vis Sci. 1985;26:68–73.
5. Shepro D, Morel NM. Pericyte physiology. FASEB J. 1993;7:1031–8.
6. Andreeva ER, Pugach IM, Gordon D, Orekhov AN. Continuous subendothelial network formed by pericyte-like cells in human vascular bed. Tissue Cell. 1998;30:127–35.
7. Bostrom K, Watson KE, Horn S, Wortham C, Herman IM, Demer LL. Bone morphogenetic protein expression in human atherosclerotic lesions. J Clin Invest. 1993;91:1800–9.
8. Ivanov D, Philippova M, Antropova J, et al. Expression of cell adhesion molecule T-cadherin in the human vasculature. Histochem Cell Biol. 2001;115:231–42.
9. Nayak RC, Berman AB, George KL, Eisenbarth GS, King GL. A monoclonal antibody (3G5)-defined ganglioside antigen is expressed on the cell surface of microvascular pericytes. J Exp Med. 1988;167:1003–15.
10. DeNofrio D, Hoock TC, Herman IM. Functional sorting of actin isoforms in microvascular pericytes. J Cell Biol. 1989;109:191–202.
11. Herman IM, D'Amore PA. Microvascular pericytes contain muscle and nonmuscle actins. J Cell Biol. 1985;101:43–52.
12. Nehls V, Drenckhahn D. Heterogeneity of microvascular pericytes for smooth muscle type alpha-actin. J Cell Biol. 1991;113:147–54.
13. Skalli O, Pelte MF, Peclet MC, et al. Alpha-smooth muscle actin, a differentiation marker of smooth muscle cells, is present in microfilamentous bundles of pericytes. J Histochem Cytochem. 1989;37:315–21.
14. Alliot F, Rutin J, Leenen PJ, Pessac B. Pericytes and periendothelial cells of brain parenchyma vessels co-express aminopeptidase N, aminopeptidase A, and nestin. J Neurosci Res. 1999;58:367–78.
15. Kunz J, Krause D, Kremer M, Dermietzel R. The 140-kDa protein of blood-brain barrier-associated pericytes is identical to aminopeptidase N. J Neurochem. 1994;62:2375–86.
16. Schlingemann RO, Oosterwijk E, Wesseling P, Rietveld FJ, Ruiter DJ. Aminopeptidase A is a constituent of activated pericytes in angiogenesis. J Pathol. 1996;179:436–42.
17. Nehls V, Denzer K, Drenckhahn D. Pericyte involvement in capillary sprouting during angiogenesis in situ. Cell Tissue Res. 1992;270:469–74.
18. Nico B, Ennas MG, Crivellato E, et al. Desmin-positive pericytes in the chick embryo chorioallantoic membrane in response to fibroblast growth factor-2. Microvasc Res. 2004;68:13–9.
19. Murfee WL, Skalak TC, Peirce SM. Differential arterial/venous expression of NG2 proteoglycan in perivascular cells along microvessels: identifying a venule-specific phenotype. Microcirculation. 2005;12:151–60.
20. Schlingemann RO, Rietveld FJ, de Waal RM, Ferrone S, Ruiter DJ. Expression of the high molecular weight melanoma-associated antigen by pericytes during angiogenesis in tumors and in healing wounds. Am J Pathol. 1990;136:1393–405.

21. Schlingemann RO, Rietveld FJ, Kwaspen F, van de Kerkhof PC, de Waal RM, Ruiter DJ. Differential expression of markers for endothelial cells, pericytes, and basal lamina in the microvasculature of tumors and granulation tissue. Am J Pathol. 1991;138:1335–47.
22. Sundberg C, Ljungstrom M, Lindmark G, Gerdin B, Rubin K. Microvascular pericytes express platelet-derived growth factor-beta receptors in human healing wounds and colorectal adenocarcinoma. Am J Pathol. 1993;143:1377–88.
23. Bondjers C, Kalen M, Hellstrom M, et al. Transcription profiling of platelet-derived growth factor-B-deficient mouse embryos identifies RGS5 as a novel marker for pericytes and vascular smooth muscle cells. Am J Pathol. 2003;162:721–9.
24. Cho H, Kozasa T, Bondjers C, Betsholtz C, Kehrl JH. Pericyte-specific expression of Rgs5: implications for PDGF and EDG receptor signaling during vascular maturation. FASEB J. 2003;17:440–2.
25. Gerhardt H, Wolburg H, Redies C. N-cadherin mediates pericytic-endothelial interaction during brain angiogenesis in the chicken. Dev Dyn. 2000;218:472–9.
26. Morikawa S, Baluk P, Kaidoh T, Haskell A, Jain RK, McDonald DM. Abnormalities in pericytes on blood vessels and endothelial sprouts in tumors. Am J Pathol. 2002;160:985–1000.
27. Majesky MW. Developmental basis of vascular smooth muscle diversity. Arterioscler Thromb Vasc Biol. 2007;27:1248–58.
28. Etchevers HC, Vincent C, Le Douarin NM, Couly GF. The cephalic neural crest provides pericytes and smooth muscle cells to all blood vessels of the face and forebrain. Development. 2001;128:1059–68.
29. Yamashita J, Itoh H, Hirashima M, et al. Flk1-positive cells derived from embryonic stem cells serve as vascular progenitors. Nature. 2000;408:92–6.
30. Hellstrom M, Gerhardt H, Kalen M, et al. Lack of pericytes leads to endothelial hyperplasia and abnormal vascular morphogenesis. J Cell Biol. 2001;153:543–53.
31. Hellstrom M, Kalen M, Lindahl P, Abramsson A, Betsholtz C. Role of PDGF-B and PDGFR-beta in recruitment of vascular smooth muscle cells and pericytes during embryonic blood vessel formation in the mouse. Development. 1999;126:3047–55.
32. Lindahl P, Johansson BR, Leveen P, Betsholtz C. Pericyte loss and microaneurysm formation in PDGF-B-deficient mice. Science. 1997;277:242–5.
33. Lindblom P, Gerhardt H, Liebner S, et al. Endothelial PDGF-B retention is required for proper investment of pericytes in the microvessel wall. Genes Dev. 2003;17:1835–40.
34. Nicosia RF, Villaschi S. Rat aortic smooth muscle cells become pericytes during angiogenesis in vitro. Lab Invest. 1995;73:658–66.
35. DeRuiter MC, Poelmann RE, VanMunsteren JC, Mironov V, Markwald RR, Gittenberger-de Groot AC. Embryonic endothelial cells transdifferentiate into mesenchymal cells expressing smooth muscle actins in vivo and in vitro. Circ Res. 1997;80:444–51.
36. Bababeygy SR, Cheshier SH, Hou LC, Higgins DM, Weissman IL, Tse VC. Hematopoietic stem cell-derived pericytic cells in brain tumor angio-architecture. Stem Cells Dev. 2008;17:11–8.
37. Kokovay E, Li L, Cunningham LA. Angiogenic recruitment of pericytes from bone marrow after stroke. J Cereb Blood Flow Metab. 2006;26:545–55.
38. Ozerdem U, Alitalo K, Salven P, Li A. Contribution of bone marrow-derived pericyte precursor cells to corneal vasculogenesis. Invest Ophthalmol Vis Sci. 2005;46:3502–6.
39. Rajantie I, Ilmonen M, Alminaite A, Ozerdem U, Alitalo K, Salven P. Adult bone marrow-derived cells recruited during angiogenesis comprise precursors for periendothelial vascular mural cells. Blood. 2004;104:2084–6.
40. Howson KM, Aplin AC, Gelati M, Alessandri G, Parati EA, Nicosia RF. The postnatal rat aorta contains pericyte progenitor cells that form spheroidal colonies in suspension culture. Am J Physiol Cell Physiol. 2005;289:C1396–407.
41. Tilton RG. Capillary pericytes: perspectives and future trends. J Electron Microsc Tech. 1991;19:327–44.
42. Richardson RL, Hausman GJ, Campion DR. Response of pericytes to thermal lesion in the inguinal fat pad of 10-day-old rats. Acta Anat (Basel). 1982;114:41–57.

43. Sato K, Urist MR. Induced regeneration of calvaria by bone morphogenetic protein (BMP) in dogs. Clin Orthop Relat Res. 1985;197:301–11.
44. Diaz-Flores L, Gutierrez R, Gonzalez P, Varela H. Inducible perivascular cells contribute to the neochondrogenesis in grafted perichondrium. Anat Rec. 1991;229:1–8.
45. Diaz-Flores L, Gutierrez R, Lopez-Alonso A, Gonzalez R, Varela H. Pericytes as a supplementary source of osteoblasts in periosteal osteogenesis. Clin Orthop Relat Res. 1992;275:280–6.
46. Brighton CT, Hunt RM. Early histologic and ultrastructural changes in microvessels of periosteal callus. J Orthop Trauma. 1997;11:244–53.
47. Decker B, Bartels H, Decker S. Relationships between endothelial cells, pericytes, and osteoblasts during bone formation in the sheep femur following implantation of tricalciumphosphate-ceramic. Anat Rec. 1995;242:310–20.
48. Schor AM, Allen TD, Canfield AE, Sloan P, Schor SL. Pericytes derived from the retinal microvasculature undergo calcification in vitro. J Cell Sci. 1990;97(Pt 3):449–61.
49. Farrington-Rock C, Crofts NJ, Doherty MJ, Ashton BA, Griffin-Jones C, Canfield AE. Chondrogenic and adipogenic potential of microvascular pericytes. Circulation. 2004;110: 2226–32.
50. Doherty MJ, Ashton BA, Walsh S, Beresford JN, Grant ME, Canfield AE. Vascular pericytes express osteogenic potential in vitro and in vivo. J Bone Miner Res. 1998;13:828–38.
51. Meyrick B, Reid L. The effect of continued hypoxia on rat pulmonary arterial circulation. An ultrastructural study. Lab Invest. 1978;38:188–200.
52. Davidoff MS, Middendorff R, Enikolopov G, Riethmacher D, Holstein AF, Muller D. Progenitor cells of the testosterone-producing Leydig cells revealed. J Cell Biol. 2004;167: 935–44.
53. Sundberg C, Ivarsson M, Gerdin B, Rubin K. Pericytes as collagen-producing cells in excessive dermal scarring. Lab Invest. 1996;74:452–66.
54. Dellavalle A, Sampaolesi M, Tonlorenzi R, et al. Pericytes of human skeletal muscle are myogenic precursors distinct from satellite cells. Nat Cell Biol. 2007;9:255–67.
55. Humphreys BD, Lin SL, Kobayashi A, et al. Fate tracing reveals the pericyte and not epithelial origin of myofibroblasts in kidney fibrosis. Am J Pathol. 2010;176:85–97.
56. Rajkumar VS, Howell K, Csiszar K, Denton CP, Black CM, Abraham DJ. Shared expression of phenotypic markers in systemic sclerosis indicates a convergence of pericytes and fibroblasts to a myofibroblast lineage in fibrosis. Arthritis Res Ther. 2005;7:R1113–23.
57. Alliot-Licht B, Bluteau G, Magne D, et al. Dexamethasone stimulates differentiation of odontoblast-like cells in human dental pulp cultures. Cell Tissue Res. 2005;321:391–400.
58. Dore-Duffy P, Katychev A, Wang X, Van Buren E. CNS microvascular pericytes exhibit multipotential stem cell activity. J Cereb Blood Flow Metab. 2006;26:613–24.
59. Goritz C, Dias DO, Tomilin N, Barbacid M, Shupliakov O, Frisen J. A pericyte origin of spinal cord scar tissue. Science 2011;333:238–42.
60. Kirton JP, Crofts NJ, George SJ, Brennan K, Canfield AE. Wnt/beta-catenin signaling stimulates chondrogenic and inhibits adipogenic differentiation of pericytes: potential relevance to vascular disease? Circ Res. 2007;101:581–9.
61. Rajashekhar G, Traktuev DO, Roell WC, et al. IFATS collection: adipose stromal cell differentiation is reduced by endothelial cell contact and paracrine communication: role of canonical Wnt signaling. Stem Cells. 2008;26:2674–81.
62. Traktuev DO, Merfeld-Clauss S, Li J, et al. A population of multipotent CD34-positive adipose stromal cells share pericyte and mesenchymal surface markers, reside in a periendothelial location, and stabilize endothelial networks. Circ Res. 2008;102:77–85.
63. Kirton JP, Wilkinson FL, Canfield AE, Alexander MY. Dexamethasone downregulates calcification-inhibitor molecules and accelerates osteogenic differentiation of vascular pericytes: implications for vascular calcification. Circ Res. 2006;98:1264–72.
64. Bianco P, Riminucci M, Gronthos S, Robey PG. Bone marrow stromal stem cells: nature, biology, and potential applications. Stem Cells. 2001;19:180–92.
65. Short B, Brouard N, Occhiodoro-Scott T, Ramakrishnan A, Simmons PJ. Mesenchymal stem cells. Arch Med Res. 2003;34:565–71.

66. Shi S, Gronthos S. Perivascular niche of postnatal mesenchymal stem cells in human bone marrow and dental pulp. J Bone Miner Res. 2003;18:696–704.
67. Seo BM, Miura M, Gronthos S, et al. Investigation of multipotent postnatal stem cells from human periodontal ligament. Lancet. 2004;364:149–55.
68. Tintut Y, Alfonso Z, Saini T, et al. Multilineage potential of cells from the artery wall. Circulation. 2003;108:2505–10.
69. Sarugaser R, Lickorish D, Baksh D, Hosseini MM, Davies JE. Human umbilical cord perivascular (HUCPV) cells: a source of mesenchymal progenitors. Stem Cells. 2005;23:220–9.
70. Zannettino AC, Paton S, Arthur A, et al. Multipotential human adipose-derived stromal stem cells exhibit a perivascular phenotype in vitro and in vivo. J Cell Physiol. 2008;214:413–21.
71. Tavazoie M, Van der Veken L, Silva-Vargas V, et al. A specialized vascular niche for adult neural stem cells. Cell Stem Cell. 2008;3:279–88.
72. Khan WS, Tew SR, Adesida AB, Hardingham TE. Human infrapatellar fat pad-derived stem cells express the pericyte marker 3G5 and show enhanced chondrogenesis after expansion in fibroblast growth factor-2. Arthritis Res Ther. 2008;10:R74.
73. Castrechini NM, Murthi P, Gude NM, et al. Mesenchymal stem cells in human placental chorionic villi reside in a vascular niche. Placenta. 2010;31:203–12.
74. Maes C, Kobayashi T, Selig MK, et al. Osteoblast precursors, but not mature osteoblasts, move into developing and fractured bones along with invading blood vessels. Dev Cell. 2010;19:329–44.
75. Campagnolo P, Cesselli D, Al Haj Zen A, et al. Human adult vena saphena contains perivascular progenitor cells endowed with clonogenic and proangiogenic potential. Circulation. 2010;121:1735–45.
76. da Silva Meirelles L, Chagastelles PC, Nardi NB. Mesenchymal stem cells reside in virtually all post-natal organs and tissues. J Cell Sci. 2006;119:2204–13.
77. Crisan M, Yap S, Casteilla L, et al. A perivascular origin for mesenchymal stem cells in multiple human organs. Cell Stem Cell. 2008;3:301–13.
78. Caplan AI. All MSCs are pericytes? Cell Stem Cell. 2008;3:229–30.
79. Covas DT, Panepucci RA, Fontes AM, et al. Multipotent mesenchymal stromal cells obtained from diverse human tissues share functional properties and gene-expression profile with CD146+ perivascular cells and fibroblasts. Exp Hematol. 2008;36:642–54.
80. Chen CW, Montelatici E, Crisan M, et al. Perivascular multi-lineage progenitor cells in human organs: regenerative units, cytokine sources or both? Cytokine Growth Factor Rev. 2009;20:429–34.
81. Corselli M, Chen CW, Crisan M, Lazzari L, Peault B. Perivascular ancestors of adult multipotent stem cells. Arterioscler Thromb Vasc Biol. 2010;30:1104–9.
82. Katare R, Riu F, Mitchell K, et al. Transplantation of Human Pericyte Progenitor Cells Improves the Repair of Infarcted Heart Through Activation of an Angiogenic Program Involving Micro-RNA-132. Circ Res 2011;109:00–00. (Published online before print August 25, 2011)
83. Paquet-Fifield S, Schluter H, Li A, et al. A role for pericytes as microenvironmental regulators of human skin tissue regeneration. J Clin Invest. 2009;119:2795–806.
84. He W, Nieponice A, Soletti L, et al. Pericyte-based human tissue engineered vascular grafts. Biomaterials. 2010;31:8235–44.

Benefits and Risks of Manipulating the HIF Hydroxylase Pathway in Ischemic Heart Disease

2

Tammie Bishop and Peter J. Ratcliffe

2.1 Introduction

Ischemic heart disease is a major cause of morbidity and mortality in the Western world. It occurs when oxygen delivery cannot meet the metabolic needs of the heart, as observed in patients with stable coronary artery disease as well as those experiencing acute myocardial infarction. Although conditions leading to myocardial injury have been well studied, and physical means of revascularization by stenting or coronary bypass surgery are well developed, there remains a need to define treatments that limit damage in the acute phase or promote revascularization by medical means. In particular, mechanisms that preserve cellular function during ischemia remain poorly understood.

Experimental models of myocardial ischemia in rodents have demonstrated that prior exposure to sublethal cycles of ischemia-reperfusion (I/R) protects tissues such as the heart from subsequent ischemia. There is compelling evidence that this ischemic preconditioning (IPC) is, at least in part, conferred through hypoxic activation of the transcription factor: hypoxia-inducible factor (HIF). HIF is a master regulator of oxygen homeostasis that induces the expression of hundreds of genes in response to hypoxia, including those that stimulate glycolysis, angiogenesis, and erythropoiesis. These changes help the organism adapt to oxygen deprivation at both the cellular and tissue levels. Pharmacological modulators of HIF are consequently being pursued as therapeutic targets for myocardial (as well as more general tissue) ischemia.

HIF is an α/β heterodimeric transcription factor, whose α subunit is regulated through posttranslational modification by HIF prolyl hydroxylases (PHDs, *p*rolyl *h*ydroxylase *d*omain): PHD1, 2 and 3 (reviewed in Kaelin and Ratcliffe [1]).

T. Bishop (✉) • P.J. Ratcliffe
Wellcome Trust Centre for Human Genetics,
University of Oxford, Oxford, UK
e-mail: tammie@well.ox.ac.uk; pjr@well.ox.ac.uk

D. Abraham et al. (eds.), *Translational Vascular Medicine*,
DOI 10.1007/978-0-85729-920-8_2, © Springer-Verlag London Limited 2012

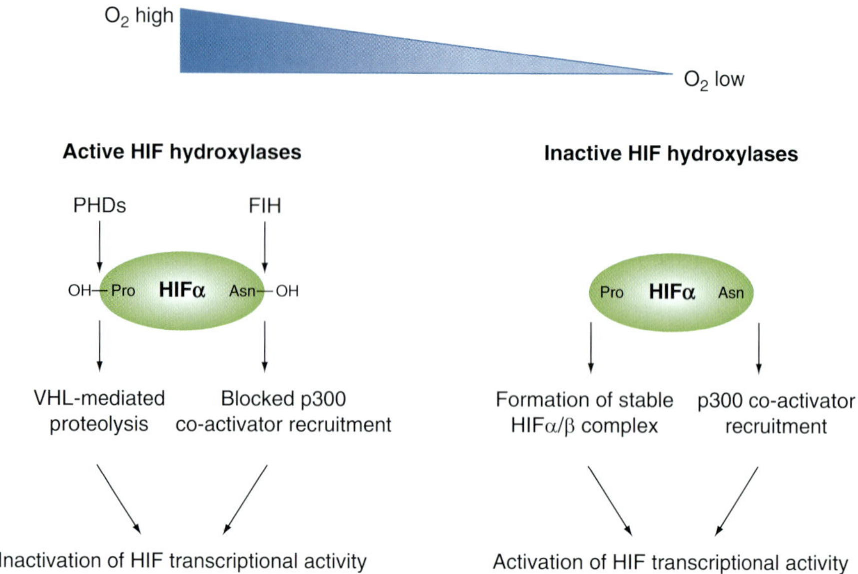

Fig. 2.1 Dual regulation of HIF-alpha subunits by prolyl and asparaginyl hydroxylation. In the presence of oxygen, active HIF prolyl hydroxylases (*PHDs*), as well as factor inhibiting HIF (*FIH*), downregulate and inactivate HIFα subunits. PHDs hydroxylate prolyl residues to promote von Hippel–Lindau tumor suppressor (*VHL*)–dependent proteolysis of HIFα subunits. FIH, on the other hand, hydroxylates an asparaginyl residue, which blocks p300 co-activator recruitment from activating HIFα-subunit transcriptional activity. In hypoxia, HIF hydroxylases (*PHDs* and *FIH*) are inactive and these processes are suppressed, which allows the formation of a transcriptionally active HIF complex

These non-heme Fe(II) and 2-oxoglutarate-dependent dioxygenase PHD enzymes are now widely regarded as cellular oxygen sensors that transduce the oxygen status to the cell via posttranslational hydroxylation of HIFα. In the presence of oxygen, PHD hydroxylates two proline residues within a central degradation domain in HIF-1α and -2α. This promotes their binding to von Hippel–Lindau tumor suppressor (VHL) E3 ubiquitin ligase, leading to proteasomal degradation. A second point of regulation involves asparaginyl hydroxylation by another non-heme Fe(II) and 2-oxoglutarate-dependent dioxygenase termed FIH (*f*actor *i*nhibiting *H*IF). During hypoxia, reduced PHD and FIH activity allows HIFα subunits to escape proteolysis and assemble into an active α/β heterodimer that induces a broad range of target genes (Fig. 2.1).

A substantial body of work indicates that despite this dual control system, activation of HIF can be achieved through inhibition of the PHD/VHL degradation pathway alone. Indeed, several PHD inhibitory drugs are in development to test whether pharmacological modulation of the HIF hydroxylase system to activate HIF protects from subsequent ischemic insult. This type of intervention may have effects in the short term through enhanced cellular metabolism (for example, stimulation of glycolysis, glucose metabolism, and reduced mitochondrial oxygen consumption) as well as in the medium to longer term through increased perfusion (for example, by stimulation of angiogenesis), giving potential applications both in the acute phase as well as in chronic ischemic heart disease.

The safety of long-term PHD inhibition/HIF activation, however, remains unclear. Given the ubiquitous distribution of the HIF hydroxylase system and wide range of processes affected by HIF, it seems unlikely that all consequences of HIF activation will be beneficial to treating myocardial ischemia; some may even impinge normal physiological function in the heart or other tissues. We consider in this review evidence relating to the benefits and risks of manipulating the HIF hydroxylase system as a therapeutic means of treating myocardial ischemia.

2.2 Benefits

2.2.1 Genetic Manipulation of HIF-1α

Evidence for the essential role of HIF-1α in IPC was obtained from transgenic mouse models, wherein haploinsufficiency of *HIF-1α* is sufficient to ablate the protective effect conferred by IPC on myocardial infarction [2, 3]. This result is similarly present in mice treated with intraventricular infusion of *HIF-1α* siRNA [4].

In agreement with this, overexpression of HIF-1α in the myocardium of mice attenuates infarct size and improves cardiac function several weeks (but not 24 h) after coronary artery occlusion [5]. This delayed protective effect is thought to be conferred, at least in part, through increased capillary density in the infarct and peri-infarct zones via transcriptional activation of pro-angiogenic HIF target genes such as vascular endothelial growth factor (VEGF) and angiopoietin-2. Together with the predicted vasodilation from HIF-mediated stimulation of inducible nitric oxide synthase, these changes are postulated to help restore delivery of blood to the heart. It should be noted that the overexpressed HIF-1α in these mice would be subject to normoxic degradation, thus limiting upregulation of the pathway in the cells that are best oxygenated. The long-term effects of more complete HIF-1α activation from blockage of the degradation pathway, therefore, cannot be readily deduced from this study.

Further, overexpression of a stable form of HIF-1α in the epidermis of mice has been shown to induce hypervascularity (in line with the predicted induction of pro-angiogenic HIF target genes) [6]. Interestingly, in contrast to transgenic mice overexpressing myocardial VEGF, in which rapid stimulation of dysregulated angiogenesis leads to fragile and immature vessel formation [7, 8], HIF-1α overexpression induces blood vessel formation without any leakage or inflammation. Most probably this is because of multiple, coordinated actions on the angiogenic process. It is also possible that effects of HIF activation at sites remote from the site of ischemia may have protective actions (for instance, by increasing circulating endothelial progenitors). This might conceivably assist perfusion of distant tissues and may underlie remote ischemic preconditioning effects, whereby IPC of, for example, the kidney can result in cardioprotection [9].

2.2.2 Pharmacological Inhibition and Genetic Manipulation of PHD Enzymes

Small molecule inhibitors of the PHD enzymes potently activate the HIF response both *in vitro* and *in vivo*. Thus, it has been proposed that administration of PHD inhibitors could mimic, at least in part, the protective effects of exposure to hypoxia. Indeed, PHD inhibition likely results in greater HIF activation than the submaximal levels achieved through ischemic insult.

Initial studies using cobalt chloride and the iron chelator desferrioxamine to inhibit PHD enzymes (by displacement of their Fe(II) center or decreasing Fe(II) availability in solution) suggested that PHD inhibition acts similarly to IPC in providing protection against myocardial infarction [10, 11]. However, such inhibitors would be predicted to target other Fe(II)-containing enzymes and likely result in side effects from dysregulation of non-HIF hydroxylase pathways.

Subsequent studies have applied more specific inhibitors of PHD activity, dimethyl-oxalylglycine (DMOG) and FG2216, to rodent models of myocardial ischemia. DMOG is a 2-oxoglutarate analogue that inhibits the 2-oxoglutarate-dependent-dioxygenase family of enzymes (which includes the PHD enzymes); FG2216, on the other hand, is a more selective analogue which is proposed to specifically target the PHD enzymes, making it attractive for therapeutic use. Both DMOG and FG2216 have been reported to minimize tissue damage 24 h to several weeks after myocardial infarction [4, 12–14].

Genetic manipulation of PHD activity has also been shown to protect from myocardial I/R. Although all three isoforms of PHD (1, 2, and 3) can hydroxylate and regulate HIFα in vitro, the ubiquitously high level of PHD2 protein across a range of cell lines is thought to account for its dominant role in setting low steady-state levels of HIF in normoxia [15]. In keeping with this, intraventricular infusion with *PHD2*, but not *PHD1* or *3*, siRNA reduced post-ischemic infarct area [4, 16, 17]. Similar results were obtained with PHD2 silencing using intramyocardial injection of *PHD2* shRNA [18].

Genetic deletion of *PHD2* (but not *PHD1* or *3*) in mice results in embryonic lethality [19]. It has been reported, however, that transgenic mice containing hypomorphic alleles for *PHD2* are viable with no obvious cardiac abnormalities. These mice have improved functional recovery, coronary flow rate, and reduced infarct size following I/R in the isolated mouse heart [20], in agreement with the dominant role of the PHD2 isoform in HIF regulation.

Interestingly, *PHD1–/–* mice, which survive until adulthood with no obvious heart defects, have also been reported to show significant protection from myocardial I/R [21]. Further, this protection against ischemic insult is observed in *PHD1–/–* skeletal muscle [22] and liver [23], indicating that the mechanisms involved are not restricted to the heart. Although the latter phenotypes are thought to involve HIF-dependent pathways, it is curious that the other hallmarks of HIF activation such as polycythemia and angiogenesis are not observed in *PHD1–/–* mice. Indeed, PHD1 has been reported to have HIF-independent functions in regulating cellular proliferation [24] and it is possible that these may contribute to the ischemic protection. Alternatively, it may be that PHD1 loss induces HIF to a lesser extent than loss of PHD2, such that there is sufficient HIF to provide protection from ischemia without activating erythropoiesis or angiogenesis. Whatever the mechanism, the findings raise the interesting possibility that PHD isoform-specific inhibitors (which have yet to be developed) could provide more targeted drug intervention.

Overall, these studies provide evidence that short-term (or mild chronic) activation of HIF, by either pharmacological inhibition of PHD enzymes or genetic manipulation of PHD/HIF, can be beneficial against myocardial I/R. The protection conferred may occur shortly after HIF induction via changes in cellular metabolism (for example, enhanced glucose uptake and metabolism through activation of HIF target genes such as GLUT-1, pyruvate dehydrogenase kinase, and 6-phosphofructokinase 1) and vasodilation (for example, by induction of nitric oxide synthases). In addition, activation of HIF may confer delayed protection via angiogenesis and vascular remodeling.

Long-term HIF activation, for example, through genetic manipulation of the HIF hydroxylase system, however, has potential detrimental effects. These are outlined below.

2.3 Risks

2.3.1 Genetic Manipulation of HIFα

Evidence for the detrimental effects of sustained HIFα activation are obtained from recent studies, whereby overexpression of a stable form of either HIF-1α or HIF-2α in cardiomyocytes results in cardiomyopathy [25, 26].

2.3.2 Genetic Manipulation of PHD Enzymes

The effects of chronic PHD inhibitor exposure are largely unknown and existing data derives from *PHD* knockout mice which may not accurately mimic the effects of catalytic inhibition (for example, because of loss of additional non-catalytic effects of the enzyme protein). It is worth noting, however, that supplementation of a certain brand of Canadian beer with cobalt sulfate was identified as a contributing etiological factor in the so-called Quebec beer-drinker's cardiomyopathy (with associated polycythemia) of the late 1960s [27]. This hints at protracted PHD inhibitor usage being potentially detrimental to cardiac function – a possibility that is supported by genetic manipulation of the PHD enzymes in mice.

Widespread, conditional inactivation of *PHD2* in adult mice results in severe polycythemia and hyperactive angiogenesis/angiectasia, in line with the predicted induction of HIFα, pro-angiogenic HIF target genes, and erythropoiesis-promoting HIF target gene erythropoietin. However, these mice also suffer from dilated cardiomyopathy and premature mortality [28–31]. The latter phenotypes may occur either as an indirect consequence of polycythemia and/or as a direct action of *PHD2* loss in cardiomyocytes. Further studies demonstrate that, in fact, cardiac-specific loss of *PHD2* is sufficient to induce dilated cardiomyopathy and premature mortality in adult mice, which is exacerbated when on a *PHD3–/–* background [25]. Thus, sustained PHD2 inactivation/HIF activation in the heart itself is detrimental to cardiac function and may even play a causal role in the pathogenesis of ischemic cardiomyopathy [25].

Aside from the risks of dysregulated erythropoiesis and angiogenesis, loss of PHD activity in other noncardiac tissues may also pose risks to both cardiovascular and other tissue functions. For instance, *PHD3–/–* mice, though viable and with no obvious cardiac abnormalities, suffer from abnormal sympathoadrenal development that is likely to be the cause of the observed reduced catecholamine secretion and systemic hypotension [32]. In humans, activating mutations in HIF-2α have been associated with pulmonary hypertension [33]. Systemic administration of PHD inhibitors may therefore result in a range of side effects from HIF activation in tissues other than the heart.

2.3.3 Genetic Manipulation of VHL

As both VHL and PHD negatively regulate HIF, and assuming a lack of divergence in the PHD/HIF/VHL oxygen-sensing pathway, one might predict loss of VHL to phenocopy loss of PHDs (in particular PHD2, given its dominant role in HIF regulation). Indeed, *VHL–/–* mice, like *PHD2–/–* mice, are embryonic lethal due to placental defects [34]. Cardiac-restricted ablation of *VHL* in adult mice leads to dilated cardiomyopathy, lipid accumulation, myocyte loss, fibrosis, and even malignant transformation, in a HIF-1α-dependent manner [35]. The cardiac phenotype after *VHL* loss is therefore more severe than observed after combined *PHD2/PHD3* inactivation, possibly because of residual PHD1 activity and/or a contribution from PHD and HIF-independent functions of VHL. However, the findings again suggest that long-term, high-level upregulation of HIF pathways is likely to entrain significant side effects.

Overall, genetic studies demonstrate that extensive HIF activation in the heart is potentially deleterious to cardiovascular function. Thus, PHD inhibitors will probably require careful dose titration to achieve the desired risk/benefit profile and/or limitation of the duration of therapy.

2.4 Summary

Current work has defined both benefits and risks associated with the manipulation of the HIF hydroxylase system as a therapeutic means of treating myocardial ischemia.

Short-term (or mild, chronic) activation of HIF, like IPC, is protective against ischemic insult. Although this has been determined using interventions that precede ischemia, two findings raise the possibility that PHD inhibitors could equally be applied post-ischemia. First, HIF activation lasts several days following ischemic insult [36]. Second, cycles of I/R applied at the onset of, rather than preceding, ischemia are still able to confer protection (a process known as ischemic post-conditioning [37]). The ability to treat myocardial ischemia by post-event drug intervention would make PHD inhibitors particularly useful in the clinical setting.

Prolonged, excessive HIF activation, on the other hand, phenocopies ischemic cardiomyopathy and is deleterious to cardiovascular function. It may also have detrimental side effects in noncardiac tissues if applied in a systemic manner. Ablation of *PHD1* in mice induces hypoxia tolerance without effect on PHD2-/HIF-regulated pathways such as erythrocytosis. In this regard, a PHD1-specific inhibitor, though not yet available, may be beneficial.

In summary, PHD inhibitors that activate HIF are an attractive therapeutic option for minimizing tissue damage from myocardial ischemia or improving perfusion by medical means. However, care will be required to avoid side effects from uncontrolled activation of hypoxia pathways. This highlights the need for time, dose, tissue, and/or PHD isoform-specific drug interventions in order to minimize the potential deleterious side effects of PHD inhibitors.

References

1. Kaelin Jr WG, Ratcliffe PJ. Oxygen sensing by metazoans: the central role of the HIF hydroxylase pathway. Mol Cell. 2008;30:393–402.
2. Cai Z, Manalo DJ, Wei G, et al. Hearts from rodents exposed to intermittent hypoxia or erythropoietin are protected against ischemia-reperfusion injury. Circulation. 2003;108:79–85.
3. Cai Z, Zhong H, Bosch-Marce M, et al. Complete loss of ischaemic preconditioning-induced cardioprotection in mice with partial deficiency of HIF-1 alpha. Cardiovasc Res. 2008;77:463–70.
4. Eckle T, Kohler D, Lehmann R, El Kasmi K, Eltzschig HK. Hypoxia-inducible factor-1 is central to cardioprotection: a new paradigm for ischemic preconditioning. Circulation. 2008;118:166–75.
5. Kido M, Du L, Sullivan CC, et al. Hypoxia-inducible factor 1-alpha reduces infarction and attenuates progression of cardiac dysfunction after myocardial infarction in the mouse. J Am Coll Cardiol. 2005;46:2116–24.
6. Elson DA, Thurston G, Huang LE, et al. Induction of hypervascularity without leakage or inflammation in transgenic mice overexpressing hypoxia-inducible factor-1alpha. Genes Dev. 2001;15:2520–32.
7. Lee RJ, Springer ML, Blanco-Bose WE, Shaw R, Ursell PC, Blau HM. VEGF gene delivery to myocardium: deleterious effects of unregulated expression. Circulation. 2000;102:898–901.
8. Carmeliet P. VEGF gene therapy: stimulating angiogenesis or angioma-genesis? Nat Med. 2000;6:1102–3.
9. Kant R, Diwan V, Jaggi AS, Singh N, Singh D. Remote renal preconditioning-induced cardioprotection: a key role of hypoxia inducible factor-prolyl 4-hydroxylases. Mol Cell Biochem. 2008;312:25–31.
10. Dendorfer A, Heidbreder M, Hellwig-Burgel T, Johren O, Qadri F, Dominiak P. Deferoxamine induces prolonged cardiac preconditioning via accumulation of oxygen radicals. Free Radic Biol Med. 2005;38:117–24.
11. Xi L, Taher M, Yin C, Salloum F, Kukreja RC. Cobalt chloride induces delayed cardiac preconditioning in mice through selective activation of HIF-1alpha and AP-1 and iNOS signaling. Am J Physiol Heart Circ Physiol. 2004;287:H2369–75.

12. Zhao HX, Wang XL, Wang YH, et al. Attenuation of myocardial injury by postconditioning: role of hypoxia inducible factor-1alpha. Basic Res Cardiol. 2010;105:109–18.
13. Ockaili R, Natarajan R, Salloum F, et al. HIF-1 activation attenuates postischemic myocardial injury: role for heme oxygenase-1 in modulating microvascular chemokine generation. Am J Physiol Heart Circ Physiol. 2005;289:H542–8.
14. Philipp S, Jurgensen JS, Fielitz J, et al. Stabilization of hypoxia inducible factor rather than modulation of collagen metabolism improves cardiac function after acute myocardial infarction in rats. Eur J Heart Fail. 2006;8:347–54.
15. Berra E, Benizri E, Ginouves A, Volmat V, Roux D, Pouyssegur J. HIF prolyl-hydroxylase 2 is the key oxygen sensor setting low steady-state levels of HIF-1alpha in normoxia. EMBO J. 2003;22:4082–90.
16. Natarajan R, Salloum FN, Fisher BJ, Kukreja RC, Fowler 3rd AA. Hypoxia inducible factor-1 activation by prolyl 4-hydroxylase-2 gene silencing attenuates myocardial ischemia reperfusion injury. Circ Res. 2006;98:133–40.
17. Natarajan R, Salloum FN, Fisher BJ, Ownby ED, Kukreja RC, Fowler 3rd AA. Activation of hypoxia-inducible factor-1 via prolyl-4 hydoxylase-2 gene silencing attenuates acute inflammatory responses in postischemic myocardium. Am J Physiol Heart Circ Physiol. 2007;293:H1571–80.
18. Huang M, Chan DA, Jia F, et al. Short hairpin RNA interference therapy for ischemic heart disease. Circulation. 2008;118:S226–33.
19. Takeda K, Ho VC, Takeda H, Duan LJ, Nagy A, Fong GH. Placental but not heart defects are associated with elevated hypoxia-inducible factor alpha levels in mice lacking prolyl hydroxylase domain protein 2. Mol Cell Biol. 2006;26:8336–46.
20. Hyvarinen J, Hassinen IE, Sormunen R, et al. Hearts of hypoxia-inducible factor prolyl 4-hydroxylase-2 hypomorphic mice show protection against acute ischemia-reperfusion injury. J Biol Chem. 2010;285:13646–57.
21. Adluri RS, Thirunavukkarasu M, Dunna NR, et al. Disruption of HIF-prolyl hydroxylase-1 (PHD-1-/-) attenuates ex vivo myocardial ischemia/reperfusion injury through HIF-1alpha transcription factor and its target genes in mice. Antiox Redox Signal 2011;15:1789–97.
22. Aragones J, Schneider M, Van Geyte K, et al. Deficiency or inhibition of oxygen sensor Phd1 induces hypoxia tolerance by reprogramming basal metabolism. Nat Genet. 2008;40: 170–80.
23. Schneider M, Van Geyte K, Fraisl P, et al. Loss or silencing of the PHD1 prolyl hydroxylase protects livers of mice against ischemia/reperfusion injury. Gastroenterology. 2010;138:1143–54. e1–2.
24. Zhang Q, Gu J, Li L, et al. Control of cyclin D1 and breast tumorigenesis by the EglN2 prolyl hydroxylase. Cancer Cell. 2009;16:413–24.
25. Moslehi J, Minamishima YA, Shi J, et al. Loss of hypoxia-inducible factor prolyl hydroxylase activity in cardiomyocytes phenocopies ischemic cardiomyopathy. Circulation. 2010;122: 1004–16.
26. Bekeredjian R, Walton CB, MacCannell KA, et al. Conditional HIF-1alpha expression produces a reversible cardiomyopathy. PLoS One. 2010;5:e11693.
27. Morin Y, Daniel P. Quebec beer-drinkers' cardiomyopathy: etiological considerations. Can Med Assoc J. 1967;97:926–8.
28. Minamishima YA, Moslehi J, Bardeesy N, Cullen D, Bronson RT, Kaelin Jr WG. Somatic inactivation of the PHD2 prolyl hydroxylase causes polycythemia and congestive heart failure. Blood. 2008;111:3236–44.
29. Minamishima YA, Moslehi J, Padera RF, Bronson RT, Liao R, Kaelin Jr WG. A feedback loop involving the Phd3 prolyl hydroxylase tunes the mammalian hypoxic response in vivo. Mol Cell Biol. 2009;29:5729–41.
30. Takeda K, Aguila HL, Parikh NS, et al. Regulation of adult erythropoiesis by prolyl hydroxylase domain proteins. Blood. 2008;111:3229–35.
31. Takeda K, Cowan A, Fong GH. Essential role for prolyl hydroxylase domain protein 2 in oxygen homeostasis of the adult vascular system. Circulation. 2007;116:774–81.

32. Bishop T, Gallagher D, Pascual A, et al. Abnormal sympathoadrenal development and systemic hypotension in PHD3-/- mice. Mol Cell Biol. 2008;28:3386–400.
33. Gale DP, Harten SK, Reid CD, Tuddenham EG, Maxwell PH. Autosomal dominant erythrocytosis and pulmonary arterial hypertension associated with an activating HIF2 alpha mutation. Blood. 2008;112:919–21.
34. Gnarra JR, Ward JM, Porter FD, et al. Defective placental vasculogenesis causes embryonic lethality in VHL-deficient mice. Proc Natl Acad Sci USA. 1997;94:9102–7.
35. Lei L, Mason S, Liu D, et al. Hypoxia-inducible factor-dependent degeneration, failure, and malignant transformation of the heart in the absence of the von Hippel-Lindau protein. Mol Cell Biol. 2008;28:3790–803.
36. Willam C, Maxwell PH, Nichols L, et al. HIF prolyl hydroxylases in the rat; organ distribution and changes in expression following hypoxia and coronary artery ligation. J Mol Cell Cardiol. 2006;41:68–77.
37. Zhao ZQ, Corvera JS, Halkos ME, et al. Inhibition of myocardial injury by ischemic postconditioning during reperfusion: comparison with ischemic preconditioning. Am J Physiol Heart Circ Physiol. 2003;285:H579–88.

3 Cytoprotective Mechanisms in the Vasculature

Justin C. Mason

3.1 Introduction

The vascular endothelium forms an essential barrier, separating blood constituents and the extravascular tissues. For a long time considered an inert semipermeable membrane, the vascular endothelium is now recognized to be multifunctional, dynamic, and heterogeneous organ. In health, the endothelium contributes to the control of vasodilatation and permeability, while maintaining an anti-thrombotic, anti-inflammatory, anti-adhesive phenotype. This is an active process controlled by intrinsic gene expression and external stimuli. As a consequence specialized endothelium is found in the blood-brain barrier, lining fenestrated capillaries in the kidney, as sinusoidal endothelium in the liver and in lung alveoli to facilitate gas exchange. The endothelium is also highly adaptable, changing phenotype in response to specific stimuli and so facilitating hemostasis and regulating the response to inflammatory stimuli. In the latter, the endothelium regulates vascular permeability, expression of cellular adhesion molecules and recruitment of leukocytes. In addition, release of growth factors such as vascular endothelial growth factor (VEGF) and subsequent endothelial proliferation are important in tissue repair.

As a consequence of its anatomic location, the vascular endothelium is continuously exposed to potentially harmful factors such as endotoxin, cytokines, advanced glycation end-products, complement components, activated leukocytes, and oxidatively modified low-density lipoproteins (ox-LDL). If uncontrolled, these noxious stimuli predispose to endothelial dysfunction, predominantly driven by reduced expression of endothelial nitric oxide synthase (eNOS) [1].

Endothelial injury is the earliest detectable event in atherogenesis [2], and induces a local inflammatory response resulting in endothelial dysfunction, characterized by

J.C. Mason
Bywaters Centre for Vascular Inflammation,
National Heart and Lung Institute, Imperial College London,
Hammersmith Hospital, London, UK
e-mail: justin.mason@imperial.ac.uk

D. Abraham et al. (eds.), *Translational Vascular Medicine*,
DOI 10.1007/978-0-85729-920-8_3,

reduced NO biosynthesis, oxidative stress, increased permeability to lipoproteins, and monocyte recruitment [3]. Moreover, apoptosis occurs preferentially at sites of endothelial injury and atherosclerosis [4], where denudation of vascular endothelium enhances the risk of thrombosis. Thus, mechanisms that control endothelial inflammation and minimize vascular injury are essential for the maintenance of vascular integrity, initiation of repair, and resistance to atherogenesis. A detailed understanding of these molecular mechanisms may in turn reveal novel therapeutic targets which will help to prevent vascular injury and allow the maintenance of vascular endothelial homeostasis and integrity [5].

3.2 Accelerated Atherosclerosis

Heart attack and stroke as a consequence of atherosclerosis remain the leading cause of death in the western world. Moreover, certain disease groups are exposed to the risk of accelerated atherogenesis, with hyperlipidemia, the metabolic syndrome, and diabetes mellitus the best recognized. Over the last decade, the increased risk of accelerated atherogenesis in patients suffering from systemic inflammatory diseases has emerged as an intense area of research.

Prolonged systemic inflammation, such as that associated with rheumatoid arthritis (RA) and systemic lupus erythematosus (SLE), may accelerate atherogenesis with cardiovascular disease responsible for 35–50% increased mortality in RA [6]. Importantly, the disease itself represents a specific risk factor [7]. Likewise, SLE is an independent risk factor and responsible for a 10–50 fold increase in myocardial infarction in a female population characteristically protected against cardiovascular disease [8]. Thus, although patients with chronic inflammatory disease commonly have more traditional risk factors than age- and sex-matched controls, these alone do not account for the increased cardiovascular risk. Additional mechanisms implicated include increased oxidative stress, pro-inflammatory cytokines, endothelial activation leading to enhanced leukocyte adhesion, and the deleterious effects of immune complexes, anti-phospholipid antibodies, homocysteinemia, hypercoagulability, $CD4^{+}CD28^{-}$ T cells, and drug toxicity [6]. The significance of chronic systemic inflammation is reinforced by evidence of accelerated atherosclerosis in patients with vasculitides and other non-rheumatic inflammatory diseases.

A current challenge is to identify early the subgroup of patients with these diseases most at risk of developing accelerated atherogenesis. The advance in novel noninvasive imaging techniques is one approach that has been adopted in recent years. For example, high-resolution ultrasound can monitor intima–media thickness and demonstrate early disease [9]. Using positron emission tomography with oxygen-15-labeled water, we have demonstrated that the increase in myocardial blood flow in response to intravenous adenosine is significantly attenuated in some patients with RA and SLE. These patients were known to have normal or minimally diseased (≤20% luminal reduction) coronary arteries and no significant difference in conventional cardiovascular risk factors when compared with age- and sex-matched controls [10]. Likewise, we have shown that an integrated method for cardiovascular magnetic resonance angiography (CMR) in patients with Takayasu's arteritis provides not only accurate delineation of arterial wall thickening, but can also identify early atherosclerotic plaques, demonstrate dynamic ventricular function and myocardial scarring [11]. These techniques may have the potential to identify patients most at risk of accelerated atherosclerosis, so allowing early preventative therapy.

However, current treatments for atherosclerosis are directed predominantly at established symptomatic lesions, with an outstanding need for new preventative therapies. Intensive management of inflammation combined with traditional risk factor modification is required to minimize cardiovascular risk in rheumatic diseases. Methotrexate and mycophenolate mofetil demonstrate anti-atherogenic effects, with methotrexate reducing cardiovascular mortality in RA by 70% [12, 13]. Anti-tumor necrosis factor-α therapy may enhance endothelial function, and the risk of myocardial infarction in patients with RA who respond to anti-TNF agents is significantly reduced when compared to non-responders [14]. An important additional approach is to target endothelial dysfunction, an end achieved to some extent by statins [15] and angiotensin-converting enzyme inhibitors [16], which also exert anti-inflammatory effects. However, efficacy needs to be established by prospective studies, and to optimize this approach, we need a detailed understanding of vascular endothelial cytoprotective signaling pathways and their downstream target genes.

3.3 Vascular Cytoprotection

Exogenous factors and intracellular mechanisms combine to control inflammatory responses, prevent bystander endothelial injury, and maintain the integrity of the vascular wall (Fig. 3.1). I will provide a brief overview of these mechanisms before dealing in more detail with some recent advances.

A variety of anti-inflammatory cytokines and growth factors play an important role in the maintenance of endothelial homeostasis, regulation of inflammation and reparative mechanisms including angiogenesis. The IL-1 receptor antagonist (IL-1ra) and soluble TNF receptors exert potent anti-inflammatory effects and are used clinically in the treatment of systemic inflammatory diseases including autoinflammatory disorders [17] and rheumatoid arthritis [18]. In murine models IL-1ra is atheroprotective, inhibiting early atherogenesis in ApoE-deficient mice [19], while IL-1ra knockouts suffer arterial inflammation and a low expressing polymorphism has been linked to coronary artery disease [20]. IL-10 is particularly important for its effects on macrophages, inhibiting pro-inflammatory cytokine synthesis and favoring a CD163hi anti-inflammatory phenotype [21]. Although its effects on vascular endothelium are less well understood, IL-10 is reported to inhibit NF-κB, vascular inflammation, and endothelial cell adhesion molecule expression [22].

Growth factors play an essential role in endothelial homeostasis with basic fibroblast growth factor (FGF-2) and VEGF (see below) capable of activating anti-apoptotic pathways. In addition to its mitogenic actions, FGF-2 induces expression of the anti-apoptotic protein Bcl-2 [23]. It may also enhance protection against complement-mediated injury through induction of decay-accelerating factor (DAF) [24], and exert

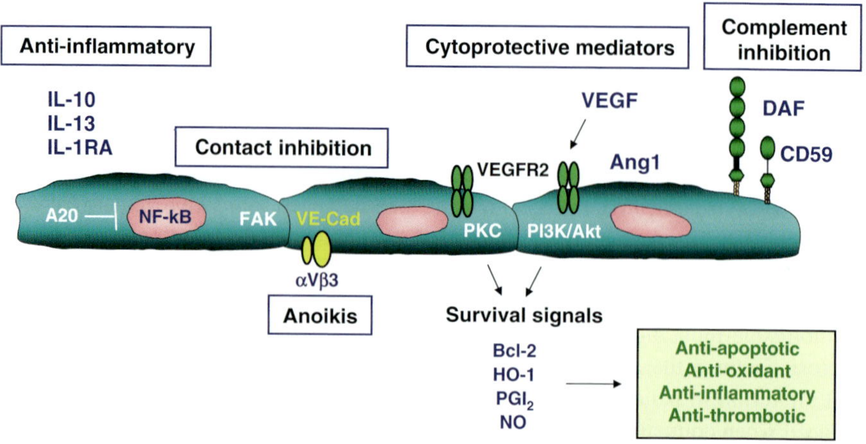

Fig. 3.1 Cytoprotective mechanisms in the vasculature. The vascular endothelium is protected by exogenous anti-inflammatory and pro-survival factors including IL-10, Ang1, and VEGF. Induction of intrinsic cytoprotective genes via PKC and PI-3 K/Akt-dependent pathways results in enhanced protection against apoptosis, oxidative stress, thrombosis, and complement-mediated injury. *IL* interleukin, *IL-1RA* IL-1 receptor antagonist, *VE-Cad* VE-cadherin, *PKC* protein kinase C, *VEGF* vascular endothelial growth factor, *HO-1* heme oxygenase-1, PGI_2 prostacyclin, *DAF* decay-accelerating factor, *Ang1* angiopoietin 1

anti-thrombotic, anti-inflammatory effects [25, 26]. TGF-β signals via two type 1 receptors, activin receptor-like kinases Alk-1 and Alk-5 which induce opposite effects, with Alk-1 driving pro-proliferative and pro-migratory gene expression, while Alk-5 signaling facilitates growth arrest, matrix synthesis, and formation of a stable vessel [27]. TGF-β may also inhibit expression of E-selectin so reducing leukocyte adhesion. Angiopoietin-1 (Ang-1) has also emerged as an important vasculoprotective growth factor, signaling predominantly via Tie2, with its actions opposed by family member Ang 2. Ang1 may exert multiple protective effects in the vasculature including anti-inflammatory, anti-apoptotic, and anti-thrombotic actions [27, 28].

3.3.1 Cytoprotective Genes

Many of the vasculoprotective properties of exogenous mediators are facilitated through induction of intrinsic cytoprotective genes. In the vascular endothelium, these include endothelial nitric oxide synthase (eNOS), superoxide dismutases, A1, A20, B cell lymphoma protein (Bcl)-2, Bcl-xL, heme oxygenase-1 (HO-1) [5], and membrane-bound complement regulatory proteins DAF and CD59 [29, 30]. A20 is an inducible ubiquitin-editing anti-inflammatory protein that negatively regulates NF-κB-dependent gene expression. A20 is induced as a consequence of NF-κB activation and exerts a negative feedback on further activation acting at multiple levels within the NF-κB pathway [31]. The importance of A20 is well illustrated by the phenotype of the knockout mice which are markedly susceptible to TNF and develop severe widespread inflammation and cachexia [32].

Bcl-2, Bcl-xL, and A1 are members of the anti-apoptotic Bcl-2 family, which may also exert important anti-inflammatory and cytoprotective effects. The balance between the pro- and anti-apoptotic members of the Bcl-2 family is critical in determining cell fate [33]. Thus, if pro-apoptotic Bim, Bid, and Bad are present in sufficient amounts to bind to and overwhelm Bcl-2 and Bcl-X_L, sequestered Bax and Bak are released allowing the escape of mitochondrial cytochrome *c*. This in turn results in the generation of the apoptosome, which cleaves and activates downstream apoptosis effector caspases 3, 6, and 7, [33]. We have recently reported that protein kinase Cε forms a signaling complex and acts co-operatively with anti-apoptotic kinase (Akt) to protect human vascular endothelial cells against apoptosis, through induction of Bcl-2 and inhibition of caspase-3 cleavage [30].

3.3.2 Resistance to Complement

Mechanisms implicated in complement deposition on the EC surface include activation of the classical pathway by immune complexes, anti-phospholipid, and anti-endothelial cell Abs, and through recognition of apoptotic cell blebs by C1q [34]. Induction of both CD59 and DAF on EC via distinct signaling pathways contributes significantly to the regulation of complement activity and protection against bystander injury [35, 36]. In particular, the propensity for DAF expression to be induced suggests it represents an important response to inflammation and injury and hence in the protection against vascular injury. We have shown DAF expression to be upregulated in response to TNFα, IFNγ, thrombin, VEGF, and bFGF, while epidermal growth factor and PlGF failed to alter expression. The increase in DAF led to enhanced protection against complement-mediated injury (Fig. 3.2), which was

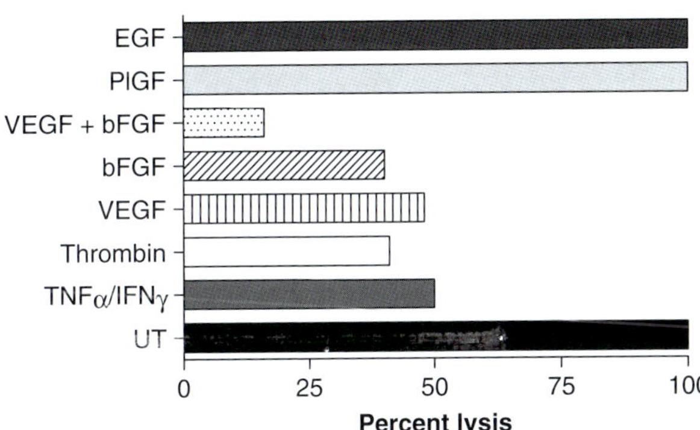

Fig. 3.2 Induction of decay-accelerating factor (DAF) protects endothelial cells against complement. Exposure of human endothelial cells to certain growth factors and pro-inflammatory mediators for 24–48 h increases DAF expression on the cell surface and resistance to complement-mediated injury following opsonization with an anti-endoglin mAb and exposure to normal human serum. Data expressed as percent cell lysis versus untreated (UT) control. *EGF* epidermal growth factor, *PlGF* placenta growth factor, *VEGF* vascular endothelial growth factor, *bFGF* basic fibroblast growth factor, *TNF* tumor necrosis factor, *IFNγ* interferon-γ

reversed by inclusion of an inhibitory DAF mAb [24, 29, 35, 37, 38]. Moreover, $DAF^{-/-}$ and $CD59^{-/-}$ mice, when crossed with atherosclerosis prone strains, suffer accelerated disease [39–41].

3.3.3 Vascular Endothelial Growth Factor

In addition to their better known roles in vasculogenesis and angiogenesis, it is increasingly recognized that VEGFs are important in adult endothelial homeostasis [42]. The five main VEGF ligands, VEGFA-D, and placenta growth factor PIGF are also found as splice variants. Thus, the isoforms of human VEGFA are VEGFA121, VEGF145, VEGFA165, VEGFA189, and VEGFA206. The VEGFs signal via the receptor tyrosine kinases VEGFR1-3 and this signaling maybe modulated by co-receptors such as neuropilin and heparan sulfate proteoglycans (see ref. [43] for a detailed review). The ability of VEGFA to induce cytoprotective gene expression in vascular endothelium is well established and the cytoprotective actions of VEGFA include induction of the anti-apoptotic genes Bcl-2 and A1 [30, 44]. In addition, VEGF increases eNOS expression and NO release [45], induces expression of HO-1 [46], and contributes to the maintenance of an anti-thrombotic endothelial surface through induction of prostacyclin synthesis [45]. VEGF also enhances protection against complement-mediated injury via upregulation of DAF expression [47]. Recent elegant studies, in which mice with an inducible podocyte-specific deletion of *vegfA* were generated, have demonstrated endothelial cell swelling, local thrombosis, and subsequent proteinuria and hypertension [48]. These abnormalities may contribute to the side effects associated with the use of the anti-VEGFA mAb bevacizumab in disseminated colonic carcinoma, which include both hypertension and thrombosis [48, 49].

3.3.4 Heme Oxygenase-1

HO-1 is an inducible cytoprotective enzyme which degrades heme, generating carbon monoxide, bilirubin, and ferrous iron which is rapidly sequestered by intracellular ferritin [50, 51]. The cytoprotective properties of HO-1 are attributed to its products and include antioxidant, anti-apoptotic, anti-thrombotic, and anti-inflammatory actions [50] (Fig. 3.3). The importance of these cytoprotective actions are reflected in the severe sequelae of HO-1 deficiency, which include intravascular hemolysis, anemia, diffuse endothelial damage, and accelerated atherosclerosis [52]. We have recently reported an additional cytoprotective action of HO-1, the regulation of complement activation, mediated via induction of DAF. Analysis of cardiac EC isolated from $Hmox1^{-/-}$ mice revealed a significant reduction in DAF expression as compared to $Hmox1^{+/+}$ EC, while the $Hmox1^{-/-}$ cells displayed enhanced sensitivity to complement-mediated lysis [53]. HO-1 expression is required for prolonged allograft survival, and both HO-1 expression and complement regulation are important in accommodation, the resistance of a transplanted organ to graft-specific antibodies and complement fixation [54]. Therefore, our data

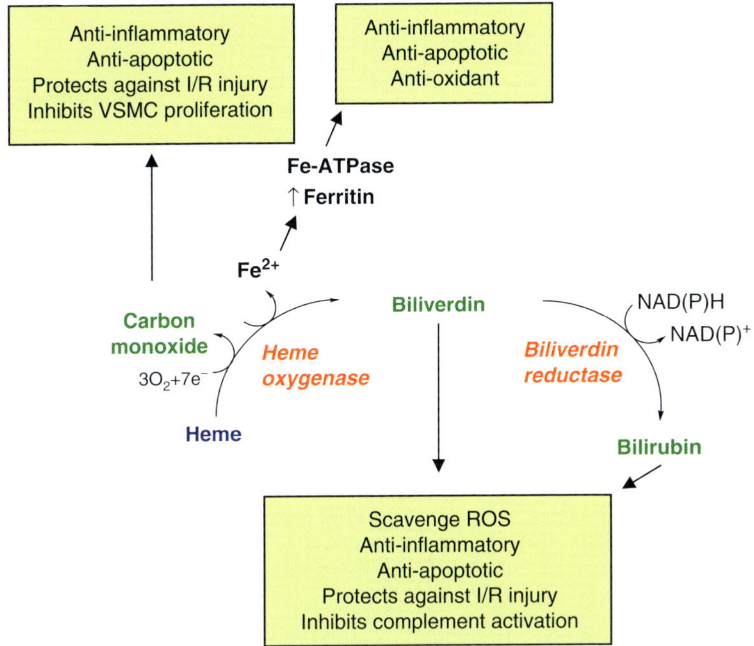

Fig. 3.3 Heme oxygenase-1-mediated degradation of heme. HO-1 catalyzes the breakdown of heme into equimolar amounts of carbon monoxide (CO), biliverdin, and free iron (Fe^{2+}). Biliverdin reductase subsequently catalyzes the conversion of biliverdin to bilirubin. The increase in intracellular Fe^{2+} induces expression of the iron-binding protein heavy chain-ferritin and the opening of Fe^{2+} export channels. The products of heme degradation exert a variety of effects on endothelial cells which are protective against atherosclerosis. *ROS* reactive oxygen species, *I/R* ischemia reperfusion

linking the activity of HO-1 and expression of DAF is likely to be important in accommodation, resistance to post-transplant vasculopathy, and prolonged graft survival [55, 56]. HO-1 and its products may also protect against atherogenesis. Inhibition of VSMC proliferation, combined with its anti-inflammatory, antioxidant actions and anti-thrombotic actions, contributes to the protective role of HO-1 against atherogenesis and its ability to stabilize the vulnerable plaque. Furthermore, epidemiological studies suggest that a mildly raised serum bilirubin significantly protects against ischemic heart disease.

3.4 Vascular Cytoprotection and Shear Stress

The geometric nature of atherosclerosis within the arterial tree led to the study of blood flow patterns as an influence in atherogenesis. These studies suggest that a disturbed flow (DF) waveform, with low shear reversing flow patterns, such as that located at arterial branch points, is pro-atherogenic, whereas unidirectional pulsatile

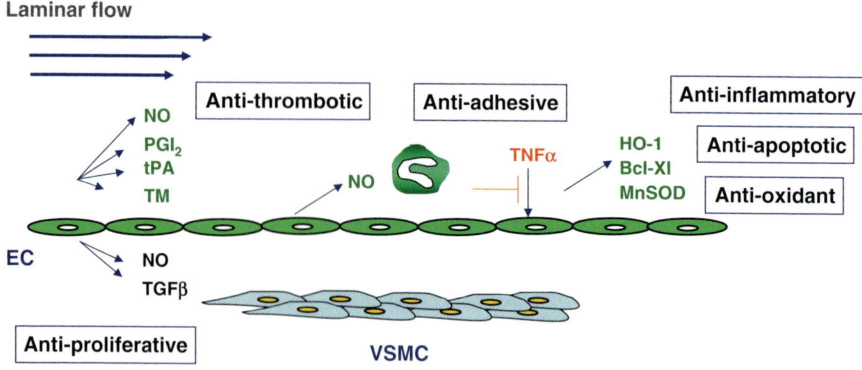

Fig. 3.4 Shear stress-induced vascular cytoprotection. Laminar shear stress is a critical component in the cytoprotection of arterial endothelium. Shear activates anti-thrombotic and cell survival genes including anti-apoptotic and antioxidant proteins. Nitric oxide exerts anti-adhesive, anti-thrombotic, and vasodilatory effects. In addition, release of nitric oxide and TGFβ inhibits vascular smooth muscle cell proliferation. *NO* nitric oxide, *PGI2* prostacyclin, *tPA* tissue plasminogen activator, *TM* thrombomodulin, *TNF* tumor necrosis factor, *HO-1* heme oxygenase-1, *MnSOD* manganese superoxide dismutase, *EC* endothelial cells, *VSMC* vascular smooth muscle cells

laminar shear stress (LSS) >10 dyn/cm^2 is atheroprotective [57]. This is reflected in the phenotype of EC exposed to LSS, typically characterized by enhanced endothelial nitric oxide synthase (eNOS) expression and nitric oxide (NO) biosynthesis, prolonged EC survival, and an anticoagulant, anti-adhesive cell surface [58, 59] (Fig. 3.4). In contrast, endothelium exposed to DF exhibits reduced levels of eNOS, increased apoptosis, generation of reactive oxygen species, permeability to LDL, and leukocyte adhesion [57].

3.4.1 Cytoprotective Transcription Factors

Considerable recent attention has been given to the investigation of LSS-inducible cytoprotective transcription factors, and in particular Kruppel-like factors (KLF) 2 and 4 and NF-E2-related factor-2 (Nrf2). KLF2 and KLF4 are members of a family of 17 zinc-finger transcription factors. In vitro, endothelial KLF2 is induced by LSS but not DF, while in vivo, KLF2 is differentially expressed in areas of the aorta exposed to LSS and DF [60]. Importantly, KLF2 activity has been shown to be an important regulator of cytoprotective genes including eNOS, thrombomodulin, and HO-1 [61–63]. An ERK5/myocyte enhancing factor 2 pathway has been identified upstream of KLF2 transcription, and this can be therapeutically activated by statins [64].

KLF4 activity has also been linked to the regulation of vasculoprotective genes including eNOS and thrombomodulin [65, 66]. A recent study has demonstrated that shear stress–induced KLF4 expression via MEK5/MEF2 pathway shared with KLF2 [66]. Subsequent microarray analysis demonstrated significant overlap in

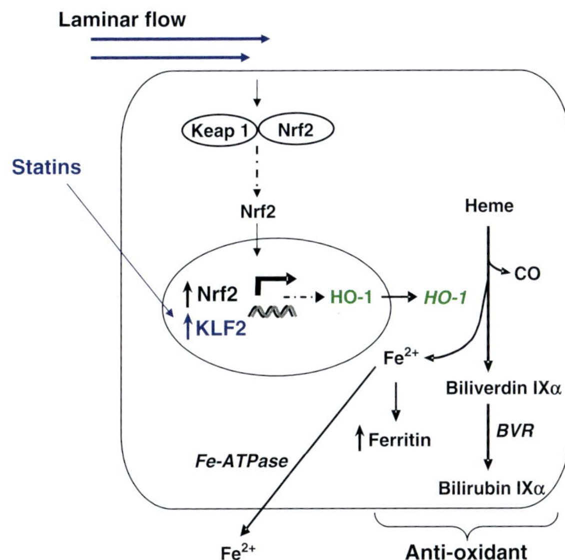

Fig. 3.5 Synergistic induction of HO-1 by statins and laminar shear stress. Endothelial cells exposed to laminar shear stress exhibit increased responsiveness to statins as evidenced by synergistic induction of HO-1 expression and activity and resistance to oxidative stress. This response is dependent upon activation of both Nrf2 and KLF-2. *KLF-2* Kruppel-like factor 2, *Nrf2* NF-E2-related factor-2, *Keap1* Kelch-like ECH-associated

target genes between the two transcription factors. Thus, further details of their precise relationship in the maintenance of endothelial homeostasis both in the resting vascular endothelium and during inflammation are awaited with interest.

Nrf2 is similarly an important flow-inducible cytoprotective transcription factor. Nrf2 is retained in the cytoplasm by kelch-like ECH-associated protein (Keap1). LSS results in dissociation of the Nrf2–Keap1 complex allowing Nrf2 translocation to the nucleus, where it controls expression of phase II detoxification enzymes and antioxidant proteins including HO-1, NAD(P)H:quinine oxidoreductase 1, ferritin heavy chain, glutathione reductase, and thioredoxin reductase 1 via the antioxidant response element [67, 68].

We have recently reported that LSS induces expression of CD59 in vascular EC via an ERK5/KLF2-dependent pathway, thereby preventing C9 insertion into the MAC and protecting against complement-mediated injury. We also demonstrated regional differences in CD59 in the murine aorta, with maximal expression of CD59 at atheroprotected sites [69]. These data combined with the observation of accelerated atherosclerosis in CD59/LDLR mice suggest CD59 contributes significantly to shear stress-mediated protection against atherosclerosis [40].

Recent data suggests that shear stress may influence endothelial responsiveness to exogenous factors including drugs. Thus, we have reported that atorvastatin-mediated HO-1-dependent antioxidant effects [63] are enhanced by LSS, demonstrating that biomechanical signaling contributes to endothelial responsiveness to pharmacological agents. This synergistic relationship between LSS and statin involved Akt phosphorylation, activation of both KLF2 and Nrf2, eNOS induction, and prolonged HO-1 mRNA stability [68] (Fig. 3.5).

This observation has potentially important implications for statin efficacy in patients with ischemic heart disease, and for the increasing use of statins in prevention of accelerated atherosclerosis in patients suffering from systemic inflammatory diseases. The data emphasize the need for novel therapies to optimize vasculoprotection, and our recent report of sulforaphane-mediated activation of Nrf2 in atheroprone sites of the murine aorta offers hope in this regard [70].

3.4.2 Peroxisome Proliferator-Activated Receptors (PPARs)

PPARα, PPARδ, and PPARγ, members of the nuclear receptor family, are ligand-activated transcription factors which regulate energy balance. Transcriptional activation of their target genes requires ligand-induced heterodimerization with the retinoid X receptor and co-factor recruitment [71]. PPARs are activated by mediators including polyunsaturated fatty acids linoleic and docosahexaenoic acid, eicosanoids such as prostayclin and 15d-PGJ_2 and components of ox-LDL [71]. Expression of the PPARs in the vasculature and reported anti-inflammatory effects of synthetic ligands has led to significant interest therapeutically. PPARα agonists (Fibrates) and PPARγ agonists (Thiazolidinediones) are in current clinical use for the treatment of hyperlipidemia and diabetes mellitus respectively, while PPARδ agonists reduce abnormalities associated with the metabolic syndrome [72]. However, the relatively disappointing data to date as regards reduction of cardiovascular events with fibrates and thiazolidinediones suggests an improved understanding of PPAR biology and the actions of PPAR ligands is required [73].

Recent findings of note include the report that LSS induces expression of the PPARγ target gene CD36 via the PPARγ-responsive element in the CD36 promoter [74]. Release of endogenous PPAR ligands, particularly prostacyclin, is likely to be important in LSS-induced protective responses, and EC-derived prostacyclin has been shown to activate PPARα and δ in vascular smooth muscle cells [75]. PPARδ appears to play a multifunctional role in the vasculature, including increased fatty acid oxidation, protection against apoptosis and antioxidant, anti-inflammatory actions such as suppression of VCAM-1 [76]. These actions may contribute to the atheroprotective effect of PPARδ ligand treatment of $ApoE^{-/-}$ mice [77]. An additional important mechanism may be the ability of PPARδ ligands to upregulate HO-1 in vitro, a response that requires the co-activator PGC-1α and can be reproduced in vivo [78].

3.5 Therapeutic Manipulation of Vascular Cytoprotection

Therapeutic induction of cytoprotective genes in the vasculature has the potential to condition the vascular endothelium against injury, so minimizing or reversing endothelial dysfunction and preventing or slowing the progress of atherogenesis. This would be of particular benefit to patients known to be at particularly high risk, such as those with diabetes mellitus, hyperlipidemias, and systemic inflammatory diseases. Current immunosuppressive drugs may achieve this to some extent; however, therapies specifically targeting the vasculature are likely to be more effective.

Among these, biologic therapies such as those targeting TNFα are of particular interest, and of note, these agents appear to reduce the rate of atherogenesis and myocardial infarction in patients with RA [18].

3.5.1 Statins

Perhaps the best studied drugs in this regard are the statins, 3-hydroxy-3-methylglutaryl coenzyme A (HMG-CoA) reductase antagonists, which inhibit cholesterol synthesis and reduce serum LDL-cholesterol. This is turn reduces morbidity and mortality from ischemic heart disease. However, clinical trial data also demonstrates that the benefits of statins are rapid, extend to patients within the accepted normal LDL-cholesterol range [79] and exceed those of other lipid-lowering drugs, despite comparable falls in cholesterol [80]. These observations suggest that statins have pleiotropic effects above and beyond LDL-cholesterol lowering [81]. These cholesterol-independent actions of statins result in significant improvement in endothelial function in both hyper- and normocholesterolemic patients with atherosclerosis [82].

Statin-mediated inhibition of isoprenoid lipid production and subsequent protein prenylation and activity of signaling proteins such as the small GTPases underlie many of the LDL-cholesterol-independent actions [81, 82]. This mechanism, initially identified in vitro, has been supported by two recent studies in which a reduction in Rho-associated protein kinase (ROCK) activity and improved endothelial function was observed in patients treated with high-dose statins when compared to low-dose statins or ezetimibe (an alternative class of lipid-lowering agent) [83]. In vascular endothelium, statins increase eNOS mRNA stability and NO biosynthesis, leading to inhibition of leukocyte trafficking, an anti-inflammatory response that is lost in eNOS-deficient mice [84]. Statins also exert anti-thrombotic, antioxidant, and immunomodulatory effects in EC [81, 85–87]. We have identified an additional cytoprotective action of statins, the regulation of complement activation. At least in vitro, atorvastatin and simvastatin induce expression of DAF [88] and under hypoxic conditions, both DAF and CD59 [36] and we propose that this response may contribute to both their atheroprotective and anti-inflammatory actions.

3.5.2 Heme Oxygenase-1

There is considerable interest in the therapeutic potential of HO-1 either through modulation of its expression or delivery of its products [50]. However, such an approach is not straightforward in light of the potential toxicity of CO, free iron, and bilirubin. In vivo animal models are encouraging and HO-1 induction favors long-term allograft survival [89] and protects against atherosclerosis [90]. Exogenous CO may substitute for HO-1, conferring protection against ischemia reperfusion [91], restenosis injury, and allograft rejection [92]. Although less well studied, biliverdin and bilirubin exert similar effects [92]. In the vascular endothelium, we and others have demonstrated HO-1 induction in vitro following treatment with statins [63, 68, 93], celecoxib [94], rapamycin [95], and probucol [96]. PPARα, PPARγ, and PPARδ agonists have also been shown to induce HO-1 in vitro [78, 97]. Moreover, we have recently reported that treatment of mice with PPARδ agonists increases aortic EC expression of HO-1 [78]. Thus, the development of approaches through which HO-1 can be induced or its products delivered safely, and subsequent clinical trials investigating the efficacy of such an approach, is awaited with interest.

3.6 Conclusion

Inflammatory reactions within the vasculature are tightly regulated, with dysregulation increasingly recognized as a significant contributory feature in a variety of disease states including atherosclerosis and chronic inflammatory auto-immune diseases. Vascular endothelial cell injury predisposes to endothelial dysfunction, a critical precursor to atherogenesis, and a potential target for preventative therapy. Significant progress has been made in dissecting the molecular mechanisms through which the vascular endothelium is protected against injury, and over the next decade, it is hoped that these insights will reveal novel cytoprotective targets that can be therapeutically manipulated. This in turn may allow early intervention in those patients known to be at particularly high risk.

References

1. Bonetti PO, Lerman LO, Lerman A. Endothelial dysfunction: a marker of atherosclerotic risk. Arterioscler Thromb Vasc Biol. 2003;23(2):168–75.
2. Lerman A. Restenosis: another "dysfunction" of the endothelium. Circulation. 2005;111: 8–10.
3. Hansson GK. Inflammation, atherosclerosis, and coronary artery disease. N Engl J Med. 2005;352(16):1685–95.
4. Stoneman VE, Bennett MR. Role of apoptosis in atherosclerosis and its therapeutic implications. Clin Sci. 2004;107:343–54.
5. Tedgui A, Mallat Z. Anti-inflammatory mechanisms in the vascular wall. Circ Res. 2001;88: 877–87.
6. Sattar N, McCarey DW, Capell H, McInnes IB. Explaining how "high-grade" systemic inflammation accelerates vascular risk in rheumatoid arthritis. Circulation. 2003;108(24):2957–63.
7. del Rincon ID, Williams K, Stern MP, Freeman GL, Escalante A. High incidence of cardiovascular events in a rheumatoid arthritis cohort not explained by traditional cardiac risk factors. Arthritis Rheum. 2001;44(12):2737–45.
8. Esdaile JM, Abrahamowicz M, Grodzicky T, et al. Traditional Framingham risk factors fail to fully account for accelerated atherosclerosis in systemic lupus erythematosus. Arthritis Rheum. 2001;44:2331–7.
9. El-Magadmi M, Bodill H, Ahmad Y, et al. Systemic lupus erythematosus: an independent risk factor for endothelial dysfunction in women. Circulation. 2004;110(4):399–404.
10. Recio-Mayoral A, Mason JC, Kaski JC, Rubens MB, Harari OA, Camici PG. Chronic inflammation and coronary microvascular dysfunction in patients without risk factors for coronary artery disease. Eur Heart J. 2009;30:1837–43.
11. Keenan NG, Mason JC, Maceira A, et al. Integrated cardiac and vascular assessment in Takayasu arteritis by cardiovascular magnetic resonance. Arthritis Rheum. 2009;60(11):3501–9.

12. Choi HK, Hernan MA, Seeger JD, Robins JM, Wolfe F. Methotrexate and mortality in patients with rheumatoid arthritis: a prospective study. Lancet. 2002;359(9313):1173–7.
13. Romero F, Rodriguez-Iturbe B, Pons H, et al. Mycophenolate mofetil treatment reduces cholesterol-induced atherosclerosis in the rabbit. Atherosclerosis. 2000;152(1):127–33.
14. Dixon WG, Watson KD, Lunt M, Hyrich KL, Silman AJ, Symmons DP. Reduction in the incidence of myocardial infarction in patients with rheumatoid arthritis who respond to anti-tumor necrosis factor alpha therapy: results from the British Society for Rheumatology Biologics Register. Arthritis Rheum. 2007;56(9):2905–12.
15. Hermann F, Forster A, Chenevard R, et al. Simvastatin improves endothelial function in patients with rheumatoid arthritis. J Am Coll Cardiol. 2005;45:461–4.
16. Flammer AJ, Sudano I, Hermann F, et al. Angiotensin-converting enzyme inhibition improves vascular function in rheumatoid arthritis. Circulation. 2008;117:2262–9.
17. Hawkins PN, Lachmann HJ, McDermott MF. Interleukin-1-receptor antagonist in the Muckle-Wells syndrome. N Engl J Med. 2003;348(25):2583–4.
18. Feldmann M, Maini SR. Role of cytokines in rheumatoid arthritis: an education in pathophysiology and therapeutics. Immunol Rev. 2008;223:7–19.
19. Elhage R, Maret A, Pieraggi MT, Thiers JC, Arnal JF, Bayard F. Differential effects of interleukin-1 receptor antagonist and tumor necrosis factor binding protein on fatty-streak formation in apolipoprotein E-deficient mice. Circulation. 1998;97(3):242–4.
20. Francis SE, Camp NJ, Dewberry RM, et al. Interleukin-1 receptor antagonist gene polymorphism and coronary artery disease. Circulation. 1999;99(7):861–6.
21. Philippidis P, Mason JC, Evans BJ, et al. Hemoglobin scavenger receptor CD163 mediates interleukin-10 release and heme oxygenase-1 synthesis: antiinflammatory monocyte-macrophage responses in vitro, in resolving skin blisters in vivo, and after cardiopulmonary bypass surgery. Circ Res. 2004;94:119–26.
22. Henke PK, DeBrunye LA, Strieter RM, et al. Viral IL-10 gene transfer decreases inflammation and cell adhesion molecule expression in a rat model of venous thrombosis. J Immunol. 2000;164(4):2131–41.
23. Karsan A, Yee E, Poirier GG, Zhou P, Craig R, Harlan JM. Fibroblast growth factor-2 inhibits endothelial cell apoptosis by Bcl-2-dependent and independent mechanisms. Am J Pathol. 1997;151(6):1775–84.
24. Mason JC, Lidington EA, Ahmad SR, Haskard DO. bFGF and VEGF synergistically enhance endothelial cytoprotection via decay-accelerating factor upregulation. Am J Physiol Cell Physiol. 2002;282:C578–87.
25. Pendurthi UR, Williams JT, Rao LV. Acidic and basic fibroblast growth factors suppress transcriptional activation of tissue factor and other inflammatory genes in endothelial cells. Arterioscler Thromb Vasc Biol. 1997;17(5):940–6.
26. Zhang H, Issekutz AC. Down-modulation of monocyte transendothelial migration and endothelial adhesion molecule expression by fibroblast growth factor: reversal by the anti-angiogenic agent SU6668. Am J Pathol. 2002;160:2219–30.
27. Gaengel K, Genove G, Armulik A, Betsholtz C. Endothelial-mural cell signaling in vascular development and angiogenesis. Arterioscler Thromb Vasc Biol. 2009;29(5):630–8.
28. Brindle NP, Saharinen P, Alitalo K. Signaling and functions of angiopoietin-1 in vascular protection. Circ Res. 2006;98(8):1014–23.
29. Mason JC, Steinberg R, Lidington EA, Kinderlerer AR, Ohba M, Haskard DO. Decay-accelerating factor induction on vascular endothelium by VEGF is mediated via a VEGF-R2 and PKCα/ε-dependent cytoprotective signaling pathway and is inhibited by cyclosporin A. J Biol Chem. 2004;279:41611–8.
30. Steinberg R, Harari OA, Lidington EA, et al. A PKCε/Akt signalling complex protects human vascular endothelial cells against apoptosis through induction of Bcl-2. J Biol Chem. 2007;282:32288–97.
31. Vereecke L, Beyaert R, van Loo G. The ubiquitin-editing enzyme A20 (TNFAIP3) is a central regulator of immunopathology. Trends Immunol. 2009;30(8):383–91.
32. Lee EG, Boone DL, Chai S, et al. Failure to regulate TNF-induced NF-kappaB and cell death responses in A20-deficient mice. Science. 2000;289(5488):2350–4.

33. Youle RJ, Strasser A. The BCL-2 protein family: opposing activities that mediate cell death. Nat Rev Mol Cell Biol. 2008;9(1):47–59.
34. Navratil JS, Watkins SC, Wisnieski JJ, Ahearn JM. The globular heads of C1q specifically recognize surface blebs of apoptotic vascular endothelial cells. J Immunol. 2001;166:3231–9.
35. Mason JC, Yarwood H, Sugars K, Morgan BP, Davies KA, Haskard DO. Induction of decay-accelerating factor by cytokines or the membrane-attack complex protects vascular endothelial cells against complement deposition. Blood. 1999;94:1673–82.
36. Kinderlerer AR, Steinberg R, Johns M, et al. Statin-induced expression of CD59 on vascular endothelium in hypoxia. A potential mechanism for the anti-inflammatory actions of statins in rheumatoid arthritis. Arthritis Res Ther. 2006;8:R130–41.
37. Lidington EA, Haskard DO, Mason JC. Induction of decay-accelerating factor by thrombin through a protease-activated receptor1 and protein kinase C-dependent pathway protects vascular endothelial cells from complement-mediated injury. Blood. 2000;96:2784–92.
38. Ahmad SR, Lidington EA, Ohta R, et al. Decay-accelerating factor induction by TNFα through a phosphatidylinositol-3 kinase and protein kinase C-dependent pathway protects murine vascular endothelial cells against complement deposition. Immunology. 2003;110: 258–68.
39. Leung VWY, Yun S, Botto M, et al. Decay-accelerating factor suppresses complement C3 activation and retards atherosclerosis in low density lipoprotein receptor deficient mice. Am J Pathol. 2009;175:1757–67.
40. Yun S, Leung V, Botto M, Boyle J, Haskard D. Accelerated atherosclerosis in low-density lipoprotein receptor-deficient mice lacking the membrane-bound complement regulator CD59. Arterioscler Thromb Vasc Biol. 2008;28:1714–6.
41. Wu G, Hu W, Shahsafaei A, et al. Complement regulator CD59 protects against atherosclerosis by restricting the formation of complement membrane attack complex. Circ Res. 2009;104(4): 550–8.
42. Lee S, Chen TT, Barber CL, et al. Autocrine VEGF signaling is required for vascular homeostasis. Cell. 2007;130(4):691–703.
43. Olsson AK, Dimberg A, Kreuger J, Claesson-Welsh L. VEGF receptor signalling – in control of vascular function. Nat Rev Mol Cell Biol. 2006;7:359–71.
44. Gerber HP, Dixit V, Ferrara N. Vascular endothelial growth factor induces expression of the antiapoptotic proteins Bcl-2 and A1 in vascular endothelial cells. J Biol Chem. 1998;273: 13313–6.
45. He H, Venema VJ, Guo XL, Venema RC, Marrero MB, Caldwell RB. Vascular endothelial growth factor signals endothelial cell production of nitric oxide and prostacyclin through Flk-1/KDR activation of c-Src. J Biol Chem. 1999;274:25130–5.
46. Bussolati B, Ahmed A, Pemberton H, et al. Bifunctional role for VEGF-induced heme oxygenase-1 in vivo: induction of angiogenesis and inhibition of leukocytic infiltration. Blood. 2004;103:761–6.
47. Mason JC, Lidington EA, Yarwood H, Lublin DM, Haskard DO. Induction of endothelial cell decay-accelerating factor by vascular endothelial growth factor – a mechanism for cytoprotection against complement-mediated injury during inflammatory angiogenesis. Arthritis Rheum. 2001;44:138–50.
48. Eremina V, Jefferson JA, Kowalewska J, et al. VEGF inhibition and renal thrombotic microangiopathy. N Engl J Med. 2008;358:1129–36.
49. Hurwitz H, Saini S. Bevacizumab in the treatment of metastatic colorectal cancer: safety profile and management of adverse events. Semin Oncol. 2006;33(5 Suppl 10):S26–34.
50. Loboda A, Jazwa A, Grochot-Przeczek A, et al. Heme oxygenase-1 and the vascular bed: from molecular mechanisms to therapeutic opportunities. Antioxid Redox Signal. 2008;10: 1767–812.
51. Ryter SW, Alam J, Choi AM. Heme oxygenase-1/carbon monoxide: from basic science to therapeutic applications. Physiol Rev. 2006;86:583–650.
52. Yachie A, Niida Y, Wada T, et al. Oxidative stress causes enhanced endothelial cell injury in human heme oxygenase-1 deficiency. J Clin Invest. 1999;103(1):129–35.

53. Kinderlerer AR, Gregoire IP, Hamdulay SS, et al. Heme-oxygenase-1 expression enhances vascular endothelial resistance to complement-mediated injury through induction of decay-accelerating factor. A role for bilirubin and ferritin. Blood. 2009;113:1598–607.
54. Soares MP, Lin Y, Anrather J, et al. Expression of heme oxygenase-1 can determine cardiac xenograft survival. Nat Med. 1998;4(9):1073–7.
55. Wehner J, Morrell CN, Reynolds T, Rodriguez ER, Baldwin 3rd WM. Antibody and complement in transplant vasculopathy. Circ Res. 2007;100(2):191–203.
56. Soares MP, Bach FH. Heme oxygenase-1 in organ transplantation. Front Biosci. 2007;12:4932–45.
57. Berk BC. Atheroprotective signaling mechanisms activated by steady laminar flow in endothelial cells. Circulation. 2008;117(8):1082–9.
58. Dimmeler S, Hermann C, Galle J, Zeiher AM. Upregulation of superoxide dismutase and nitric oxide synthase mediates the apoptosis-suppressive effects of shear stress on endothelial cells. Arterioscler Thromb Vasc Biol. 1999;19:656–64.
59. Sheikh S, Rainger G, Gale Z, Rahman M, Nash G. Exposure to fluid shear stress modulates the ability of endothelial cells to recruit neutrophils in response to tumor necrosis factor -alpha: a basis for local variations in vascular sensitivity to inflammation. Blood. 2003;102:2828–34.
60. Dekker RJ, van Soest S, Fontijn RD, et al. Prolonged fluid shear stress induces a distinct set of endothelial cell genes, most specifically lung Kruppel-like factor (KLF2). Blood. 2002;100(5):1689–98.
61. SenBanerjee S, Lin Z, Atkins GB, et al. KLF2 is a novel transcriptional regulator of endothelial proinflammatory activation. J Exp Med. 2004;199:1305–15.
62. Parmar KM, Larman HB, Dai G, et al. Integration of flow-dependent endothelial phenotypes by Kruppel-like factor 2. J Clin Invest. 2006;116:49–58.
63. Ali F, Hamdulay SS, Kinderlerer AR, et al. Statin-mediated cytoprotection of human vascular endothelial cells: a role for Kruppel-like factor 2-dependent induction of heme oxygenase-1. J Thromb Haemost. 2007;5:2537–46.
64. Parmar KM, Nambudiri V, Dai G, Larman HB, Gimbrone Jr MA, Garcia-Cardena G. Statins exert endothelial atheroprotective effects via the KLF2 transcription factor. J Biol Chem. 2005;280(29):26714–9.
65. Hamik A, Lin Z, Kumar A, et al. Kruppel-like factor 4 regulates endothelial inflammation. J Biol Chem. 2007;282(18):13769–79.
66. Villarreal Jr G, Zhang Y, Larman HB, Gracia-Sancho J, Koo A, Garcia-Cardena G. Defining the regulation of KLF4 expression and its downstream transcriptional targets in vascular endothelial cells. Biochem Biophys Res Commun. 2010;391(1):984–9.
67. Dai G, Vaughn S, Zhang Y, Wang ET, Garcia-Cardena G, Gimbrone Jr MA. Biomechanical forces in atherosclerosis-resistant vascular regions regulate endothelial redox balance via phosphoinositol 3-kinase/Akt-dependent activation of Nrf2. Circ Res. 2007;101(7):723–33.
68. Ali F, Zakkar M, Karu K, et al. Induction of the cytoprotective enzyme heme oxygenase-1 by statins is enhanced in vascular endothelium exposed to laminar shear stress. J Biol Chem. 2009;284:18882–92.
69. Kinderlerer AR, Ali F, Johns M, et al. KLF-2-dependent, shear stress-induced expression of CD59: a novel cytoprotective mechanism against complement-mediated injury in the vasculature. J Biol Chem. 2008;283:14636–44.
70. Zakkar M, Van der Heiden K, Luong LA, et al. Activation of Nrf2 in endothelial cells protects arteries from exhibiting a proinflammatory state. Arterioscler Thromb Vasc Biol. 2009;29: 1851–7.
71. Brown JD, Plutzky J. Peroxisome proliferator-activated receptors as transcriptional nodal points and therapeutic targets. Circulation. 2007;115(4):518–33.
72. Riserus U, Sprecher D, Johnson T, et al. Activation of peroxisome proliferator-activated receptor (PPAR)δ promotes reversal of multiple metabolic abnormalities, reduces oxidative stress, and increases fatty acid oxidation in moderately obese men. Diabetes. 2008;57(2):332–9.
73. Hamblin M, Chang L, Fan Y, Zhang J, Chen YE. PPARs and the cardiovascular system. Antioxid Redox Signal. 2009;11(6):1415–52.

74. Liu Y, Zhu Y, Rannou F, et al. Laminar flow activates peroxisome proliferator-activated receptor-gamma in vascular endothelial cells. Circulation. 2004;110(9):1128–33.
75. Tsai MC, Chen L, Zhou J, et al. Shear stress induces synthetic-to-contractile phenotypic modulation in smooth muscle cells via peroxisome proliferator-activated receptor alpha/delta activations by prostacyclin released by sheared endothelial cells. Circ Res. 2009;105(5):471–80.
76. Fan Y, Wang Y, Tang Z, et al. Suppression of pro-inflammatory adhesion molecules by PPAR-δ in human vascular endothelial cells. Arterioscler Thromb Vasc Biol. 2008;28:315–21.
77. Barish GD, Atkins AR, Downes M, et al. PPARδ regulates multiple proinflammatory pathways to suppress atherosclerosis. Proc Natl Acad Sci USA. 2008;105(11):4271–6.
78. Ali F, Ali NS, Bauer A, et al. PPARδ and PGC1α act cooperatively to induce haem oxygenase-1 and enhance vascular endothelial cell resistance to stress. Cardiovasc Res. 2010;85(4): 701–10.
79. Group HPCS. MRC/BHF heart protection study of cholesterol lowering with simvastatin in 20536 high-risk individuals: a randomised placebo-controlled trial. Lancet. 2002;360:7–22.
80. Landmesser U, Bahlmann F, Mueller M, et al. Simvastatin versus ezetimibe: pleiotropic and lipid-lowering effects on endothelial function in humans. Circulation. 2005;111:2356–63.
81. Greenwood J, Mason JC. Statins and the vascular endothelial inflammatory response. Trends Immunol. 2007;28:88–98.
82. Mason JC. Statins and their role in vascular protection. Clin Sci (Lond). 2003;105:251–66.
83. Liu PY, Liu YW, Lin LJ, Chen JH, Liao JK. Evidence for statin pleiotropy in humans: differential effects of statins and ezetimibe on rho-associated coiled-coil containing protein kinase activity, endothelial function, and inflammation. Circulation. 2009;119(1):131–8.
84. Endres M, Laufs U, Huang Z, et al. Stroke protection by 3-hydroxy-3-methylglutaryl (HMG)-CoA reductase inhibitors mediated by endothelial nitric oxide synthase. PNAS. 1998;95(15): 8880–5.
85. Bourcier T, Libby P. HMG CoA reductase inhibitors reduce plasminogen activator inhibitor-1 expression by human vascular smooth muscle and endothelial cells. Arterioscler Thromb Vasc Biol. 2000;20(2):556–62.
86. Kwak B, Mulhaupt F, Myit S, Mach F. Statins as a newly recognized type of immunomodulator. Nat Med. 2000;6(12):1399–402.
87. Wagner AH, Kohler T, Ruckschloss U, Just I, Hecker M. Improvement of nitric oxide-dependent vasodilatation by HMG-CoA reductase inhibitors through attenuation of endothelial superoxide anion formation. Arterioscler Thromb Vasc Biol. 2000;20(1):61–9.
88. Mason JC, Ahmed Z, Mankoff R, et al. Statin-induced expression of decay-accelerating factor protects vascular endothelium against complement-mediated injury. Circ Res. 2002;91: 696–703.
89. Hancock WW, Buelow R, Sayegh MH, Turka LA. Antibody-induced transplant arteriosclerosis is prevented by graft expression of anti-oxidant and anti-apoptotic genes. Nat Med. 1998;4(12):1392–6.
90. Cheng C, Noordeloos AM, Jeney V, et al. Heme oxygenase-1 determines atherosclerotic lesion progression into a vulnerable plaque. Circulation. 2009;119(23):3017–27.
91. Fujita T, Toda K, Karimova A, et al. Paradoxical rescue from ischemic lung injury by inhaled carbon monoxide driven by derepression of fibrinolysis. Nat Med. 2001;7(5):598–604.
92. Otterbein LE, Zuckerbraun BS, Haga M, et al. Carbon monoxide suppresses arteriosclerotic lesions associated with chronic graft rejection and with balloon injury. Nat Med. 2003;9(2): 183–90.
93. Lee TS, Chang CC, Zhu Y, Shyy JY. Simvastatin induces heme oxygenase-1: a novel mechanism of vessel protection. Circulation. 2004;110(10):1296–302.
94. Hamdulay SS, Wang B, Birdsey GM, et al. Celecoxib activates PI-3 K/Akt and mitochondrial redox signaling to enhance heme oxygenase-1-mediated anti-inflammatory activity in vascular endothelium. Free Radic Biol Med. 2010;48(8):1013–23. 2010 Jan 18.
95. Visner GA, Lu F, Zhou H, Liu J, Kazemfar K, Agarwal A. Rapamycin induces heme oxygenase-1 in human pulmonary vascular cells: implications in the antiproliferative response to rapamycin. Circulation. 2003;107:911–6.

96. Wu BJ, Kathir K, Witting PK, et al. Antioxidants protect from atherosclerosis by a heme oxygenase-1 pathway that is independent of free radical scavenging. J Exp Med. 2006;203: 1117–27.
97. Kronke G, Kadl A, Ikonomu E, et al. Expression of heme oxygenase-1 in human vascular cells is regulated by peroxisome proliferator-activated receptors. Arterioscler Thromb Vasc Biol. 2007;27:1276–82.

4 Notch Signaling in Vascular Development

Shalini Jadeja and Marcus Fruttiger

4.1 Introduction

During vascular development, numerous cell fate decisions must occur. The hematopoietic, endothelial, and mural cell lineages are all derived from the mesoderm, and during early embryonic development precursor cells must "decide" which of the three lineages to enter. Also, within each of these lineages, more decision making is needed to generate additional sub-specification of cells. For instance, the vascular tree is split into different caliber vessels, arteries, veins, and capillaries; therefore, endothelial cells building this complex network are far from a uniform cell population. Instead they differentiate into vessel-specific and even tissue-specific phenotypes. The main function of the Notch signaling pathway is to generate cell diversity by mediating cell fate decision, and it is therefore no surprise that this signaling pathway participates critically on many levels throughout vascular development (Fig. 4.1).

4.1.1 Development of Hematopoietic and Vascular Cells

During early embryonic development, mesodermal progenitors give rise to blood cells and primitive vascular networks. In amniotes, this occurs in two areas: in the embryo proper and extra-embryonically, in the yolk sac. Fish and amphibians do not

S. Jadeja
Medical and Developmental Genetics, MRC Human Genetics Unit,
Western General Hospital, Edinburgh, UK
e-mail: shalini.jadeja@hgu.mrc.ac.uk

M. Fruttiger (✉)
Cell Biology, UCL Institute of Ophthalmology,
London, UK
e-mail: m.fruttiger@ucl.ac.uk

D. Abraham et al. (eds.), *Translational Vascular Medicine*,
DOI 10.1007/978-0-85729-920-8_4,

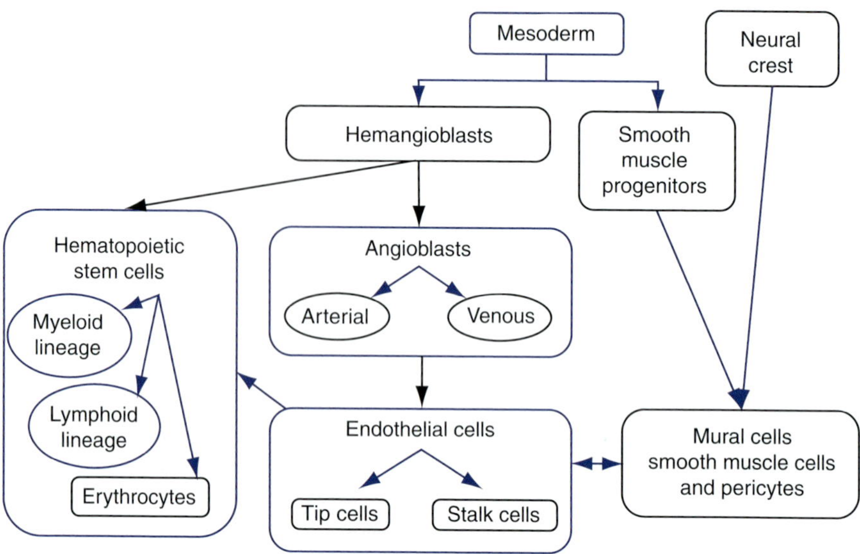

Fig. 4.1 Notch signaling is implicated extensively in hematopoietic and vascular development, from the early mesoderm progenitor stage through to endothelial cell differentiation. *Blue coloring* indicates Notch signaling involvement

have a yolk sac and therefore have no extra-embryonic source of vessels and blood. However, the precise lineage relationship of blood cells, vascular cells, and their mesodermal precursors is complex and only partially understood. During gastrulation, mesoderm cells emerge from the posterior primitive streak and migrate to the proximal region of the yolk sac [1, 2]. There they give rise to so-called blood islands containing hematopoietic and endothelial precursors. In mice by circa embryonic day (E) 7.5, the first blood islands can be detected in the yolk sac [3]. These morphologically distinct cell clusters segregate into blood cells and ensheathing endothelial cells, then remodel into smaller channels and eventually into blood filled vascular networks [4]. Due to the close spatiotemporal relationship between hematopoietic and endothelial precursor cells, the existence of a common mesodermal progenitor (the hemangioblast) has been proposed [5, 6]. Indeed, detailed mapping studies have shown that early mesodermal precursors are already committed to the hemangioblast lineage when they are still in the primitive streak [7]. Furthermore, genetic deletion of the Vascular endothelial growth factor receptor 2 (*Vegfr2*) in mice results in a lack of hematopoietic and endothelial cells and blood islands do not form [8]. However, it has also been suggested that, as the mesodermal precursors emerge from the primitive streak, allocation to the hematopoietic lineage may occur before, and independently of, the bulk of vascular commitment [9].

Furthermore, mesodermal precursor cells not only give rise to the blood and endothelial lineage, but they also generate smooth muscle cells; the cells that participate in building the vascular wall (therefore referred to as "mural" cells). *In vitro* experiments have shown that cell colonies cultured from the primitive streak have hematopoietic, endothelial, and smooth muscle potential [7]. It is also possible to generate endothelial and smooth muscle cells from embryonic stem cell–derived cells that are Vegfr2-positive [10, 11]. In addition, embryonic stem cell–derived embryoid bodies contain blast colony–forming cells (BL-CFC) that have the potential to generate hematopoietic, endothelial, and smooth muscle cells depending on culture conditions [5, 12]. Alternatively, the three lineages may simply be derivatives of ventral mesoderm that can give rise to a broader array of cell types [13]. It is therefore not entirely clear yet whether the three cell types are derived from a single common precursor or even whether the hemangioblast exists *in vivo*.

Once endothelial precursor cells (also known as angioblasts) have been generated, they migrate, coalesce, and differentiate into endothelial cells, which form a primitive vascular plexus de novo. This is the classic definition of "vasculogenesis" [14], the process responsible for forming blood vessels in the yolk sac and, intraembryonically, the endocardial tube, dorsal aortae, and cardinal veins. Subsequently, the primitive vessel networks recruit mural cells, remodel, and form further vessels in a process termed "angiogenesis" [15, 16].

4.2 Notch Signaling Mechanisms

In mammals, there are four Notch receptors (Notch1-4) and two types of ligands: The Delta-like ligands (Dll1, Dll2 and Dll4) and the Jagged ligands (Jag1 and Jag2). Flies only have one Notch receptor and two ligands, Delta and Serrate (homologue to Jagged). Receptors and ligands are both transmembrane proteins and as a result Notch signaling is mediated between neighboring cells. Ligand–receptor binding induces proteolytic receptor cleavage, first by the Adam metalloproteases, and then for a second time by a γ-secretase complex (containing Presenilin) to release the Notch intracellular domain (NICD), which then translocates to the nucleus. There it forms a complex with the transcription factor "Recombination signal binding protein for immunoglobulin kappa J region" (Rbpj, also known as CSL) and relieves the repression of downstream target genes. This mechanism is known as the "canonical" Notch pathway (reviewed by Kopan and Ilagan [17]). Non-canonical Notch signaling has also been observed [18, 19], but is not discussed here.

Despite the pleiotropic function of Notch signaling in numerous cell types in all metazoa, to date, only a few downstream target genes have been identified. The best characterized target genes are basic Helix-loop-helix (bHLH) transcription factors from the *Hes* and *Hey* gene families (in vertebrates) and the related *Hairy* and *E(spl)* genes (in Drosophila). In classic examples, in flies, it has been shown that these transcription factors are part of a feedback mechanism that allows initially identical (or very similar) neighboring cells to take on different identities [20, 21]. Notch signaling–mediated activation of the bHLH transcription factors causes increased expression of Notch receptor and decreased expression of Notch ligand. In the absence of Notch signaling, the opposite occurs; an increase of the ligand and a decrease of the receptor. Such a bi-stable system amplifies small, initial differences and simultaneously forces neighboring cells into opposite states regarding Notch expression. In this model, referred to as the "lateral inhibition model," a cell that expresses Notch ligand usually suppresses a particular differentiation outcome in its neighbors. However, some Notch signaling is also mediated by "lateral induction," where a ligand-expressing cell induces a specific cell fate in its neighbors [21, 22].

Superficially, the mode of action of the canonical Notch pathway appears relatively simple, but there is a remarkable array of posttranslational and biological processes that modulate signaling strength and add significant complexity to the system (reviewed by [22–24]). For instance, ligand expression in receptor positive cells may have inhibitory function under certain circumstances (so-called cis-inhibition). Furthermore, in Drosophila, it has been shown that ubiquitination and subsequent endocytosis of Notch ligands is essential for their activity [23]. Similarly, Notch receptor function also critically depends on posttranslational modifications. Glycosylation of Notch receptors is initiated by *O*-fucosyl transferase by adding a fucose molecule. The carbohydrate chains are then extended by glycosyl transferases such as the Fringe family. This can modify the responsiveness of Notch receptors to specific ligands. For instance, Fringe-modified Notch becomes more responsive to Delta-like/Delta and less responsive to Jagged/Serrate ligands [25, 26]. It has also been shown that receptor endocytosis and subsequent trafficking can influence Notch activity. In addition, the multiple proteolytic cleavages and the various proteases involved to generate the NICD add more possibilities to regulate Notch signaling and complicate things further. The bewildering complexity of these regulatory mechanisms seems to suggest that Notch signaling is so fundamental and important for biological function that it requires sophisticated and tight regulation.

4.3 Vascular Notch Expression and Knockout (KO) Mice

Most Notch receptors and their ligands are expressed in the developing vasculature (Table 4.1), whereas in the adult vasculature, expression becomes more restricted. It is remarkable that in the developing mammalian vascular system, with the exception of Notch 2 and Dll3, all Notch receptors and ligands are expressed. Interestingly, endothelial cells are usually not polarized into receptor- and ligand-expressing cells (as during lateral inhibition) but often express Notch receptors and ligands simultaneously. In line with the prominent expression of Notch genes in the vasculature, most Notch gene deletions tend to cause embryonic lethality due to disturbed vascular and cardiac development (Table 4.1).

4.3.1 Notch Receptors

Notch1 and 4 are expressed in endothelial cells. While Notch1 expression is widespread in numerous other cells types, Notch4 is largely restricted to the vascular endothelium [27, 28]. Genetic deletion of *Notch1* in mice is lethal by embryonic day (E)10.5. In these mice, development proceeds normally until E9.5 but subsequently somite condensation fails and cell death is apparent in the nervous

Table 4.1 Expression of Notch signaling mechanisms and mouse knockouts

Gene	Expression	Knockout phenotype	Reference
Notch1	Endothelial cells (widespread in many other tissues)	Lethal by E10.5 Defective remodeling of the vasculature	[28]
Notch3	Smooth muscle cells	Viable Artery differentiation defects in adults	[37, 38]
Notch4	Endothelial cells	Viable	[30]
Dll1	Endothelial cells (arterial)	Lethal at E12 Segmentation defects	[40]
Dll4	Endothelial cells	Embryonic lethal at E10.5 Vascular development defects	[43–45]
Jag1	Endothelial cells and smooth muscle cells	Lethal between E10.5 & E11.5 Defects in remodeling yolk sac and embryonic vasculature	[51]
Jag2	Endothelial cells and hematopoietic precursors	Perinatal lethal Craniofacial malformations	[53]

systems. Although embryonic lethality was not attributed to primary defects in the vasculature, the vessels that did form were anastomosing [28]. Endothelial specific loss of *Notch1* results in embryonic death by E10.5, with severe vascular defects [29].

In contrast, *Notch4* KO mice are viable and fertile [30]. It is possible that Notch1 can compensate for the loss of Notch4 because compound knockouts for both *Notch1* and *Notch4* exhibit a more severe phenotype than the single *Notch1* KO embryos. In these *Notch1/4* double KO embryos, vascular remodeling and sprouting is disturbed, as in *Notch1* KO mutants; but whereas in the *Notch1* KO mutants, the dorsal aorta is collapsed with a closed lumen, in the *Notch1/4* double mutants, both the dorsal aorta and the anterior cardinal vein are collapsed [30]. Notch signaling is usually activated transiently with a high signal turnover. Switching on the expression of constitutively active *Notch4* in endothelial cells during early development also leads to severely abnormal vascular remodeling [31]. Overactive *Notch4* postnatally causes brain arteriovenous malformations resulting in hemorrhage, neurological damage, and perinatal death [32]. In summary, it appears that both under- and overactive Notch signaling can both severely disturb vascular remodeling [33] and, although Notch4 is redundant during embryonic development, it may still play a role in this process.

Notch3 is expressed predominantly in vascular smooth muscle cells late in development and in adults, with particular high levels in arteries [34–36]. Notch3 function is not needed for viability and fertility in mice [37], but adult *Notch3* KO mice display structural defects in arteries [38]. Vascular smooth muscle cells are recruited to the arteries in these KO mice but display abnormal differentiation and morphology, resulting in a thinner coating of the arteries with smooth muscle cells.

4.3.2 Delta-Like Ligands

Dll1 is expressed in many tissues including the endothelium of major blood vessels during late development [39]; in the postnatal vasculature, its expression is limited to endothelial cells of arteries [40]. Dll4, on the other hand, is relatively vascular endothelium specific (apart from a few exceptions). It is most strongly expressed in developing arteries and is downregulated in mature vessels [30, 34, 41]. Both, *Dll1* and *Dll4* KO mice are embryonic lethal, due to severe hemorrhages, although the *Dll1* KOs also suffer from abnormal somite formation [42]. However, while heterozygous *Dll1* deletion results in only a minor vascular phenotype (impaired arteriogenesis in adults), heterozygous *Dll4* KO mutants also died in early embryogenesis [43–45]. Although, subsequent studies have shown that in certain genetic backgrounds heterozygous Dll4 KO mutants are viable and fertile [46–48].

4.3.3 Jagged Ligands

Jagged1 is found in blood vessels throughout development into adulthood, where it is expressed by endothelial cells, vascular smooth muscle cells, and other cell types (e.g., neurons in certain brain nuclei) [49, 50]. Null mutants exhibit early embryonic lethality due to hemorrhaging and defects in remodeling of the embryonic and yolk sac vasculature [51]. Endothelial cell–specific knockouts of *Jag1* exhibit the same phenotype with initial vascular patterning unperturbed, but remodeling of the blood vessels and vascular smooth muscle development affected [52]. In contrast to Jagged1, Jagged2 is expressed in virtually all postnatal neurons but appears only transiently in the developing vasculature [49]. Perhaps unsurprisingly, *Jag2* null mutants die perinatally with severe craniofacial defects, but they do not display vascular defects [53], suggesting that Jagged2 function in the vasculature is redundant.

4.4 Notch Signaling in Hematopoiesis

In zebrafish, it has been shown that Notch signaling influences the balance between the endothelia and hematopoietic lineage [54]. In mammals and birds, the situation is more complicated because in these animals, the first wave of hematopoiesis occurs in the yolk sac (termed "primitive hematopoiesis") and only generates primitive erythrocytes and some macrophage progenitors [3, 55]. Hematopoiesis then shifts to intra-embryonic sites such as the aorta-gonad mesonephros (AGM) region where hematopoietic stem cells (HSC) bud off from the ventral wall of the dorsal aorta [56–58], giving rise to all hematopoietic lineages (termed "definite hematopoiesis"). In contrast to extra-embryonic hematopoiesis, where endothelial and blood cells may have a common mesodermal precursor, in the AGM the hematopoietic lineage is derived from endothelial cells. Deleting *Notch1* or *Rbpj* in mice severely impairs intra-embryonic hematopoiesis but, interestingly, has no effect on yolk sac hematopoiesis [59, 60]. Furthermore, deletion of *Jag1* (but not *Jag2*) also disrupts the generation of HSCs in the AGM [61], demonstrating that Jag1–Notch1 interactions are required for definitive hematopoiesis. Notch signaling also plays a major role in the immune system, where it is critically involved in T and B cell development and cell fate decisions in the myeloid lineage, which is reviewed elsewhere [62, 63].

4.5 Notch Signaling in Vascular Wall Development

As discussed previously, early ventral mesoderm does not only generate hematopoietic and endothelial precursors, but is also a source of vascular smooth muscle cells [64]. Notch signaling has been implicated by several studies in the specification of the vascular smooth muscle lineage. In vivo electroporation of chicken ventral mesoderm cells with constitutively active Notch1 led to a strong bias toward smooth muscle cell differentiation, whereas application of a Notch inhibitor N-[N-(3,5-Difluorophenacetyl)-L-alanyl]-S-phenylglycine t-butyl ester (DAPT) skewed the balance toward the hematopoietic/vascular lineage [65]. Similarly, constitutively active Notch1 favoured the generation of mural cells from cultured embryonic stem cells [66]. This suggests that Notch signaling pushes mesodermal precursor cells toward the mural cell lineage.

Notch signaling has also been implicated in smooth muscle cell differentiation, which occurs later, once mural precursors have been generated, and when they start to invest primitive endothelial networks. At this stage, bi-directional signaling between mural and endothelial cells mediates vascular remodeling and initiates vascular network maturation. Signaling through Notch3 (among other signaling systems) is part of this cross-talk. This is demonstrated by the abnormal artery differentiation in *Notch3* KO mice (see above).

Similarly, human mutations in *Notch3* cause cerebral arteriopathy with subcortical infarcts and leukoencephalopathy (CADASIL), which is an autosomal dominant disorder of small arterial vessels in the brain [67].

Jagged1 is a likely ligand for Notch3 signaling during artery differentiation and maturation, as endothelial-specific deletion of *Jag1* in mice causes deficits in vascular smooth muscle cell differentiation [52]. Co-culture experiments have also shown that endothelial cells can induce and activate Notch3 via a positive feedback loop that includes endothelial-derived Jagged1 [68]. The role of this pathway in mural cell differentiation is further supported by the observation that in vitro stimulation of a mesenchymal cell line (C3H10T1/2) with Jagged1 (but not with Dll4) induced multiple smooth muscle marker genes [69]. However, how the role of Notch in mural cell differentiation relates to its earlier function in mural cell generation is not fully understood yet.

4.6 Notch Signaling in Vasculogenesis

During early embryogenesis, the dorsal aortae and cardinal veins are formed de novo from migrating and coalescing angioblasts [14]. Although arteries and veins are structurally and functionally distinct, for many years, it was thought that this distinction was due to differences in blood flow and pressure. However, it is now clear that endothelial cells start to express either arterial or venous markers before the onset of circulation, suggesting a genetic influence [70].

Lineage tracing in zebrafish has shown that individual angioblasts may already be restricted to an arterial or venous fate before the first embryonic vessels are fully established [71]. Angioblasts migrating toward the midline contribute either to the dorsal aorta or the cardinal vein. Exposure of these cells to Vascular endothelial growth factor (Vegf) activates Notch signaling and arterial specification [72]. Loss of Notch signaling leads to reduced arterial markers and ectopic expression of venous markers in the dorsal aorta [73]. In addition, the dorsal aorta fails to form in fish that lack the bHLH transcription factor gridlock (*grl*), the zebrafish orthologue of the mammalian *Hey2* and a downstream target of Notch signaling [71, 74].

Between fish and mammals, the role of Notch signaling in vasculogenesis is remarkably well conserved. Transgenic overexpression of *Dll4* in mouse embryos leads to grossly enlarged dorsal aortae and embryonic lethality before E10.5 [75]. Conversely, mice lacking *Dll4* display reduced dorsal aorta calibers and increased endothelial cell migration away from the dorsal aorta [76]. Similarly, experiments using constitutively active *Notch4* have shown that overactive Notch signaling results in enlarged dorsal aortae and underdeveloped cardinal veins, whereas the opposite (small aortae and enlarged veins) was found in endothelial-specific *Notch1* KO mice [77].

Interestingly, this study also found that the overall number of endothelial cells was unchanged by the manipulation of Notch signaling, suggesting that Notch signaling reciprocally balances the size of the dorsal aortae and cardinal veins by modulating how angioblasts are allocated to either the developing aortae or the veins.

Dll4 and Jag1 are both expressed during early dorsal aorta vasculogenesis [39–41], but it appears that only Dll4 is essential for early arterial cell fate specification. In *Jag1* KO mice, the primitive vascular plexus is initially established but then fails to remodel properly [51], suggesting a non-essential role of Jag1 in vasculogenesis. Dll1 is not expressed in the vasculature until after the first blood vessels are formed and therefore also not required for vasculogenesis [78]. However, Dll1 is needed to stabilize arteries after they have formed and, in comparison to Dll4, seems to act at a later stage of artery differentiation [40, 78].

4.7 Notch Signaling in Angiogenesis

After the first blood vessels have been generated, the vascular system is further expanded by angiogenesis (vascular growth from existing vessels). For instance, the developing retinal vasculature in mice is formed by angiogenesis [79]. The mouse retina is avascular at birth and its vasculature develops in the first 3 weeks postnatally. During the first postnatal week, a primary vascular plexus emerges from the optic nerve head and uses a template of retinal astrocytes as a substrate to spread across the inner surface of the retina. Angiogenic sprouting activity occurs at the growing edge, and vascular remodeling and differentiation can be observed more centrally.

Furthermore, the plexus consists of radially alternating and easily identifiable arteries and veins. In the second week, a secondary, deeper plexus sprouts from the superficial, primary network into the retina; and in the third week, vessels fully mature. Because of its stereotypical development and its 2-dimensional topology in the first week, the developing mouse retinal vasculature has become a popular model system to study sprouting angiogenesis [80].

In particular, the so-called tip cells at the sprouting edge can easily be identified in the growing retinal vasculature. These specialized endothelial cells at the leading tip of angiogenic sprouts have pronounced filopodia and respond to angiogenic growth factors such as Vegf by migration. The endothelial cells that follow the tip cells are called "stalk cells" and proliferate in response to Vegf [81]. Endothelial cells can switch rapidly between the tip or stalk state and compete between each other for tip cell status [82]. This competition is mediated by a transcriptional feedback loop that is based on lateral inhibition via Dll4-Notch signaling and regulates the sensitivity of endothelial cells toward Vegf. High levels of Vegf in tip cells promote the expression of Dll4 and Vegfr2. The strong Dll4 expression in tip cells then activates Notch signaling and suppresses Dll4 and Vegfr2 transcription in adjacent stalk cells.

However, if stalk cells are exposed to high Vegf concentrations, they can upregulate Dll4 and Vegfr2 and turn themselves into tip cells [82]. Inhibition of Dll4-Notch signaling by chemical or genetic manipulation in mice or zebrafish disturbs the balance between tip and stalk cells and results in more tip cells and excessive vascular branching [46–48, 83, 84]. Interestingly, there is also a Notch ligand that can inhibit this signaling axis. This is based on the fact that glycosylation of Notch receptors leads to stronger binding of Delta-like ligands versus Jagged ligands. Experiments studying the retinal vasculature of *Jag1* KO mice have shown that Jag1 can compete with Dll4 for receptor binding/activation and therefore act as an endogenous Dll4 antagonist [85]. There are also other signaling pathways that can regulate Notch activity, such as the TGFβ or Wnt signaling pathways [86, 87]. How these signaling pathways precisely interact with Notch signaling is one of the current challenges in vascular biology.

4.8 Conclusions

In summary, Notch signaling plays a critical role at several, distinct stages of vascular development. Because a properly functioning vascular system is strictly required for the survival of embryos once they grow beyond the size of 1–2 mm, complete loss of function mutations in the Notch signaling pathway is therefore often embryonic lethal. This might explain, at least in part, why as yet only two genetic Notch mutations have been characterized in humans, the Alagille Syndrome (*JAG1*) and CADASIL (*NOTCH3*) [67, 88]. The fact that they are both autosomal dominant conditions further highlights the importance of Notch signaling in development. Because Notch signaling is used at multiple times throughout vascular development, it is usually difficult to interpret phenotypes in KO mice. Sophisticated model systems that allow for conditional mutations in a cell- and time-specific manner will therefore play a particularly important role in future research of Notch signaling in the vascular system.

References

1. Kinder SJ, Tsang TE, Wakamiya M, et al. The organizer of the mouse gastrula is composed of a dynamic population of progenitor cells for the axial mesoderm. Development. 2001; 128(18):3623–34.
2. Lawson KA, Meneses JJ, Pedersen RA. Clonal analysis of epiblast fate during germ layer formation in the mouse embryo. Development. 1991;113(3):891–911.
3. Baron MH. Embryonic origins of mammalian hematopoiesis. Exp Hematol. 2003;31(12):1160–9.
4. Ferkowicz MJ, Starr M, Xie X, et al. CD41 expression defines the onset of primitive and definitive hematopoiesis in the murine embryo. Development. 2003;130(18):4393–403.
5. Choi K, Kennedy M, Kazarov A, Papadimitriou JC, Keller G. A common precursor for hematopoietic and endothelial cells. Development. 1998;125(4):725–32.
6. Sabin F. Studies on the origin of blood vessels and of red blood corpuscles as seen in the living blastoderm of chicks during the second day of incubation. Contrib Embryol Carnegie Inst Wash. 1920;9:214–62.
7. Huber TL, Kouskoff V, Fehling HJ, Palis J, Keller G. Haemangioblast commitment is initiated in the primitive streak of the mouse embryo. Nature. 2004;432(7017):625–30.
8. Shalaby F, Rossant J, Yamaguchi TP, et al. Failure of blood-island formation and vasculogenesis in Flk-1-deficient mice. Nature. 1995;376(6535):62–6.
9. Kinder SJ, Tsang TE, Quinlan GA, Hadjantonakis AK, Nagy A, Tam PP. The orderly allocation of mesodermal cells to the extraembryonic structures and the anteroposterior axis during gastrulation of the mouse embryo. Development. 1999;126(21):4691–701.
10. Kattman SJ, Huber TL, Keller GM. Multipotent flk-1+ cardiovascular progenitor cells give rise to the cardiomyocyte, endothelial, and vascular smooth muscle lineages. Dev Cell. 2006;11(5):723–32.
11. Yamashita J, Itoh H, Hirashima M, et al. Flk1-positive cells derived from embryonic stem cells serve as vascular progenitors. Nature. 2000;408(6808):92–6.
12. Ema M, Rossant J. Cell fate decisions in early blood vessel formation. Trends Cardiovasc Med. 2003;13(6):254–9.
13. Furuta C, Ema H, Takayanagi S, et al. Discordant developmental waves of angioblasts and hemangioblasts in the early gastrulating mouse embryo. Development. 2006;133(14):2771–9.
14. Risau W, Flamme I. Vasculogenesis. Annu Rev Cell Dev Biol. 1995;11:73–91.
15. Poole TJ, Coffin JD. Vasculogenesis and angiogenesis: two distinct morphogenetic mechanisms establish embryonic vascular pattern. J Exp Zool. 1989;251(2):224–31.
16. Risau W. Mechanisms of angiogenesis. Nature. 1997;386(6626):671–4.
17. Kopan R, Ilagan MX. The canonical Notch signaling pathway: unfolding the activation mechanism. Cell. 2009;137(2):216–33.
18. Talora C, Campese AF, Bellavia D, et al. Notch signaling and diseases: an evolutionary journey from a simple beginning to complex outcomes. Biochim Biophys Acta. 2008;1782(9):489–97.
19. Sanalkumar R, Dhanesh SB, James J. Non-canonical activation of Notch signaling/target genes in vertebrates. Cell Mol Life Sci. 2010;67(17):2957–68.
20. Artavanis-Tsakonas S, Rand MD, Lake RJ. Notch signaling: cell fate control and signal integration in development. Science. 1999;284(5415):770–6.
21. Bray S. Notch signalling in Drosophila: three ways to use a pathway. Semin Cell Dev Biol. 1998;9(6):591–7.
22. Bray SJ. Notch signalling: a simple pathway becomes complex. Nat Rev Mol Cell Biol. 2006;7(9):678–89.

23. Fortini ME. Notch signaling: the core pathway and its posttranslational regulation. Dev Cell. 2009;16(5):633–47.
24. D'Souza B, Miyamoto A, Weinmaster G. The many facets of Notch ligands. Oncogene. 2008;27(38):5148–67.
25. Yang LT, Nichols JT, Yao C, Manilay JO, Robey EA, Weinmaster G. Fringe glycosyltransferases differentially modulate Notch1 proteolysis induced by Delta1 and Jagged1. Mol Biol Cell. 2005;16(2):927–42.
26. Haines N, Irvine KD. Glycosylation regulates Notch signalling. Nat Rev Mol Cell Biol. 2003;4(10):786–97.
27. Uyttendaele H, Marazzi G, Wu G, Yan Q, Sassoon D, Kitajewski J. Notch4/int-3, a mammary proto-oncogene, is an endothelial cell-specific mammalian Notch gene. Development. 1996;122(7):2251–9.
28. Swiatek PJ, Lindsell CE, del Amo FF, Weinmaster G, Gridley T. Notch1 is essential for postimplantation development in mice. Genes Dev. 1994;8(6):707–19.
29. Limbourg FP, Takeshita K, Radtke F, Bronson RT, Chin MT, Liao JK. Essential role of endothelial Notch1 in angiogenesis. Circulation. 2005;111(14):1826–32.
30. Krebs LT, Xue Y, Norton CR, et al. Notch signaling is essential for vascular morphogenesis in mice. Genes Dev. 2000;14(11):1343–52.
31. Uyttendaele H, Ho J, Rossant J, Kitajewski J. Vascular patterning defects associated with expression of activated Notch4 in embryonic endothelium. Proc Natl Acad Sci USA. 2001;98(10):5643–8.
32. Murphy PA, Lam MT, Wu X, et al. Endothelial Notch4 signaling induces hallmarks of brain arteriovenous malformations in mice. Proc Natl Acad Sci USA. 2008;105(31):10901–6.
33. Alva JA, Iruela-Arispe ML. Notch signaling in vascular morphogenesis. Curr Opin Hematol. 2004;11(4):278–83.
34. Claxton S, Fruttiger M. Periodic Delta-like 4 expression in developing retinal arteries. Gene Expr Patterns. 2004;5(1):123–7.
35. Villa N, Walker L, Lindsell CE, Gasson J, Iruela-Arispe ML, Weinmaster G. Vascular expression of Notch pathway receptors and ligands is restricted to arterial vessels. Mech Dev. 2001;108(1–2):161–4.
36. Joutel A, Andreux F, Gaulis S, et al. The ectodomain of the Notch3 receptor accumulates within the cerebrovasculature of CADASIL patients. J Clin Invest. 2000;105(5):597–605.
37. Krebs LT, Xue Y, Norton CR, et al. Characterization of Notch3-deficient mice: normal embryonic development and absence of genetic interactions with a Notch1 mutation. Genesis. 2003;37(3):139–43.
38. Domenga V, Fardoux P, Lacombe P, et al. Notch3 is required for arterial identity and maturation of vascular smooth muscle cells. Genes Dev. 2004;18(22):2730–5.
39. Beckers J, Clark A, Wunsch K, De Hrabe AM, Gossler A. Expression of the mouse Delta1 gene during organogenesis and fetal development. Mech Dev. 1999;84(1–2):165–8.
40. Limbourg A, Ploom M, Elligsen D, et al. Notch ligand Delta-like 1 is essential for postnatal arteriogenesis. Circ Res. 2007;100(3):363–71.
41. Shutter JR, Scully S, Fan W, et al. Dll4, a novel Notch ligand expressed in arterial endothelium. Genes Dev. 2000;14(11):1313–8.
42. De Hrabe AM, McIntyre J, Gossler A. Maintenance of somite borders in mice requires the Delta homologue DII1. Nature. 1997;386(6626):717–21.
43. Duarte A, Hirashima M, Benedito R, et al. Dosage-sensitive requirement for mouse Dll4 in artery development. Genes Dev. 2004;18(20):2474–8.
44. Gale NW, Dominguez MG, Noguera I, et al. Haploinsufficiency of delta-like 4 ligand results in embryonic lethality due to major defects in arterial and vascular development. Proc Natl Acad Sci USA. 2004;101(45):15949–54.
45. Krebs LT, Shutter JR, Tanigaki K, Honjo T, Stark KL, Gridley T. Haploinsufficient lethality and formation of arteriovenous malformations in Notch pathway mutants. Genes Dev. 2004;18(20):2469–73.

46. Hellstrom M, Phng LK, Hofmann JJ, et al. Dll4 signalling through Notch1 regulates formation of tip cells during angiogenesis. Nature. 2007;445(7129):776–80.
47. Suchting S, Freitas C, le Noble F, et al. The Notch ligand Delta-like 4 negatively regulates endothelial tip cell formation and vessel branching. Proc Natl Acad Sci USA. 2007;104(9): 3225–30.
48. Lobov IB, Renard RA, Papadopoulos N, et al. Delta-like ligand 4 (Dll4) is induced by VEGF as a negative regulator of angiogenic sprouting. Proc Natl Acad Sci USA. 2007;104(9):3219–24.
49. Irvin DK, Nakano I, Paucar A, Kornblum HI. Patterns of Jagged1, Jagged2, Delta-like 1 and Delta-like 3 expression during late embryonic and postnatal brain development suggest multiple functional roles in progenitors and differentiated cells. J Neurosci Res. 2004;75(3):330–43.
50. Loomes KM, Underkoffler LA, Morabito J, et al. The expression of Jagged1 in the developing mammalian heart correlates with cardiovascular disease in Alagille syndrome. Hum Mol Genet. 1999;8(13):2443–9.
51. Xue Y, Gao X, Lindsell CE, et al. Embryonic lethality and vascular defects in mice lacking the Notch ligand Jagged1. Hum Mol Genet. 1999;8(5):723–30.
52. High FA, Lu MM, Pear WS, Loomes KM, Kaestner KH, Epstein JA. Endothelial expression of the Notch ligand Jagged1 is required for vascular smooth muscle development. Proc Natl Acad Sci USA. 2008;105(6):1955–9.
53. Jiang R, Lan Y, Chapman HD, et al. Defects in limb, craniofacial, and thymic development in Jagged2 mutant mice. Genes Dev. 1998;12(7):1046–57.
54. Lee CY, Vogeli KM, Kim SH, et al. Notch signaling functions as a cell-fate switch between the endothelial and hematopoietic lineages. Curr Biol. 2009;19(19):1616–22.
55. Sheng G. Primitive and definitive erythropoiesis in the yolk sac: a bird's eye view. Int J Dev Biol. 2010;54(6–7):1033–43.
56. Taoudi S, Medvinsky A. Functional identification of the hematopoietic stem cell niche in the ventral domain of the embryonic dorsal aorta. Proc Natl Acad Sci USA. 2007;104(22):9399–403.
57. Medvinsky A, Dzierzak E. Definitive hematopoiesis is autonomously initiated by the AGM region. Cell. 1996;86(6):897–906.
58. Garcia-Porrero JA, Godin IE, Dieterlen-Lievre F. Potential intraembryonic hemogenic sites at pre-liver stages in the mouse. Anat Embryol (Berl). 1995;192(5):425–35.
59. Kumano K, Chiba S, Kunisato A, et al. Notch1 but not Notch2 is essential for generating hematopoietic stem cells from endothelial cells. Immunity. 2003;18(5):699–711.
60. Robert-Moreno A, Espinosa L, de la Pompa JL, Bigas A. RBPjkappa-dependent Notch function regulates Gata2 and is essential for the formation of intra-embryonic hematopoietic cells. Development. 2005;132(5):1117–26.
61. Robert-Moreno A, Guiu J, Ruiz-Herguido C, et al. Impaired embryonic haematopoiesis yet normal arterial development in the absence of the Notch ligand Jagged1. EMBO J. 2008;27(13): 1886–95.
62. Radtke F, Fasnacht N, Macdonald HR. Notch signaling in the immune system. Immunity. 2010;32(1):14–27.
63. Yuan JS, Kousis PC, Suliman S, Visan I, Guidos CJ. Functions of notch signaling in the immune system: consensus and controversies. Annu Rev Immunol. 2010;28:343–65.
64. Majesky MW. Developmental basis of vascular smooth muscle diversity. Arterioscler Thromb Vasc Biol. 2007;27(6):1248–58.
65. Shin M, Nagai H, Sheng G. Notch mediates Wnt and BMP signals in the early separation of smooth muscle progenitors and blood/endothelial common progenitors. Development. 2009;136(4):595–603.
66. Schroeder T, Meier-Stiegen F, Schwanbeck R, et al. Activated Notch1 alters differentiation of embryonic stem cells into mesodermal cell lineages at multiple stages of development. Mech Dev. 2006;123(7):570–9.
67. Joutel A, Corpechot C, Ducros A, et al. Notch3 mutations in CADASIL, a hereditary adult-onset condition causing stroke and dementia. Nature. 1996;383(6602):707–10.
68. Liu H, Kennard S, Lilly B. NOTCH3 expression is induced in mural cells through an autoregulatory loop that requires endothelial-expressed JAGGED1. Circ Res. 2009;104(4):466–75.

69. Doi H, Iso T, Sato H, et al. Jagged1-selective notch signaling induces smooth muscle differentiation via a RBP-Jkappa-dependent pathway. J Biol Chem. 2006;281(39):28555–64.
70. Wang HU, Chen ZF, Anderson DJ. Molecular distinction and angiogenic interaction between embryonic arteries and veins revealed by ephrin-B2 and its receptor Eph-B4. Cell. 1998;93(5): 741–53.
71. Zhong TP, Childs S, Leu JP, Fishman MC. Gridlock signalling pathway fashions the first embryonic artery. Nature. 2001;414(6860):216–20.
72. Lawson ND, Vogel AM, Weinstein BM. Sonic hedgehog and vascular endothelial growth factor act upstream of the Notch pathway during arterial endothelial differentiation. Dev Cell. 2002;3(1):127–36.
73. Lawson ND, Scheer N, Pham VN, et al. Notch signaling is required for arterial-venous differentiation during embryonic vascular development. Development. 2001;128(19):3675–83.
74. Zhong TP, Rosenberg M, Mohideen MA, Weinstein B, Fishman MC. Gridlock, an HLH gene required for assembly of the aorta in zebrafish. Science. 2000;287(5459):1820–4.
75. Trindade A, Kumar SR, Scehnet JS, et al. Overexpression of delta-like 4 induces arterialization and attenuates vessel formation in developing mouse embryos. Blood. 2008;112(5):1720–9.
76. Benedito R, Trindade A, Hirashima M, et al. Loss of Notch signalling induced by Dll4 causes arterial calibre reduction by increasing endothelial cell response to angiogenic stimuli. BMC Dev Biol. 2008;8:117.
77. Kim YH, Hu H, Guevara-Gallardo S, Lam MT, Fong SY, Wang RA. Artery and vein size is balanced by Notch and ephrin B2/EphB4 during angiogenesis. Development. 2008;135(22): 3755–64.
78. Sorensen I, Adams RH, Gossler A. DLL1-mediated Notch activation regulates endothelial identity in mouse fetal arteries. Blood. 2009;113(22):5680–8.
79. Fruttiger M. Development of the mouse retinal vasculature: angiogenesis versus vasculogenesis. Invest Ophthalmol Vis Sci. 2002;43(2):522–7.
80. Fruttiger M. Development of the retinal vasculature. Angiogenesis. 2007;10(2):77–88.
81. Gerhardt H, Golding M, Fruttiger M, et al. VEGF guides angiogenic sprouting utilizing endothelial tip cell filopodia. J Cell Biol. 2003;161(6):1163–77.
82. Phng LK, Gerhardt H. Angiogenesis: a team effort coordinated by notch. Dev Cell. 2009;16(2):196–208.
83. Leslie JD, Ariza-McNaughton L, Bermange AL, McAdow R, Johnson SL, Lewis J. Endothelial signalling by the Notch ligand Delta-like 4 restricts angiogenesis. Development. 2007;134(5): 839–44.
84. Siekmann AF, Lawson ND. Notch signalling limits angiogenic cell behaviour in developing zebrafish arteries. Nature. 2007;445(7129):781–4.
85. Benedito R, Roca C, Sorensen I, et al. The notch ligands Dll4 and Jagged1 have opposing effects on angiogenesis. Cell. 2009;137(6):1124–35.
86. Holderfield MT, Hughes CC. Crosstalk between vascular endothelial growth factor, notch, and transforming growth factor-beta in vascular morphogenesis. Circ Res. 2008;102(6):637–52.
87. Franco CA, Liebner S, Gerhardt H. Vascular morphogenesis: a Wnt for every vessel? Curr Opin Genet Dev. 2009;19(5):476–83.
88. Oda T, Elkahloun AG, Pike BL, et al. Mutations in the human Jagged1 gene are responsible for Alagille syndrome. Nat Genet. 1997;16(3):235–42.

Section II

Novel Molecular Mediators Regulating Cardiovascular System

5 The Therapeutic Potential of Dimethylarginine Dimethylaminohydrolase-Mediated Regulation of Nitric Oxide Synthesis

James Leiper, Francesca Arrigoni, and Bierina Ahmetaj

5.1 Introduction

The establishment and progression of cardiovascular disease is associated with endothelial dysfunction. It is widely accepted that nitric oxide production from the vascular endothelium plays a key role in regulation of vascular function in normal health and during disease. Therefore, mechanisms that regulate vascular nitric oxide production have become the focus of significant attention from both vascular biologists and the pharmaceutical industry. The inhibition of nitric oxide synthase activity by endogenously produced competitive inhibitors has recently been linked to reduced nitric oxide synthesis in numerous animal models of disease and several human disease states. In this chapter, we will review the current literature describing these relationships and briefly focus on the pharmacological effects that some of the current therapies for treating these diseases might have on this pathway.

5.2 ADMA Synthesis

Asymmetric dimethylarginine (ADMA) is an amino acid that is constitutively produced following the posttranslational modification of Arginine residues. A complex process, this methylation is carried out by protein arginine methytransferases (PRMTs) that can catalyze monomethylation, producing mono-methylated Arginines such as L-NG-monomethyl Arginine (L-NMMA) [1].The PRMTs themselves exist

J. Leiper (✉)
MRC Clinical Sciences Centre, Hammersmith Hospital, Imperial College London, London, UK
e-mail: james.leiper@csc.mrc.ac.uk

F. Arrigoni • B. Ahmetaj
School of Pharmacy and Chemistry, Kingston University, Kingston-Upon-Thames, Surrey, UK

D. Abraham et al. (eds.), *Translational Vascular Medicine*,
DOI 10.1007/978-0-85729-920-8_5,

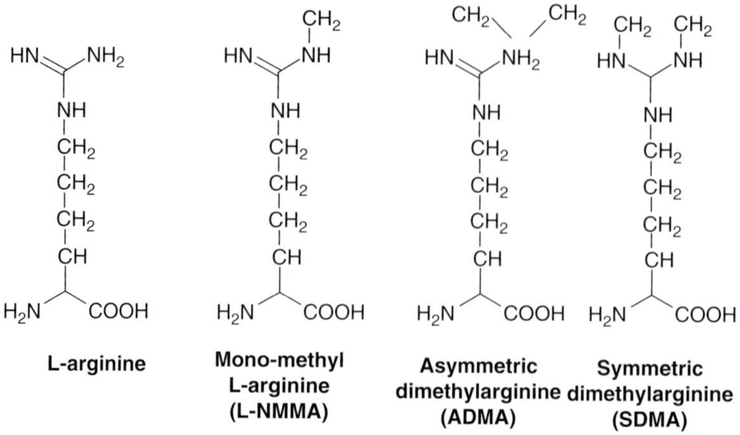

Fig. 5.1 Structures of L-arginine and the endogenous methylarginines L-NMMA, ADMA, and SDMA. The methylarginines L-NMMA and ADMA are both inhibitors of NOS, whereas SDMA is not

as different isoforms, and are classified according to their enzyme activity (Type I enzymes and Type II enzymes) and substrate specificity. PRMT 1 produces ADMA and PRMT2 the symmetrical isomer, SDMA (Fig. 5.1) [2]. The activity of PRMT enzymes is regulated by many factors including cellular stresses [3–5].

As there is no evidence to suggest that ADMA can be made from the methylation of free arginines [6, 7], the proteolysis of methylated arginines appears to be the sole source of ADMA [8] that correlates with elevated levels of ADMA in induced cardiovascular diseases [3, 5].

The intracellular pool size of ADMA and other monomethylarginines is believed to be controlled predominantly by the hydrolysis of ADMA to citrulline and dimethylamine by dimethylarginine dimethylaminohydrolase (DDAH) [9, 10], while SDMA is left unhydrolyzed [10]. Alternative pathways can also metabolize DMAs into derivatives of alpha-ketoacids (renal dimethyl pyruvate transferase DPT) and acetylated metabolites; however, these low-capacity pathways are minor and not thought to provide any major metabolic changes [11].

The movement of the methylated arginines into and out of the cell is regulated by the cationic amino acid transporter family (CAT) [12]. These transport amino acids, which include ADMA and SDMA, into and out of the cell in a one-to-one exchange for another amino acid via an antiporter mechanism [13]. ADMA has a high affinity for both the CAT type 1 and 2 transporters [12, 14] demonstrating a greater affinity for them than L-arginine [15]; both ADMA and SDMA are considered to have an equal affinity to other members of the CAT transporter family such as the CAT2B isoform [12]. The overall effect of competition between methylated and non-methylated arginine at these transporters is not fully clear; however, at very high, superphysiological, concentrations, methylarginines may prevent the uptake of L-Arginine into the cell, and promote Arginine efflux [16].

Due to the cationic nature of the transporter, the potential across the cell membrane can regulate its activity, both positively and negatively [17]. This can provide a driving force of cationic amino acids into the cells so that by inducing membrane hyperpolarization, vasoactive agonists like Acetylcholine and Bradykinin can increase the driving force of CAT-mediated amino acid entry into the cell [18, 19], potentially altering cellular Arginine and ADMA uptake [17].

Following its export into the plasma, all the free methylated Arginines are ultimately cleared from the body by renal excretion and hepatic metabolism [7].

5.3 ADMA-Mediated Regulation of Nitric Oxide Synthesis

ADMA is a potent reversible inhibitor of all three Nitric Oxide Synthase (NOS) isoforms (nNOS, eNOS, iNOS) [20, 21] by behaving as an L-arginine analogue. L-Arginine is the substrate of the Nitric Oxide Synthase (NOS) enzyme family that produces Nitric Oxide (NO) and L-citrulline. The enzymes, eNOS (endothelial), iNOS (inducible), and nNOS (neuronal), coded by different genes [22] were originally classified according to their cellular distribution: in the endothelium, inducibly in most cells, and neuronally but this has since been shown not to be exclusive. The isoforms are classified according to their dependence on intracellular Ca^{2+} [23], duration of action, and the fact that iNOS is inducible and nNOS and eNOS are constitutively produced [24, 25].

Optimal NOS activity requires the presence of a number of cofactors including NADPH and BH4, substrate L-Arginine, and, depending on the isoform, Ca^{2+}. NO diffuses from the cell of origin into the target cells. NO reversibly binds to the heme group in soluble guanylate cyclase to form nitrosyl complexes in the target tissue that leads to cGMP production [26]. cGMP in turn activates Protein Kinase G (PKG) [27] that in the smooth muscle lowers intracellular Ca^{2+}, resulting in the dephosphorylation of myosin light chains causing a decrease in vascular tone. The actions of cGMP are terminated upon hydrolysis by a family of phosphodiesterases or prolonged pharmacologically by phosphodiesterase inhibitors [28].

When produced in large enough quantities, such as following the induction of iNOS, NO can feedback negatively to inhibit NOS activity by s-nitrosylating the enzyme [29, 30].

Other end products of NOS activity include nitrate and nitrite that can be reduced to NO, a reactive oxygen species, under conditions of low oxygen tension. If it reacts with superoxide, peroxynitrite (ONOO) forms, leading to cellular damage and death [31]. Other actions of peroxynitrite that will affect NO production are thought to occur indirectly through tyrosine nitration of the CAT transporters [32] that may lead to increases in intracellular ADMA and reductions in L-arginine [33].

As a result of its actions, Nitric Oxide (NO) is involved in a wide variety of regulatory mechanisms that in the cardiovascular system include: the maintenance of vascular tone, anti-thrombotic effects, control of smooth muscle cell proliferation, leukocyte adhesion, endothelial cell proliferation, motility and survival [34–37], the promotion and expression of VEGF and pro-angiogenic factors [37, 38], and the inhibition of anti-angiogenic factors [34].

Consequently, a reduction in NO bioavailability in vivo and in vitro results in numerous alterations in vascular function. The endothelial dysfunction that is directly linked to a decrease in the production of NO from eNOS is a risk factor associated with atherogenesis.

All NOS isoforms are inhibited by ADMA [20, 21, 39, 40] with IC_{50} values ranging from 1 to 10 μm [41] depending on the prevailing substrate concentration. Inhibition of eNOS by methylated Arginine analogues has been comprehensively measured by Cardounel and colleagues [15]. For endothelial NOS, the K_m for eNOS = 3.14 μmol/l. The Ki for eNOS by ADMA is 0.9 μmol/l and by L-NMMA is 1.1 μmol/l. Under normal physiological conditions, these methylated arginine levels inhibit only 10% of eNOS activity [15].

Under pathophysiological conditions with plasma concentrations of ADMA increasing threefold to ninefold, cellular NO output can be inhibited by 30–70% [7, 42, 43]. This effect may be amplified by the action of CAT transporters that are able to concentrate methylarginines by up to 10× more inside than outside the cell. Some of these transporters have a higher affinity for methylarginines than arginine and would therefore tend to increase the intracellular methylarginine:arginine ratio [15]. Thus, small changes in plasma levels of ADMA can result in large changes intracellularly [14].

Other effects of increasing intracellular ADMA levels is the uncoupling of eNOS, which leads to superoxide production and subsequent increases in oxidative stress [44, 45] that underlie the pathologies of many cardiovascular diseases [44, 46, 47].

In healthy humans, the plasma levels of ADMA range from 0.35 to 0.7 μmol/l [17, 20, 21, 48]. These levels are due to the PRMT activity, hydrolysis by DDAH, and removal from the plasma by the kidneys [49].

The selective nature of the metabolism of ADMA by DDAH means that ADMA/SDMA ratios effectively demonstrate DDAH metabolism since the CAT transporters are not selective for DMAs. Consequently, SDMA levels are determined by PRMT activity and renal clearance and may therefore be a useful marker of renal function [50]. Indeed, SDMA correlates with creatinine clearance, while ADMA levels do not correlate with glomerular filtration rate [51–53] as a consequence of metabolism by DDAH.

5.4 DDAH-Catalyzed ADMA Hydrolysis

In mammals, there are two DDAH isoforms encoded by different genes [54, 55]. While ADMA is ubiquitously expressed in all cells, DDAH is selectively expressed to varying degrees in different organs, cellular and subcellular structures with some similarities to NOS isoforms. DDAH-1 is expressed in the pancreas, forebrain, aorta, peritoneal neutrophils and macrophages [55, 56], and in the liver and kidney at sites of NOS expression [57–59]. Using murine DDAH-1 knockouts, decreased expression of DDAH-1 independent of DDAH-2 can be found in skeletal muscle, lung, brain, and heart [60].

DDAH-2 expression is high in fetal tissue, vascular endothelium (in cytosol), smooth muscle, heart, placenta, spleen, thymus, peripheral leukocytes, lymph nodes, and bone marrow [55]. In the kidney, the selective structural distribution of DDAH includes the proximal tubule [61], macula densa, distal convoluted tubule, the thick ascending limb of the loop of Henle, and the collecting ducts of the cortex and the medulla [61].

DDAH isoforms are highly conserved at the amino acid level particularly in residues important for substrate binding and hydrolysis. Across species, DDAH isoforms are highly conserved with homology in the murine, bovine, and human gene sequences of DDAH-1 (92%) and DDAH-2 (95%).

DDAH catalyzes the metabolism of one molecule of ADMA to one molecule of Dimethylamine and L-Citrulline and does not hydrolyze SDMA [62]. The K_m for ADMA metabolism by DDAH is 180 μmol/l [62]. Interestingly, recombinantly expressed DDAH-2 has a greater K_m for L-NMMA of 0.51 mmol/l compared to 0.36 mmol/l for rat DDAH-1 [62]. K_m values for ADMA and L-NMMA have been reported ranging from 69 to 170 μmol/l and 53.6 to 90 μmol/l respectively for native and recombinant DDAH1 [63, 64]. All investigations have demonstrated that the K_m values for DDAH are greater than intracellular concentrations of ADMA, which suggests that the DDAH enzyme active site is never fully saturated, allowing ADMA metabolism to be proportional to its concentration.

It has been estimated that more than 70% of ADMA can be metabolized by DDAH [65] with global heterozygous deletion of DDAH1 in the mouse increasing ADMA in the plasma, brain, and lung by 20% [60].

There are a wide variety of factors that regulate DDAH activity and expression, some are isoform specific. DDAH can be competitively inhibited by L-Arginine, although the required Ki is relatively high (Ki of 2.5 mM). This causes inhibition of ADMA metabolism in HepG2 liver cells increasing intracellular ADMA levels [66, 67]. This may explain not only the inability of supplemental L-Arginine to improve vascular function but also the adverse effects that have been observed following administration [68].

NO itself also regulates DDAH activity and expression. It is known that excess NO production found following iNOS stimulation often leads to inhibition of activity of constitutively expressed NOS isozymes by s-nitrosylation [29, 30]. NO can also reversibly inhibit recombinant DDAH in vitro and in mammalian DDAH extracts in a similar fashion via s-nitrosylation of cys-249 in the DDAH active site. This occurs after cytokine induced expression of the inducible NOS isoforms [69]. Interestingly, in IL-1β-stimulated smooth muscle cells, the induction of iNOS is associated with increased DDAH activity and expression, causing ADMA levels to decrease [40]. One can assume that any consequence of nitrosylation of the enzyme that occurs with high NO is negligible in this case, and further investigations into the precise mechanisms of DDAH regulation need to be undertaken. One putative mechanism might be via a cGMP-dependent pathway that can increase expression of the DDAH-2 isoform following increases in NO levels [70] maintaining intracellular levels of NO.

Other regulators of DDAH are: estradiol that increases DDAH activity [71] and expression [72]; insulin that increases DDAH activity [73], by inducing SIRT-1, an enzyme associated with the prevention of premature senescence [74]; and all-trans-retinoic acid that can influence angiogenesis and is a transcriptional regulator of DDAH2. All-trans-retinoic acid targets the promoter region of the DDAH-2 gene, which also contains the PPAR/RXR site, [65] and various PPAR ligands have been shown to increase the expression and activity of DDAH [74, 75].

DDAH activity can be downregulated by factors that induce reactive oxygen species as part of their mechanistic actions. These include: CF6, a component of mitochondrial ATP synthase that inhibits phospholipase A2 [76], LPS [77, 78], and TNF α [73, 77]. Sensitive to oxidation, peroxynitrite, and H_2O_2 [63], DDAH has been reported by some to be less sensitive to in vitro inactivation by the potent oxidizer H_2O_2 [64], because the active site may be protected from direct oxidation [79], perhaps because of the high pKa of the active site [64].

5.5 ADMA, DDAH, and the Regulation of Vascular Function

Increased plasma ADMA concentrations as a consequence of the regulatory mechanisms described above are linked to numerous vascular diseases alongside new and classical cardiovascular risk factors and are all associated with low NO output and endothelial dysfunction [80–84]. Patients with pro-atherogenic cardiovascular diseases such as hypercholesterolemia, hyperhomocystinemia, and hypertriglyceridemia demonstrate reduced endothelium-dependent flow-mediated vasodilatation in association with elevated plasma ADMA and reduced L-Arginine/ADMA ratios [85–87].

The contribution of DDAH to NO-mediated dilatation has been demonstrated using experimental models in DDAH1$^{+/-}$ mice, where in vitro vasorelaxation to Acetylcholine (ACh) and the calcium ionophore is reduced [60]. Using small inhibitory RNA (siRNA) constructs targeted to DDAH-1 and DDAH-2 in rats, Wang and colleagues demonstrated that while DDAH-1 appeared to be responsible for regulating serum levels of ADMA, NO-mediated vasodilatation was regulated primarily through the DDAH-2 isoform [88].

The angiogenic capabilities of endothelial cells are also affected by DDAH, improving following transfection of DDAH-2 by enhancing VEGF mRNA expression [9, 89]. Overexpression of DDAH1 increases neovascularization of tumor cells in vivo [90] and results in improved endothelial regeneration following femoral artery injury in DDAH1 transgenic mice [91].

Conversely, in DDAH1$^{+/-}$ mice, with elevated levels of ADMA, endothelium-mediated angiogenesis is inhibited [89, 92, 93].These pro-apoptotic and anti-proliferative effects of ADMA are thought to occur via an increase in reactive oxygen species (ROS) and a p38 MAPK pathway in endothelial cells [44] that induces apoptotic responses [94].

Increased levels of ADMA in patients with stable angina have also been shown to be associated with a decrease in myeloid endothelial progenitor cell number [95], suggesting a role for DDAH in vascular repair mechanisms of the endothelium.

5.6 Nitric Oxide Synthase–Independent Actions of Methylarginines

Not all actions of L-Arginine analogues are related directly to the activity of NOS and NO production. Other effects of L-Arginine analogues include: inhibition of cytochrome C [96], antagonism of muscarininc ACh receptors [97], impairment of the urea cycle [98], and induction of cytokines [99]. Rats overexpressing DDAH suppress the gene and protein expression of the cytokine TGF-β in a rat model of chronic kidney disease [100].

The importance of ADMA in non-NO-related pathology is demonstrated in eNOS$^{-/-}$ mice, where following long-term administration of ADMA, coronary vascular lesions are found, typified by medial thickening and a perivascular fibrosis in the coronary microvessels [101, 102]. As this group found no expression of iNOS or nNOS in the thickened microvessels, the lesions occurring as a consequence of ADMA inhibition of other NOS isoforms were ruled out. Indeed, while NOS triple knockout mice are viable, homozygous null mice for DDAH-1 are embryonically lethal [60] supporting the proposition of significant non-NO-dependent effect of ADMA.

Several studies have investigated the NO-independent relationship of ADMA with angiotensin. Angiotensin II is used to artificially induce hypertension and renal injury, and in these cases, ADMA is elevated perhaps as a result of increased PRMT synthesis [5]. The augmented levels of ADMA further upregulate angiotensin-converting enzyme [47, 102] that converts angiotensin I to angiotensin II.

Angiotensin II can however reduce ADMA levels by acting on AT-1 receptors, causing an increase in the mRNA expression of arginases, DDAH-2 [103] and CAT transporter expression and activity in the healthy kidney of angiotensin II hypertensive rats [103, 104]. This feedback system may consequently contribute to the paradoxically stable ADMA levels observed in rat models of angiotensin II hypertension when the kidneys are healthy [5].

5.7 ADMA Clearance: The Liver and Kidneys

Free methylarginines are cleared from the plasma by renal excretion and hepatic metabolism [7, 105].

Hepatocytes take up large amounts of particular amino acids from the hepatic circulation that include Arginine and ADMA [57, 58] and regulate the circulating levels of ADMA by expressing high levels of Arginases [57, 58] and DDAH [55]. Consequently, in liver failure, the plasma levels of Arginine, ADMA, SDMA, and other amino acids are elevated [57, 58, 106].

In the kidney, SDMA and ADMA are excreted equally. The kidney is also very sensitive to circulating levels of L-arginine and plays a major role in Arginine metabolism. ADMA can be both generated and metabolized by the kidney as ADMA is taken up from the circulation via CAT transporters [17].

In chronic kidney disease, there is reduced nitric oxide production [107] and increased ADMA and SDMA levels. SDMA levels are associated with high levels of creatinine, a marker of kidney dysfunction [108] and high SDMA levels are suggestive of an increased expression of PRMTs [109].

Relatively small increases in ADMA concentrations that occur in early-stage renal failure are associated with large increases in cardiovascular event rates. Dysfunctional kidneys excrete less ADMA, the severity of the renal disease correlating to increased ADMA concentration and reduced NO bioavailability [110]. High ADMA levels in turn cause a decrease in renal plasma flow contributing to further progression of kidney damage that will raise ADMA to pathophysiological levels [20, 21, 111] and contribute to the progression of cardiovascular dysfunction [107]. Experimentally induced chronic NOS inhibition can result in: systemic and glomerular hypertension; tubulointerstitial injury; proteinuria, glomerular ischemia, and glomerulosclerosis [112]; and chronic renal disease [113]. Consequently, plasma ADMA concentration is a strong independent predictor of disease progression in patients with kidney failure [111, 114] with elevated plasma ADMA strongly associated with mortality in patients with renal failure [84] and an increased morbidity and mortality in renal transplant patients [115]. Interestingly, in end-stage renal disease, the frequency of hemodialysis has very little effect on ADMA levels [116].

5.8 Associations Between DDAH/ADMA and Disease

Clearly, ADMA has the potential to exert significant effects on nitric oxide synthesis and DDAH is a key regulator of ADMA levels in vivo (Fig. 5.2). In the following sections, we will review the literature implicating dysregulation of ADMA levels in several major human diseases.

5.8.1 Cardiovascular Disease

The pathologies of most patients with renal disease are characterized by cardiovascular morbidity and mortality due to complications and premature atherosclerosis [117].

Atherosclerosis is the leading cause of death and disability in North America [117], and in 2003, the World Health Organisation (WHO) estimated that approximately 16.7 million people die annually of cardiovascular disease [118].

Atherogenesis proceeds as a result of continuing endothelial dysfunction that is associated with cardiovascular risks. These include: aging, hyperhomocysteinemia, postmenopausal state, smoking, diabetes, hypercholesterolemia, and hypertension

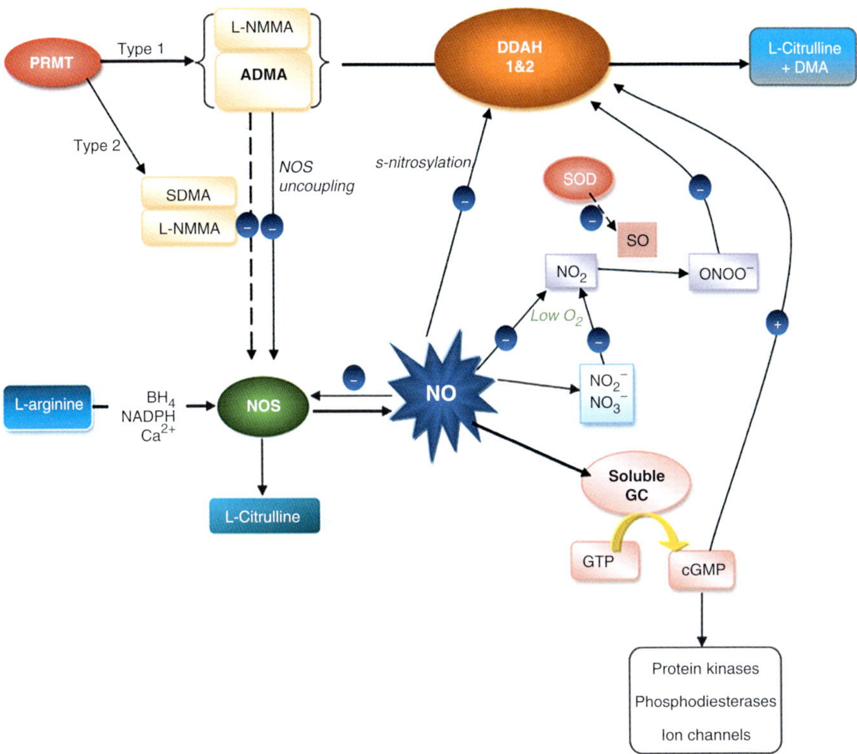

Fig. 5.2 The DDAH/ADMA/NOS pathway. The methylation of protein incorporated arginine by PRMTs and subsequent proteolysis of arginine methylated proteins leads to a production of the methylarginines ADMA, L-NMMA, and SDMA. ADMA and L-NMMA (but not SDMA) inhibit the enzyme NOS which is essential for the production of NO in the presence of tetrahydrobiopterin (BH_4), NADPH, and Ca^{2+}. NO is converted to nitrite (NO_2^-) and nitrate (NO_3^-) and under low oxygen conditions can be converted to the superoxide (SO•) to form the peroxynitrite ($ONOO^-$), leading to cellular damage and death and the suppression of DDAH 1 and 2 expression. NO binds to, and activates, soluble guanylate cyclase (sGC) forming cGMP. cGMP may enhance DDAH expression/activity. The methylarginines ADMA and L-NMMA are converted to citrulline and dimethylamine (DMA)

[119] and can be assessed according to the Framingham Risk score [120], although the Framingham Risk currently underestimates event rates of chronic Kidney disease [121, 122].

The relationship between the progression of atherosclerosis and the NO pathway is a close one. Plasma ADMA is elevated in established atherosclerosis [81], peripheral vascular disease [85], and coronary artery disease [66, 67].

Behaving as an independent predictor of cardiovascular disease in patients with coronary artery disease [42, 123], elevated ADMA is associated with what are considered to be classical risk factors for cardiovascular disease that include: hypercholesterolemia [85, 124, 125], raised low-density lipoproteins [77], triglycerides [86], raised C-reactive protein [123], ageing [81], hypertension, pulmonary hypertension[126], diabetes, hyperlipidemia [124, 127–129], and hyperhomocysteinemia [130, 131]. High levels of homocysteine, which are associated with coronary and peripheral vascular disease [132], result in elevated oxidative stress and increased ADMA levels in both animal [124] and human experiments [133] by directly increasing ADMA accumulation [131, 134] causing endothelial dysfunction [130].

5.8.2 Hypertension

Hypertension involves an interaction of multiple underlying mechanisms that include: the renin-angiotensin system [135, 136], oxidative stress [137], and nitric oxide synthesis.

Nitric oxide plays an integral role in the regulation of vascular tone and blood pressure [138, 139], and in both human and animal experiments, there is an increasing body of evidence associating the development of hypertension with NO deficiency. Blood pressure is associated with plasma levels of ADMA in healthy subjects [81] and the infusion of ADMA into healthy subjects will moderately elevate blood pressure, offset by decreases in cardiac output and cardiac dysfunction [127]. In patients with essential hypertension, plasma ADMA has been reported by some groups to be elevated [140–143] and by others to remain unaffected [144, 145].

However, increased ADMA levels in hypertensive patients correspond with impaired flow-mediated vasodilatation [146] and experiments performed ex vivo on resistance vessels taken from patients with essential hypertension demonstrate elevated ADMA levels that correlate with endothelial dysfunction and a reduced NOS activity [147]. Experimentally increased plasma ADMA concentrations have been shown to result in hypertension. Genetic or pharmacological inhibition of ADMA metabolism causes elevated systemic and pulmonary pressures [60], while genetic overexpression of DDAH1 produces the opposite effects.

ADMA may also regulate blood pressure by affecting the kidney excretion of Na^+ ions. Mice lacking eNOS are salt sensitive [148] and the effects of endogenous NOS inhibitors can induce kidney-mediated salt-sensitive hypertension in rats [149].

5.8.3 Metabolic Syndrome

Metabolic syndrome is a cluster of the most dangerous risk factors, characterized by obesity, dyslipidemia, hypertension, and insulin resistance. It has reached epidemic proportions globally primarily due to an increased sedentary lifestyle and dietary habits. It is associated with an approximate twofold increased risk of cardiovascular morbidity and mortality in the European population [150]. This increased association is due in part to vascular complications contributing to an increase in cardiovascular risk [151, 152] with impaired NO-mediated vasorelaxation [153].

In rat models of metabolic syndrome, associations with either reduced NO [154] or endothelial dysfunction alongside oxidative stress [155] have been demonstrated.

Hypertension in the rat model of metabolic syndrome used by Roberts and co-workers was associated with NO downregulation and dysfunction of the pathway downstream from NO [156]. Patients with metabolic syndrome show correlations between NO and BMI, systolic blood pressure and triglyceride levels [157], and decreased vascular reactivity to ACh compared to age-matched healthy controls [158].

Decreased NO bioavailability in metabolic syndrome is associated with increased levels of ADMA [159]. Insulin resistance, which is pivotal to this syndrome [160], positively correlates with ADMA levels in nondiabetic, normotensive individuals [159]. Increased ADMA levels are most likely the result of a decrease in DDAH activity that is associated with obesity, hypercholesterolemia, and oxidative stress [161, 162] and an upregulation of PRMT1 expression that has been shown in the presence of low-density lipoprotein or oxidized LDL in cultured endothelial cells [2, 44].

5.8.4 Diabetes

In approximately 70% of all deaths in patients with diabetes, cardiovascular disease is responsible [163, 164]. Insulin signaling pathways in the vascular endothelium share similarities with metabolic insulin signaling pathways in adipose tissue and skeletal muscle [165]. In skeletal muscle, insulin can stimulate an increase in NO production, resulting in increased blood flow [166]. In this situation, increased NO production is the result of eNOS phosphorylation and activation that is downstream of Akt signaling. [167, 168]. When insulin-mediated glucose uptake is defective, it has been suggested that the MAP-kinase pathway can also regulate insulin-dependent NO production [169, 170]. Insulin can also increase arginine bioavailability via improved CAT transport, upregulate eNOS expression and activity in cultured Human Umbilical endothelial cells (HUVEC) [171], and increase DDAH activity [73].

Predictably, in animal models of diabetes [172, 173] and in patients with impaired glucose tolerance [81], insulin resistance [159], and both type 1 and 2 diabetes [174–176], plasma ADMA levels are elevated.

5.8.5 Insulin-Resistant Type II Diabetes

Insulin resistance is typically defined as "decreased sensitivity and/or responsiveness to the metabolic actions of insulin that promote glucose disposal" [177]. Type 2 diabetes is strongly linked to the metabolic syndrome and cardiovascular disease [160], and obese patients with insulin resistance have higher plasma ADMA levels than obese patients without insulin resistance [178] with a decline in ADMA plasma levels reported only with weight loss in patients with insulin resistance and not those without insulin resistance [179].

Strong associations between the duration of the disease, smoking, nephropathy, and diabetic retinopathy were significantly associated with ADMA levels in a cohort of 343 patients with type 2 diabetes [180]. Genetic variations in both DDAH-1 and DDAH-2 genes in this cohort associated with the elevated levels of ADMA [180].

Underlying the acquired insulin resistance present in type 2 diabetes is a degree of glucotoxicity, lipotoxicity, and inflammation which are responsible for the increased levels of oxidative stress and inflammatory molecules which contribute to endothelial dysfunction [181].

Elevation of intracellular glucose levels is associated with an increase in ADMA [182, 183] that is secondary to increased oxidative stress, reduced DDAH activity [3] and eNOS expression [183, 184], and increased PRMT activity [3]. Hyperglycemia has also been shown to downregulate DDAH activity in rat models of critical illness [182], cultured endothelial cells [185], and rat models of type 2 diabetes [172].

Further consequences of high plasma glucose levels and oxidative stress are the production of Advanced End Glycation (AGE) products that are associated with elevated plasma ADMA in both type 2 diabetes [186] and in hypercholesterolemia [187]. One putative mechanism for the action of AGE is via the inhibition of eNOS [184] and/or a decrease in DDAH activity that can attenuate NO-dependent vasorelaxation in rat aortic rings [189].

5.9 Pharmacotherapy of ADMA

The importance of ADMA as an independent marker of cardiovascular disease risk [81] and a marker for atherosclerotic change [190] [110] suggests that the outcomes of certain cardiovascular diseases might be improved by pharmacologically manipulating the ADMA/DDAH pathway. Here we discuss some of the more common therapies used in treatment of cardiovascular diseases that have been shown to modulate the activity of the ADMA/DDAH pathway (Table 5.1).

5.9.1 ACE and ARB Inhibitors

Numerous studies have shown a link between ADMA and the renin-angiotensin system (RAS). Angiotensin-converting enzyme inhibitors (ACEIs) and angiotensin AT_1 receptor blockers (ARBs) prevent eNOS uncoupling and oxidative stress via inhibitory effects on the activity of free radical–producing enzymes [87, 191, 192]. ACEIs work by inhibiting angiotensin II which increases ROS formation by vascular NADPH oxidase. The production of ROS leads to inactivation of DDAH and also upregulates activity of PRMTs, consequently contributing to increased levels of ADMA.

Delles et al. first demonstrated the link between an activated renin-angiotensin system and the ADMA pathway by showing that (independent of blood pressure lowering effects) the monotherapy or combination therapy of an ACEI and ARB reduced ADMA plasma concentrations in young, mildly hypertensive men [193]. The effects of such treatments were later confirmed by Suda et al., who revealed that vascular lesions and superoxide production in both wild-type and endothelial

Table 5.1 Pharmacological treatments and their effects on ADMA levels, CAT/y^+ transport, L-arginine levels and DDAH activity and/or protein expression

Pharmaceutical reagents	ADMA	CAT/y^+ transport	L-arginine	DDAH activity and/or protein expression
ACEIs and ARBs				
Valsartan	↓	–	–	↑ DDAH 2 expression
Telmirsartan	↓	–	–	↑ DDAH 2 activity and expression via activation of PPARγ signaling
Eprosartan	↓	–	–	–
Statins + LDLs				
Rosuvastatin	↓	–	–	–
Pravastatin	–/↓	–	–	↓ Inhibition of DDAH activity
Simvastatin	↓	–	–	↑ DDAH 1 mRNA expression via the knockdown of SREBP2
Atorvastatin	–	–	–	–
Probucol	↓			↓ PRMT 1 expression/↑ DDAH activity
Antioxidants				
Kaempferol	↓	–	–	↑ DDAH 2 expression
Taurine	↓	–	↑	↑ DDAH activity via reduction of lipid peroxidation
Vitamin E	↓	–	–	
Acetylcholine/Bradykinin	↓	↑	↑	–
Anti-inflammatory Drugs				
Aspirin/Fenofibrates	↓	–	–	↓ TNFα
GW4046	↓	↑ CAT-1	–	–
Antidiabetics				
Rosiglitazone	–/↑	–	–	↑ PRMT 1 expression
Pioglitazone [194]	↓	–	–	↑ DDAH 2 expression
Metformin	↓	↑	–	–
Insulin	↓	↑	↑	↑ DDAH activity when induced by TNFα/ Upregulates eNOS activity
L-arginine	↑	–	↑	↓ DDAH activity
Vitamin A	↓	–	–	↑ DDAH 2 mRNA and protein expression

NOS-deficient mice were caused by chronic treatment with ADMA, and treatment by either ACEI or ARBs prevented these changes [102].

In addition, treatment with Telmisartan, an ARB commonly used in the management of hypertension as well as a selective modulator of PPAR-γ, delayed endothelial cell senescence, decreased oxidative stress, and upregulated the activity and protein expression of DDAH II. Importantly, Telmisartan was also shown to decrease the concentration of ADMA in endothelial cells, thereby inducing NO synthesis [74]. Studies comparing Telmisartan to another ACEI (Valsartan) in hypertensive patients with type 2 diabetes and overt nephropathy have shown renoprotection but no significant difference in ADMA levels over a course of 12 months [195].

While a significant number of data indicates the positive effects of ACEIs and ARBs on lowering ADMA, the effects of these agents on PRMT activity and (methyl)-arginine transport remain unclear [196].

5.9.2 Statins

Statins are commonly prescribed for adults with clinical evidence of cardiovascular disease. Different statin types exist and are prescribed on an individual basis according to their difference in ability to reduce cholesterol levels. These Hydroxymethylglutaryl Co-enzyme A reductase inhibitors decrease plasma cholesterol but can also inhibit platelet and leukocyte adherence to the endothelium, block proliferation of vascular smooth muscle, and stimulate eNOS expression [197]. They also improve oxidative shear stress by reducing the activity and/or expression of NAD(P) oxidase that leads to a reduction in vascular superoxide production [198].

The suggestion that native or oxidized-LDL may cause ADMA accumulation via increases in PRMT activity or by oxidative inhibition of DDAH activity [2] has meant that the effects exhibited by drugs such as statins might potentially lead to improvement of endothelial dysfunction.

However, studies investigating the effects of statins have shown an improvement in endothelial function in cardiovascular diseases independently of ADMA levels and the L-arginine/ADMA ratio [199]. Young and co-workers showed that a double-blinded, placebo-controlled crossover study of 40 mg Atorvastatin administered once daily for 6 weeks on patients with non-ischemic left ventricular dysfunction did improve lipid profiles, and endothelium-dependent vasodilatory responses of both the microvascular and macrovascular circulation, however, did not influence ADMA levels [199].

Similarly, a 24-month study in patients with mild-to-moderate Chronic Kidney Disease (CKD) showed that plasma ADMA concentrations, which did not alter over time, were not influenced by Pravastatin, or homocysteine-lowering therapy [200].

Of the major trials involving statins, the treatment of patients with hypercholesterolemia by the administration of Rosuvastatin (10 mg/day for 6 weeks) was the only one shown to decrease plasma ADMA levels significantly. Reduction in ADMA levels and low-density lipoprotein cholesterol corresponded with increases in flow-mediated dilatation [162]. Rosuvastatin has also been shown to be potent in lowering plasma cholesterol levels in patients with hypercholesterolemia [201] as well as decreasing vascular endothelial NO production in mice subjected to myocardial ischemia reperfusion injury [202]. Interestingly, recent experiments on cultured endothelial cells have shown that another statin, Simvastatin, decreases ADMA concentration by increasing DDAH1 mRNA expression via an SREBP2-dependent mechanism [203]. Overall, the effects of statins on ADMA levels remain unclear, dependent on the specific Statin prescribed and the type of cardiovascular disease.

5.9.3 Antioxidants and DDAH

Cardiovascular disease is linked to inflammation and oxidative stress [137, 204], induced by elevating levels of superoxide anions and peroxynitrite [205, 206] which are known to inhibit DDAH and induce ADMA [161, 162]. Numerous synthetic antioxidants have been shown to reduce the formation of ADMA and prevent a decrease in DDAH activity. Probucol, a potent antioxidant drug which inhibits the oxidation of cholesterol in LDL, significantly reduces levels of ADMA and improves endothelium-dependent relaxation by inhibiting PRMT 1 expression and enhancing the activity of DDAH [44, 207].

Studies on the sulfur-containing semi-essential amino acid Taurine, which has shown to be a potent antioxidant with the potential to inhibit lipid peroxidation and lower production of oxidant free radical [208, 209] significantly decreased ADMA levels in vivo by increasing DDAH activity via the reduction of lipid peroxidation [210].

More recently, Xiao et al. [211] showed that Kaempferol, a naturally occurring flavonoid with antioxidant properties, increased DDAH2 expression, decreased plasma ADMA levels, and increased plasma NO in ApoE$^{-/-}$ mice. This effect was accompanied by a significant decrease in ROS production levels [211].

In patients with mild-to-moderate Chronic Kidney Disease (CKD) administration of Vitamin E decreased plasma ADMA concentrations, perhaps as a result of improved DDAH activity [210], but neither Pravastatin nor other antioxidant therapy that included vitamin B6, B12, and folic acid affected ADMA levels [200]. This suggests that not all antioxidant therapies produce the same anti-inhibitory mechanisms for DDAH. The variations of the antioxidant effects on DDAH activity could in part be due to the protection of the active site by oxidized proteins that release zinc ions [212].

5.9.4 Antidiabetic Drugs

The rise in obesity in developed countries has led to an increase in the prevalence of associated diseases such as type 2 diabetes and metabolic syndrome. These diseases are closely associated to hypercholesterolemia as well as oxidative stress which lead to increased plasma ADMA levels by reducing DDAH activity [161, 162]. Pharmacological antidiabetic drugs include PPAR-γ agonists, sulfonylureas, insulin mimetics, and biguanides.

5.9.5 PPAR Agonists

Thiazolidinediones (TZDs) are PPAR-γ agonists which have shown beneficial effects for the glycemic management of type 2 diabetes mellitus. By acting directly on the vascular wall and peripheral tissues, they are thought to improve vascular structure and function, improve flow-mediated dilation, and have antiatherogenic effects among others [213, 214, 215]. Disadvantages in the use of TZDs have been shown to be an increased risk of fractures, particularly in women aged 65 years and over, particularly as individuals with type 2 diabetes are prone to rapid bone loss and the TZDs decrease bone formation; thus, therapy must be tailored appropriately to suit the patient's requirements [216].

Studies of the effects of the TZD Rosiglitazone have shown a variety of effects, with some studies showing positive ADMA-lowering effects by up to 30% in seven insulin-resistant patients with hypertension [159] as well as reduced plasma ADMA levels in patients with metabolic syndrome [217]. However, a recent study in a mouse model of high cardiovascular risk has shown that although Rosiglitazone can prevent carotid remodeling, a subsequent increase in superoxide and ADMA production and oxidative stress impairs endothelial dilatation of carotid arteries in response to ACh [218].

A randomized 6-month study comparing Rosiglitazone with Glyburide, a commonly prescribed sulfonylurea used to treat type 2 diabetes mellitus, showed that compared to Glyburide, Rosiglitazone significantly decreased c-reactive protein, c-peptides, improved arterial flow mediators, and showed trends toward improvements in carotoid artery distension. However, ADMA levels and other markers of oxidative stress remained unchanged in both groups, suggesting that ADMA was not associated with the improvements obtained by Rosiglitazone in this study [219]. This was confirmed by Richer et al. [220] and Mittermayer et al. who showed no effect in ADMA lowering by Rosiligtazone in critically ill patients [220, 221].

5.9.6 Biguanides

Metformin is a biguanide antidiabetic drug which can be transported to cells by the CAT transporter system due to its similar structure to ADMA [222] and unlike other antidiabetic drugs does not cause hypoglycemia. Asagami et al. [223] first looked at the effect of Metformin, either as monotherapy or in combination with sulfonylurea treatment, on ADMA, glucose, and L-Arginine levels in patients with type 2 diabetes. The study revealed that metformin (1–2 g/day for 3 months) decreased plasma ADMA concentrations by 30% in association with improved glycemic control in patients and this occurred regardless of single or combination use. Metformin did not have any effect on L-arginine levels [223]. Several other studies have confirmed this by showing that metformin treatment in women with polycystic ovaries (PCOS) reduced plasma ADMA levels as well as improved hormonal and metabolic parameters [224, 225].

5.9.7 Insulin

In 2007, Eid et al. showed that co-stimulation of HUVECs and HCAECs with insulin (10 nM) or adiponectin (20 μg/ml) for 48 h inhibited dose-dependent TNF-induced ADMA. A reduction in ADMA was a result of an increase in DDAH activity [73]. A study on young people with type 1 diabetes confirmed these findings by showing that ADMA levels were not affected by acute change in glycemia but were significantly reduced by insulin infusion [226]. Insulin sensitivity was shown to be augmented by a decrease in ADMA and an overexpression of DDAH [227].

5.9.8 L-Arginine Supplementation

The supplementation of L-arginine should in theory provide increased substrate for NOS and therefore increase the levels of NO released by cells. In patients with hypertension [228, 229], diabetes [220], and hypercholesterolemia [230, 231], short-term effects of L-arginine infusion do demonstrate improvements in vasodilation and lower levels of ADMA.

However, in 17 human studies on oral L-arginine supplementation, five of them have demonstrated no benefits at all [232].

Two studies by Blum showed that oral L-arginine supplementation (9 g/daily for one month) did not enhance NO synthesis and release in postmenopausal women [233, 234], or improve NO bioavailability in coronary artery disease patients [233, 234]. Chin-Dusting and colleagues [235, 236] measured forearm blood flow and showed that in normal healthy patients, endothelial function was not improved by oral L-arginine supplementation (20 g/day for 28 days) [235], as was the case in patients with heart failure [236] and in some cases was associated with their death [68]. Such results could be explained by the fact that L-arginine supplementation leads to increases in intracellular levels of L-arginine and ADMA and consequently impairs activity of DDAH which is required for the metabolism of ADMA [237].

Increasing evidence associates cardiovascular disease with endothelial dysfunction and dysregulation of the DDAH/ADMA/NO pathway. A number of currently used cardiovascular drugs reduce plasma ADMA concentrations and enhance NO-mediated vascular function. A greater understanding of the regulation of DDAH gene expression and enzyme activity may provide novel therapeutic opportunities for the treatment of cardiovascular diseases.

References

1. McBride AE, Silver PA. State of the arg: protein methylation at arginine comes of age. Cell. 2001;106(1):5–8.
2. Boger RH, Sydow K, et al. LDL cholesterol upregulates synthesis of asymmetrical dimethylarginine in human endothelial cells: involvement of S-adenosylmethionine-dependent methyltransferases. Circ Res. 2000;87(2):99–105.

3. Chen Y, Xu X, et al. PRMT-1 and DDAHs-induced ADMA upregulation is involved in ROS- and RAS-mediated diabetic retinopathy. Exp Eye Res. 2009;89(6):1028–34.
4. Osanai T, Saitoh M, et al. Effect of shear stress on asymmetric dimethylarginine release from vascular endothelial cells. Hypertension. 2003;42(5):985–90.
5. Sasser JM, Moningka NC, et al. Asymmetric dimethylarginine in angiotensin II-induced hypertension. Am J Physiol Regul Integr Comp Physiol. 2010;298(3):R740–6.
6. Leiper J, Vallance P. New tricks from an old dog: nitric oxide-independent effects of dimethylarginine dimethylaminohydrolase. Arterioscler Thromb Vasc Biol. 2006;26(7):1419–20.
7. Vallance P, Leiper J. Cardiovascular biology of the asymmetric dimethylarginine:dimethylargini ne dimethylaminohydrolase pathway. Arterioscler Thromb Vasc Biol. 2004;24(6):1023–30.
8. Bedford MT, Richard S. Arginine methylation an emerging regulator of protein function. Mol Cell. 2005;18(3):263–72.
9. Hasegawa K, Wakino S, et al. Dimethylarginine dimethylaminohydrolase 2 increases vascular endothelial growth factor expression through Sp1 transcription factor in endothelial cells. Arterioscler Thromb Vasc Biol. 2006;26(7):1488–94.
10. Murray-Rust J, Leiper J, et al. Structural insights into the hydrolysis of cellular nitric oxide synthase inhibitors by dimethylarginine dimethylaminohydrolase. Nat Struct Biol. 2001;8(8):679–83.
11. Cooke JP. Does ADMA cause endothelial dysfunction? Arterioscler Thromb Vasc Biol. 2000;20(9):2032–7.
12. Closs EI, Basha FZ, et al. Interference of L-arginine analogues with L-arginine transport mediated by the y+carrier hCAT-2B. Nitric Oxide. 1997;1(1):65–73.
13. Broer A, Wagner CA, et al. The heterodimeric amino acid transporter 4F2hc/y+LAT2 mediates arginine efflux in exchange with glutamine. Biochem J. 2000;349(Pt 3):787–95.
14. Bogle RG, MacAllister RJ, et al. Induction of NG-monomethyl-L-arginine uptake: a mechanism for differential inhibition of NO synthases? Am J Physiol. 1995;269(3 Pt 1):C750–6.
15. Cardounel AJ, Cui H, et al. Evidence for the pathophysiological role of endogenous methylarginines in regulation of endothelial NO production and vascular function. J Biol Chem. 2007;282(2):879–87.
16. Boger RH. Asymmetric dimethylarginine, an endogenous inhibitor of nitric oxide synthase, explains the "L-arginine paradox" and acts as a novel cardiovascular risk factor. J Nutr. 2004;134(10 Suppl):2842S–7. discussion 2853S.
17. Teerlink T, Luo Z, et al. Cellular ADMA: regulation and action. Pharmacol Res. 2009;60(6): 448–60.
18. Mann GE, Yudilevich DL, et al. Regulation of amino acid and glucose transporters in endothelial and smooth muscle cells. Physiol Rev. 2003;83(1):183–252.
19. Parnell MM, Chin-Dusting JP, et al. In vivo and in vitro evidence for ACh-stimulated L-arginine uptake. Am J Physiol Heart Circ Physiol. 2004;287(1):H395–400.
20. Vallance P, Leone A, et al. Accumulation of an endogenous inhibitor of nitric oxide synthesis in chronic renal failure. Lancet. 1992;339(8793):572–5.
21. Vallance P, Leone A, et al. Endogenous dimethylarginine as an inhibitor of nitric oxide synthesis. J Cardiovasc Pharmacol. 1992;20 Suppl 12:S60–2.
22. Sessa WC, Harrison JK, et al. Genomic analysis and expression patterns reveal distinct genes for endothelial and brain nitric oxide synthase. Hypertension. 1993;21(6 Pt 2):934–8.
23. Busse R, Mulsch A. Calcium-dependent nitric oxide synthesis in endothelial cytosol is mediated by calmodulin. FEBS Lett. 1990;265(1–2):133–6.
24. Moncada S, Higgs EA. Nitric oxide and the vascular endothelium. Handb Exp Pharmacol. 2006;176(Pt 1):213–54.
25. Radomski MW, Palmer RM, et al. Glucocorticoids inhibit the expression of an inducible, but not the constitutive, nitric oxide synthase in vascular endothelial cells. Proc Natl Acad Sci USA. 1990;87(24):10043–7.
26. Murad F, Mittal CK, et al. Guanylate cyclase: activation by azide, nitro compounds, nitric oxide, and hydroxyl radical and inhibition by hemoglobin and myoglobin. Adv Cyclic Nucleotide Res. 1978;9:145–58.

27. Clementi E. Role of nitric oxide and its intracellular signalling pathways in the control of Ca2+ homeostasis. Biochem Pharmacol. 1998;55(6):713–8.
28. Eardley I. The role of phosphodiesterase inhibitors in impotence. Expert Opin Investig Drugs. 1997;6(12):1803–10.
29. Assreuy J, Cunha FQ, et al. Feedback inhibition of nitric oxide synthase activity by nitric oxide. Br J Pharmacol. 1993;108(3):833–7.
30. Yasinska IM, Kozhukhar AV, et al. S-nitrosation of thioredoxin in the nitrogen monoxide/superoxide system activates apoptosis signal-regulating kinase 1. Arch Biochem Biophys. 2004;428(2):198–203.
31. Stamler JS, Singel DJ, et al. Biochemistry of nitric oxide and its redox-activated forms. Science. 1992;258(5090):1898–902.
32. Lee JR, Kim JK, et al. Role of protein tyrosine nitration in neurodegenerative diseases and atherosclerosis. Arch Pharm Res. 2009;32(8):1109–18.
33. Venardos K, Zhang WZ, et al. Effect of peroxynitrite on endothelial L-arginine transport and metabolism. Int J Biochem Cell Biol. 2009;41(12):2522–7.
34. Cooke JP. NO and angiogenesis. Atheroscler Suppl. 2003;4(4):53–60.
35. Goligorsky MS, Abedi H, et al. Nitric oxide modulation of focal adhesions in endothelial cells. Am J Physiol. 1999;276(6 Pt 1):C1271–81.
36. Murohara TAT. Nitric oxide and angiogenesis in cardiovascular disease. Antioxid Redox Signal. 2002;4(5):825–31.
37. Olsson AK, Dimberg A, et al. VEGF receptor signalling – in control of vascular function. Nat Rev Mol Cell Biol. 2006;7(5):359–71.
38. Murohara T, Asahara T. Nitric oxide and angiogenesis in cardiovascular disease. Antioxid Redox Signal. 2002;4(5):825–31.
39. Dowling RB, Newton R, et al. Effect of inhibition of nitric oxide synthase on pseudomonas aeruginosa infection of respiratory mucosa in vitro. Am J Respir Cell Mol Biol. 1998;19(6):950–8.
40. Ueda S, Kato S, et al. Regulation of cytokine-induced nitric oxide synthesis by asymmetric dimethylarginine: role of dimethylarginine dimethylaminohydrolase. Circ Res. 2003;92(2):226–33.
41. Boger RH. The emerging role of asymmetric dimethylarginine as a novel cardiovascular risk factor. Cardiovasc Res. 2003;59(4):824–33.
42. Schnabel R, Blankenberg S, et al. Asymmetric dimethylarginine and the risk of cardiovascular events and death in patients with coronary artery disease: results from the AtheroGene Study. Circ Res. 2005;97(5):e53–9.
43. Tanaka M, Sydow K, et al. Dimethylarginine dimethylaminohydrolase overexpression suppresses graft coronary artery disease. Circulation. 2005;112(11):1549–56.
44. Jiang DJ, Jia SJ, et al. Asymmetric dimethylarginine induces apoptosis via p38 MAPK/caspase-3-dependent signaling pathway in endothelial cells. J Mol Cell Cardiol. 2006;40(4):529–39.
45. Wells SM, Holian A. Asymmetric dimethylarginine induces oxidative and nitrosative stress in murine lung epithelial cells. Am J Respir Cell Mol Biol. 2007;36(5):520–8.
46. Furukawa Y, Kimura T. Hypertension in patients with ischemic heart disease. Nippon Rinsho. 2004;62 Suppl 3:465–70.
47. Grattagliano I, Portincasa P, et al. Experimental observations and clinical implications of fasting and diet supplementation in fatty livers. Eur Rev Med Pharmacol Sci. 2003;7(1):1–7.
48. Horowitz JD, Heresztyn T. An overview of plasma concentrations of asymmetric dimethylarginine (ADMA) in health and disease and in clinical studies: methodological considerations. J Chromatogr B Analyt Technol Biomed Life Sci. 2007;851(1–2):42–50.
49. Billecke SS, D'Alecy LG, et al. Blood content of asymmetric dimethylarginine: new insights into its dysregulation in renal disease. Nephrol Dial Transplant. 2009;24(2):489–96.
50. Fleck C, Schweitzer F, et al. Serum concentrations of asymmetric (ADMA) and symmetric (SDMA) dimethylarginine in patients with chronic kidney diseases. Clin Chim Acta. 2003;336(1–2):1–12.

51. Marescau B, Nagels G, et al. Guanidino compounds in serum and urine of nondialyzed patients with chronic renal insufficiency. Metabolism. 1997;46(9):1024–31.
52. Nijveldt RJ, Van Leeuwen PA, et al. Net renal extraction of asymmetrical (ADMA) and symmetrical (SDMA) dimethylarginine in fasting humans. Nephrol Dial Transplant. 2002;17(11): 1999–2002.
53. Schmidt RJ, Baylis C. Total nitric oxide production is low in patients with chronic renal disease. Kidney Int. 2000;58(3):1261–6.
54. Leiper JM, Santa Maria J, et al. Identification of two human dimethylarginine dimethylaminohydrolases with distinct tissue distributions and homology with microbial arginine deiminases. Biochem J. 1999;343(Pt 1):209–14.
55. Tran CT, Fox MF, et al. Chromosomal localization, gene structure, and expression pattern of DDAH1: comparison with DDAH2 and implications for evolutionary origins. Genomics. 2000;68(1):101–5.
56. Kimoto M, Tsuji H, et al. Detection of NG, NG-dimethylarginine dimethylaminohydrolase in the nitric oxide-generating systems of rats using monoclonal antibody. Arch Biochem Biophys. 1993;300(2):657–62.
57. Nijveldt RJ, Teerlink T, et al. The liver is an important organ in the metabolism of asymmetrical dimethylarginine (ADMA). Clin Nutr. 2003;22(1):17–22.
58. Nijveldt RJ, Teerlink T, et al. Asymmetrical dimethylarginine (ADMA) in critically ill patients: high plasma ADMA concentration is an independent risk factor of ICU mortality. Clin Nutr. 2003;22(1):23–30.
59. Tojo A, Welch WJ, et al. Colocalization of demethylating enzymes and NOS and functional effects of methylarginines in rat kidney. Kidney Int. 1997;52(6):1593–601.
60. Leiper J, Nandi M, et al. Disruption of methylarginine metabolism impairs vascular homeostasis. Nat Med. 2007;13(2):198–203.
61. Onozato ML, Tojo A, et al. Expression of NG, NG-dimethylarginine dimethylaminohydrolase and protein arginine N-methyltransferase isoforms in diabetic rat kidney: effects of angiotensin II receptor blockers. Diabetes. 2008;57(1):172–80.
62. Ogawa T, Kimoto M, et al. Purification and properties of a new enzyme, NG, NG-dimethylarginine dimethylaminohydrolase, from rat kidney. J Biol Chem. 1989;264(17):10205–9.
63. Forbes SP, Druhan LJ, et al. Mechanism of 4-HNE mediated inhibition of hDDAH-1: implications in no regulation. Biochemistry. 2008;47(6):1819–26.
64. Hong L, Fast W. Inhibition of human dimethylarginine dimethylaminohydrolase-1 by S-nitroso-L-homocysteine and hydrogen peroxide. Analysis, quantification, and implications for hyperhomocysteinemia. J Biol Chem. 2007;282(48):34684–92.
65. Achan V, Tran CT, et al. All-trans-retinoic acid increases nitric oxide synthesis by endothelial cells: a role for the induction of dimethylarginine dimethylaminohydrolase. Circ Res. 2002;90(7):764–9.
66. Wang J, Sim AS, et al. Relations between plasma asymmetric dimethylarginine (ADMA) and risk factors for coronary disease. Atherosclerosis. 2006;184(2):383–8.
67. Wang J, Sim AS, et al. L-arginine regulates asymmetric dimethylarginine metabolism by inhibiting dimethylarginine dimethylaminohydrolase activity in hepatic (HepG2) cells. Cell Mol Life Sci. 2006;63(23):2838–46.
68. Schulman SP, Becker LC, et al. L-arginine therapy in acute myocardial infarction: the Vascular Interaction with Age in Myocardial Infarction (VINTAGE MI) randomized clinical trial. JAMA. 2006;295(1):58–64.
69. Leiper J, Murray-Rust J, et al. S-nitrosylation of dimethylarginine dimethylaminohydrolase regulates enzyme activity: further interactions between nitric oxide synthase and dimethylarginine dimethylaminohydrolase. Proc Natl Acad Sci USA. 2002;99(21):13527–32.
70. Sakurada M, Shichiri M, et al. Nitric oxide upregulates dimethylarginine dimethylaminohydrolase-2 via cyclic GMP induction in endothelial cells. Hypertension. 2008;52(5):903–9.
71. Holden DP, Cartwright JE, et al. Estrogen stimulates dimethylarginine dimethylaminohydrolase activity and the metabolism of asymmetric dimethylarginine. Circulation. 2003;108(13): 1575–80.

72. Monsalve E, Oviedo PJ, et al. Estradiol counteracts oxidized LDL-induced asymmetric dimethylarginine production by cultured human endothelial cells. Cardiovasc Res. 2007;73(1): 66–72.
73. Eid HM, Lyberg T, et al. Insulin and adiponectin inhibit the TNFalpha-induced ADMA accumulation in human endothelial cells: the role of DDAH. Atherosclerosis. 2007;194(2): e1–8.
74. Scalera F, Martens-Lobenhoffer J, et al. Effect of telmisartan on nitric oxide–asymmetrical dimethylarginine system: role of angiotensin II type 1 receptor gamma and peroxisome proliferator activated receptor gamma signaling during endothelial aging. Hypertension. 2008;51(3): 696–703.
75. Wakino S, Hayashi K. Anti-hypertensive effects of PPARgamma ligands through the inhibition of Rho/Rho kinase pathway. Nippon Rinsho. 2005;63(4):693–9.
76. Tanaka M, Osanai T, et al. Effect of vasoconstrictor coupling factor 6 on gene expression profile in human vascular endothelial cells: enhanced release of asymmetric dimethylarginine. J Hypertens. 2006;24(3):489–97.
77. Ito A, Tsao PS, et al. Novel mechanism for endothelial dysfunction: dysregulation of dimethylarginine dimethylaminohydrolase. Circulation. 1999;99(24):3092–5.
78. Xin HY, Jiang DJ, et al. Regulation by DDAH/ADMA pathway of lipopolysaccharide-induced tissue factor expression in endothelial cells. Thromb Haemost. 2007;97(5):830–8.
79. Wadham C, Mangoni AA. Dimethylarginine dimethylaminohydrolase regulation: a novel therapeutic target in cardiovascular disease. Expert Opin Drug Metab Toxicol. 2009;5(3): 303–19.
80. Boger RH, Bode-Boger SM, et al. Biochemical evidence for impaired nitric oxide synthesis in patients with peripheral arterial occlusive disease. Circulation. 1997;95(8):2068–74.
81. Miyazaki H, Matsuoka H, et al. Endogenous nitric oxide synthase inhibitor: a novel marker of atherosclerosis. Circulation. 1999;99(9):1141–6.
82. Valkonen VP, Paiva H, et al. Risk of acute coronary events and serum concentration of asymmetrical dimethylarginine. Lancet. 2001;358(9299):2127–8.
83. Yoo JH, Lee SC. Elevated levels of plasma homocyst(e)ine and asymmetric dimethylarginine in elderly patients with stroke. Atherosclerosis. 2001;158(2):425–30.
84. Zoccali C, Bode-Boger S, et al. Plasma concentration of asymmetrical dimethylarginine and mortality in patients with end-stage renal disease: a prospective study. Lancet. 2001;358(9299): 2113–7.
85. Boger RH, Bode-Boger SM, et al. Asymmetric dimethylarginine (ADMA): a novel risk factor for endothelial dysfunction: its role in hypercholesterolemia. Circulation. 1998;98(18): 1842–7.
86. Lundman P, Eriksson MJ, et al. Mild to-moderate hypertriglyceridemia in young men is associated with endothelial dysfunction and increased plasma concentrations of asymmetric dimethylarginine. J Am Coll Cardiol. 2001;38(1):111–6.
87. Sydow K, Munzel T. ADMA and oxidative stress. Atheroscler Suppl. 2003;4(4):41–51.
88. Wang D, Gill PS, et al. Isoform-specific regulation by N(G), N(G)-dimethylarginine dimethylaminohydrolase of rat serum asymmetric dimethylarginine and vascular endothelium-derived relaxing factor/NO. Circ Res. 2007;101(6):627–35.
89. Smith CL, Birdsey GM, et al. Dimethylarginine dimethylaminohydrolase activity modulates ADMA levels, VEGF expression, and cell phenotype. Biochem Biophys Res Commun. 2003;308(4):984–9.
90. Kostourou V, Robinson SP, et al. Dimethylarginine dimethylaminohydrolase I enhances tumour growth and angiogenesis. Br J Cancer. 2002;87(6):673–80.
91. Konishi H, Sydow K, et al. Dimethylarginine dimethylaminohydrolase promotes endothelial repair after vascular injury. J Am Coll Cardiol. 2007;49(10):1099–105.
92. Achan V, Ho HK, et al. ADMA regulates angiogenesis: genetic and metabolic evidence. Vasc Med. 2005;10(1):7–14.
93. Wojciak-Stothard B, Torondel B, et al. The ADMA/DDAH pathway is a critical regulator of endothelial cell motility. J Cell Sci. 2007;120(Pt 6):929–42.

94. Hoefen RJ, Berk BC. The role of MAP kinases in endothelial activation. Vascul Pharmacol. 2002;38(5):271–3.
95. Thum T, Tsikas D, et al. Suppression of endothelial progenitor cells in human coronary artery disease by the endogenous nitric oxide synthase inhibitor asymmetric dimethylarginine. J Am Coll Cardiol. 2005;46(9):1693–701.
96. Peterson DA, Peterson DC, et al. The non specificity of specific nitric oxide synthase inhibitors. Biochem Biophys Res Commun. 1992;187(2):797–801.
97. Buxton IL, Cheek DJ, et al. NG-nitro L-arginine methyl ester and other alkyl esters of arginine are muscarinic receptor antagonists. Circ Res. 1993;72(2):387–95.
98. Brusilow SW, Horwich AL, Urea cycle enzymes, Scribres C, Beardet A, slyw, Valle D, editors. The metabolic basis of inherited disease (6th), Mc Graw-Hill, New York 1989; p. 629–63
99. Juretić ASG, Hörig H, Gross T, Gallati H, Samija M, Eljuga D, et al. Nitric oxide-independent inhibitory effects of L-arginine analog NG-monomethy-L-arginine on the generation of interleukin-2 activated cytotoxic activity in humans. Clin Nutr. 1996;15(1):16–20.
100. Matsumoto Y, Ueda S, et al. Dimethylarginine dimethylaminohydrolase prevents progression of renal dysfunction by inhibiting loss of peritubular capillaries and tubulointerstitial fibrosis in a rat model of chronic kidney disease. J Am Soc Nephrol. 2007;18(5):1525–33.
101. Suda O, Tsutsui M, et al. Long-term treatment with N(omega)-nitro-L-arginine methyl ester causes arteriosclerotic coronary lesions in endothelial nitric oxide synthase-deficient mice. Circulation. 2002;106(13):1729–35.
102. Suda O, Tsutsui M, et al. Asymmetric dimethylarginine produces vascular lesions in endothelial nitric oxide synthase-deficient mice: involvement of renin-angiotensin system and oxidative stress. Arterioscler Thromb Vasc Biol. 2004;24(9):1682–8.
103. Hultstrom M, Helle F, et al. AT(1) receptor activation regulates the mRNA expression of CAT1, CAT2, arginase-1, and DDAH2 in preglomerular vessels from angiotensin II hypertensive rats. Am J Physiol Renal Physiol. 2009;297(1):F163–8.
104. Helle F, Hultstrom M, et al. Angiotensin II-induced contraction is attenuated by nitric oxide in afferent arterioles from the nonclipped kidney in 2K1C. Am J Physiol Renal Physiol. 2009;296(1):F78–86.
105. Tran CT, Leiper JM, et al. The DDAH/ADMA/NOS pathway. Atheroscler Suppl. 2003;4(4):33–40.
106. Mookerjee RP, Malaki M, et al. Increasing dimethylarginine levels are associated with adverse clinical outcome in severe alcoholic hepatitis. Hepatology. 2007;45(1):62–71.
107. Baylis C. Arginine, arginine analogs and nitric oxide production in chronic kidney disease. Nat Clin Pract Nephrol. 2006;2(4):209–20.
108. Kielstein JT, Salpeter SR, et al. Symmetric dimethylarginine (SDMA) as endogenous marker of renal function – a meta-analysis. Nephrol Dial Transplant. 2006;21(9):2446–51.
109. Matsuguma K, Ueda S, et al. Molecular mechanism for elevation of asymmetric dimethylarginine and its role for hypertension in chronic kidney disease. J Am Soc Nephrol. 2006;17(8):2176–83.
110. Kielstein JT, Boger RH, et al. Marked increase of asymmetric dimethylarginine in patients with incipient primary chronic renal disease. J Am Soc Nephrol. 2002;13(1):170–6.
111. Fliser D, Kronenberg F, et al. Asymmetric dimethylarginine and progression of chronic kidney disease: the mild to moderate kidney disease study. J Am Soc Nephrol. 2005;16(8): 2456–61.
112. Zatz R, Baylis C. Chronic nitric oxide inhibition model six years on. Hypertension. 1998;32(6):958–64.
113. Kang DH, Nakagawa T, et al. Nitric oxide modulates vascular disease in the remnant kidney model. Am J Pathol. 2002;161(1):239–48.
114. Ravani P, Tripepi G, et al. Asymmetrical dimethylarginine predicts progression to dialysis and death in patients with chronic kidney disease: a competing risks modeling approach. J Am Soc Nephrol. 2005;16(8):2449–55.
115. Abedini S, Meinitzer A, et al. Asymmetrical dimethylarginine is associated with renal and cardiovascular outcomes and all-cause mortality in renal transplant recipients. Kidney Int. 2010;77(1):44–50.

116. Chan CT, Harvey PJ, et al. Dissociation between the short-term effects of nocturnal hemodialysis on endothelium dependent vasodilation and plasma ADMA. Arterioscler Thromb Vasc Biol. 2005;25(12):2685–6.
117. Baigent C, Burbury K, et al. Premature cardiovascular disease in chronic renal failure. Lancet. 2000;356(9224):147–52.
118. Brunner H, Cockcroft JR, et al. Endothelial function and dysfunction. Part II: association with cardiovascular risk factors and diseases. A statement by the Working Group on Endothelins and Endothelial Factors of the European Society of Hypertension. J Hypertens. 2005;23(2):233–46.
119. D'Agostino Sr RB, Vasan RS, et al. General cardiovascular risk profile for use in primary care: the Framingham Heart Study. Circulation. 2008;117(6):743–53.
120. Ducloux D, Kazory A, et al. Predicting coronary heart disease in renal transplant recipients: a prospective study. Kidney Int. 2004;66(1):441–7.
121. Kasiske BL, Chakkera HA, et al. Explained and unexplained ischemic heart disease risk after renal transplantation. J Am Soc Nephrol. 2000;11(9):1735–43.
122. Zoccali C, Benedetto FA, et al. Asymmetric dimethylarginine, C-reactive protein, and carotid intima-media thickness in end-stage renal disease. J Am Soc Nephrol. 2002;13(2):490–6.
123. Boger RH, Bode-Boger SM, et al. Plasma concentration of asymmetric dimethylarginine, an endogenous inhibitor of nitric oxide synthase, is elevated in monkeys with hyperhomocyst(e)inemia or hypercholesterolemia. Arterioscler Thromb Vasc Biol. 2000;20(6):1557–64.
124. Lentz SR, Rodionov RN, et al. Hyperhomocysteinemia, endothelial dysfunction, and cardiovascular risk: the potential role of ADMA. Atheroscler Suppl. 2003;4(4):61–5.
125. Achan V, Broadhead M, et al. Asymmetric dimethylarginine causes hypertension and cardiac dysfunction in humans and is actively metabolized by dimethylarginine dimethylaminohydrolase. Arterioscler Thromb Vasc Biol. 2003;23(8):1455–9.
126. Maas R, Schulze F, et al. Asymmetric dimethylarginine, smoking, and risk of coronary heart disease in apparently healthy men: prospective analysis from the population-based Monitoring of Trends and Determinants in Cardiovascular Disease/Kooperative Gesundheitsforschung in der Region Augsburg study and experimental data. Clin Chem. 2007;53(4):693–701.
127. Schulze F, Lenzen H, et al. Asymmetric dimethylarginine is an independent risk factor for coronary heart disease: results from the multicenter Coronary Artery Risk Determination investigating the Influence of ADMA Concentration (CARDIAC) study. Am Heart J. 2006;152(3):493 e1–8.
128. Stuhlinger MC, Oka RK, et al. Endothelial dysfunction induced by hyperhomocyst(e)inemia: role of asymmetric dimethylarginine. Circulation. 2003;108(8):933–8.
129. Stuhlinger MC, Tsao PS, et al. Homocysteine impairs the nitric oxide synthase pathway: role of asymmetric dimethylarginine. Circulation. 2001;104(21):2569–75.
130. Mato JM, Lu SC. Homocysteine, the bad thiol. Hepatology. 2005;41(5):976–9.
131. Boger RH, Lentz SR, et al. Elevation of asymmetrical dimethylarginine may mediate endothelial dysfunction during experimental hyperhomocyst(e)inaemia in humans. Clin Sci (Lond). 2001;100(2):161–7.
132. Abdelwhab S, Lotfy G, et al. Relation between asymmetric dimethylarginine (ADMA) and hearing loss in patients with renal impairment. Ren Fail. 2008;30(9):877–83.
133. Sarafidis PA, Khosla N, et al. Antihypertensive therapy in the presence of proteinuria. Am J Kidney Dis. 2007;49(1):12–26.
134. Stefanadi E, Tousoulis D, et al. Inflammatory biomarkers predicting events in atherosclerosis. Curr Med Chem. 2010;17(16):1690–707.
135. Arnal JF, Michel JB, Harrison DG. Nitric oxide in the pathogenesis of hypertension. Curr Opin Nephrol Hypertens. 1995;4(2):182–8.
136. Dominiczak AF, Bohr DF. Nitric oxide and its putative role in hypertension. Hypertension. 1995;25(6):1202–11.
137. Curgunlu A, Uzun H, et al. Increased circulating concentrations of asymmetric dimethylarginine (ADMA) in white coat hypertension. J Hum Hypertens. 2005;19(8):629–33.

138. Ito A, Egashira K, et al. Renin-angiotensin system is involved in the mechanism of increased serum asymmetric dimethylarginine in essential hypertension. Jpn Circ J. 2001;65(9):775–8.
139. Perticone F, Sciacqua A, et al. Asymmetric dimethylarginine, L-arginine, and endothelial dysfunction in essential hypertension. J Am Coll Cardiol. 2005;46(3):518–23.
140. Surdacki A, Nowicki M, et al. Reduced urinary excretion of nitric oxide metabolites and increased plasma levels of asymmetric dimethylarginine in men with essential hypertension. J Cardiovasc Pharmacol. 1999;33(4):652–8.
141. Paiva H, Laakso J, et al. Asymmetric dimethylarginine and hemodynamic regulation in middle-aged men. Metabolism. 2006;55(6):771–7.
142. Schulze F, Maas R, et al. Determination of a reference value for N(G), N(G)-dimethyl-L-arginine in 500 subjects. Eur J Clin Invest. 2005;35(10):622–6.
143. Xia W, Feng W, et al. Increased levels of asymmetric dimethylarginine and C-reactive protein are associated with impaired vascular reactivity in essential hypertension. Clin Exp Hypertens. 2010;32(1):43–8.
144. Wang D, Strandgaard S, et al. Asymmetric dimethylarginine, oxidative stress, and vascular nitric oxide synthase in essential hypertension. Am J Physiol Regul Integr Comp Physiol. 2009;296(2):R195–200.
145. Leonard AM, Chafe LL, et al. Increased salt-sensitivity in endothelial nitric oxide synthase-knockout mice. Am J Hypertens. 2006;19(12):1264–9.
146. Carlstrom M, Brown RD, et al. Role of nitric oxide deficiency in the development of hypertension in hydronephrotic animals. Am J Physiol Renal Physiol. 2008;294(2):F362–70.
147. Dekker JM, Girman C, et al. Metabolic syndrome and 10-year cardiovascular disease risk in the Hoorn Study. Circulation. 2005;112(5):666–73.
148. Reaven GM. Insulin resistance/compensatory hyperinsulinemia, essential hypertension, and cardiovascular disease. J Clin Endocrinol Metab. 2003;88(6):2399–403.
149. Reaven GM, Chen YD. Role of abnormal free fatty acid metabolism in the development of non-insulin-dependent diabetes mellitus. Am J Med. 1988;85(5A):106–12.
150. Koh KK, Han SH, et al. Inflammatory markers and the metabolic syndrome: insights from therapeutic interventions. J Am Coll Cardiol. 2005;46(11):1978–85.
151. Frisbee JC, Samora JB, et al. Exercise training blunts microvascular rarefaction in the metabolic syndrome. Am J Physiol Heart Circ Physiol. 2006;291(5):H2483–92.
152. Roberts CK, Barnard RJ, et al. A high-fat, refined-carbohydrate diet induces endothelial dysfunction and oxidant/antioxidant imbalance and depresses NOS protein expression. J Appl Physiol. 2005;98(1):203–10.
153. Roberts CK, Vaziri ND, et al. Enhanced NO inactivation and hypertension induced by a high-fat, refined-carbohydrate diet. Hypertension. 2000;36(3):423–9.
154. Sun YX, Hu SJ, et al. Plasma levels of vWF and NO in patients with metabolic syndrome and their relationship with metabolic disorders. Zhejiang Da Xue Xue Bao Yi Xue Ban. 2006;35(3):315–8.
155. Tesauro M, Schinzari F, et al. Ghrelin improves endothelial function in patients with metabolic syndrome. Circulation. 2005;112(19):2986–92.
156. Stuhlinger MC, Abbasi F, et al. Relationship between insulin resistance and an endogenous nitric oxide synthase inhibitor. JAMA. 2002;287(11):1420–6.
157. Facchini FS, Hua N, et al. Insulin resistance as a predictor of age-related diseases. J Clin Endocrinol Metab. 2001;86(8):3574–8.
158. Annuk M, Zilmer M et al. Endothelium-dependent vasodilation and oxidative stress in chronic renal failure: impact on cardiovascular disease. Kidney Int. 2003;Suppl(84):S50–3.
159. Lu TM, Ding YA, et al. Effect of rosuvastatin on plasma levels of asymmetric dimethylarginine in patients with hypercholesterolemia. Am J Cardiol. 2004;94(2):157–61.
160. Feher MD, Elkeles RS. Lipid modification and coronary heart disease in type 2 diabetes: different from the general population? Heart. 1999;81(1):10–1.
161. Laakso M. Hyperglycemia as a risk factor for cardiovascular disease in type 2 diabetes. Prim Care. 1999;26(4):829–39.

162. Choi JW, Pai SH, et al. Increases in nitric oxide concentrations correlate strongly with body fat in obese humans. Clin Chem. 2001;47(6):1106–9.
163. Baron AD, Clark MG. Role of blood flow in the regulation of muscle glucose uptake. Annu Rev Nutr. 1997;17:487–99.
164. Montagnani M, Chen H, et al. Insulin-stimulated activation of eNOS is independent of Ca2+ but requires phosphorylation by Akt at Ser(1179). J Biol Chem. 2001;276(32):30392–8.
165. Muniyappa R, Lee S, et al. Current approaches for assessing insulin sensitivity and resistance in vivo: advantages, limitations, and appropriate usage. Am J Physiol Endocrinol Metab. 2008;294(1):E15–26.
166. Hsueh WA, Quinones MJ. Role of endothelial dysfunction in insulin resistance. Am J Cardiol. 2003;92(4A):10J–7.
167. Zeng G, Quon MJ. Insulin-stimulated production of nitric oxide is inhibited by wortmannin. Direct measurement in vascular endothelial cells. J Clin Invest. 1996;98(4):894–8.
168. Gonzalez M, Flores C, et al. Cell signalling-mediating insulin increase of mRNA expression for cationic amino acid transporters-1 and -2 and membrane hyperpolarization in human umbilical vein endothelial cells. Pflugers Arch. 2004;448(4):383–94.
169. Lin KY, Ito A, et al. Impaired nitric oxide synthase pathway in diabetes mellitus: role of asymmetric dimethylarginine and dimethylarginine dimethylaminohydrolase. Circulation. 2002;106(8):987–92.
170. Xiong Y, Fu YF, et al. Elevated levels of the serum endogenous inhibitor of nitric oxide synthase and metabolic control in rats with streptozotocin-induced diabetes. J Cardiovasc Pharmacol. 2003;42(2):191–6.
171. Abbasi F, Asagmi T, et al. Plasma concentrations of asymmetric dimethylarginine are increased in patients with type 2 diabetes mellitus. Am J Cardiol. 2001;88(10):1201–3.
172. Paiva H, Lehtimaki T, et al. Plasma concentrations of asymmetric-dimethyl-arginine in type 2 diabetes associate with glycemic control and glomerular filtration rate but not with risk factors of vasculopathy. Metabolism. 2003;52(3):303–7.
173. Tarnow L, Hovind P, et al. Elevated plasma asymmetric dimethylarginine as a marker of cardiovascular morbidity in early diabetic nephropathy in type 1 diabetes. Diabetes Care. 2004;27(3):765–9.
174. Jiang JL, Zhang XH, et al. Probucol decreases asymmetrical dimethylarginine level by alternation of protein arginine methyltransferase I and dimethylarginine dimethylaminohydrolase activity. Cardiovasc Drugs Ther. 2006;20(4):281–94.
175. McLaughlin T, Stuhlinger M, et al. Plasma asymmetric dimethylarginine concentrations are elevated in obese insulin-resistant women and fall with weight loss. J Clin Endocrinol Metab. 2006;91(5):1896–900.
176. Krzyzanowska K, Mittermayer F, et al. Weight loss reduces circulating asymmetrical dimethylarginine concentrations in morbidly obese women. J Clin Endocrinol Metab. 2004;89(12): 6277–81.
177. Abhary S, Burdon KP, et al. Sequence variation in DDAH1 and DDAH2 genes is strongly and additively associated with serum ADMA concentrations in individuals with type 2 diabetes. PLoS One. 2010;5(3):e9462.
178. Kim JA, Montagnani M, et al. Reciprocal relationships between insulin resistance and endothelial dysfunction: molecular and pathophysiological mechanisms. Circulation. 2006;113(15):1888–904.
179. Ellger B, Debaveye Y, et al. Survival benefits of intensive insulin therapy in critical illness: impact of maintaining normoglycemia versus glycemia-independent actions of insulin. Diabetes. 2006;55(4):1096–105.
180. Sorrenti V, Mazza F, et al. High glucose-mediated imbalance of nitric oxide synthase and dimethylarginine dimethylaminohydrolase expression in endothelial cells. Curr Neurovasc Res. 2006;3(1):49–54.
181. Ellger B, Richir MC, et al. Glycemic control modulates arginine and asymmetrical-dimethylarginine levels during critical illness by preserving dimethylarginine-dimethylaminohydrolase activity. Endocrinology. 2008;149(6):3148–57.

182. Devangelio E, Santilli F, et al. Soluble RAGE in type 2 diabetes: association with oxidative stress. Free Radic Biol Med. 2007;43(4):511–8.
183. Santilli F, Bucciarelli L, et al. Decreased plasma soluble RAGE in patients with hypercholesterolemia: effects of statins. Free Radic Biol Med. 2007;43(9):1255–62.
184. Lai YL, Aoyama S, et al. Inhibition of L-arginine metabolizing enzymes by L-arginine-derived advanced glycation end products. J Clin Biochem Nutr. 2010;46(2):177–85.
185. Yin QF, Xiong Y. Pravastatin restores DDAH activity and endothelium-dependent relaxation of rat aorta after exposure to glycated protein. J Cardiovasc Pharmacol Res. 2005;45(6): 525–32.
186. Kielstein JT, Frolich JC, et al. ADMA (asymmetric dimethylarginine): an atherosclerotic disease mediating agent in patients with renal disease? Nephrol Dial Transplant. 2001;16(9): 1742–5.
187. Munzel T, Keaney Jr JF. Are ACE inhibitors a "magic bullet" against oxidative stress? Circulation. 2001;104(13):1571–4.
188. Delles C, Schneider MP, et al. Angiotensin converting enzyme inhibition and angiotensin II AT1-receptor blockade reduce the levels of asymmetrical N(G), N(G)-dimethylarginine in human essential hypertension. Am J Hypertens. 2002;15(7 Pt 1):590–3.
189. Galle J, Schwedhelm E, et al. Antiproteinuric effects of angiotensin receptor blockers: telmisartan versus valsartan in hypertensive patients with type 2 diabetes mellitus and overt nephropathy. Nephrol Dial Transplant. 2008;23(10):3174–83.
190. Laufs U, La Fata V, et al. Upregulation of endothelial nitric oxide synthase by HMG CoA reductase inhibitors. Circulation. 1998;97(12):1129–35.
191. Wagner AH, Kohler T, et al. Improvement of nitric oxide-dependent vasodilatation by HMG-CoA reductase inhibitors through attenuation of endothelial superoxide anion formation. Arterioscler Thromb Vasc Biol. 2000;20(1):61–9.
192. Young JM, Strey CH, et al. Effect of atorvastatin on plasma levels of asymmetric dimethylarginine in patients with non-ischaemic heart failure. Eur J Heart Fail. 2008;10(5): 463–6.
193. Nanayakkara PW, Kiefte-de Jong JC, et al. Randomized placebo-controlled trial assessing a treatment strategy consisting of pravastatin, vitamin E, and homocysteine lowering on plasma asymmetric dimethylarginine concentration in mild to moderate CKD. Am J Kidney Dis. 2009;53(1):41–50.
194. Olsson AG, Pears J, et al. Effect of rosuvastatin on low-density lipoprotein cholesterol in patients with hypercholesterolemia. Am J Cardiol. 2001;88(5):504–8.
195. Jones SP, Gibson MF, et al. Direct vascular and cardioprotective effects of rosuvastatin, a new HMG-CoA reductase inhibitor. J Am Coll Cardiol. 2002;40(6):1172–8.
196. Ivashchenko CY, Bradley BT, et al. Regulation of the ADMA-DDAH system in endothelial cells: a novel mechanism for the sterol response element binding proteins, SREBP1c and −2. Am J Physiol Heart Circ Physiol. 2010;298(1):H251–8.
197. Katagiri H, Yamada T, et al. Adiposity and cardiovascular disorders: disturbance of the regulatory system consisting of humoral and neuronal signals. Circ Res. 2007;101(1): 27–39.
198. Jun T, Ke-yan F, et al. Increased superoxide anion production in humans: a possible mechanism for the pathogenesis of hypertension. J Hum Hypertens. 1996;10(5):305–9.
199. Lacy F, O'Connor DT, et al. Plasma hydrogen peroxide production in hypertensives and normotensive subjects at genetic risk of hypertension. J Hypertens. 1998;16(3): 291–303.
200. Jiang JL, Li Ns NS, et al. Probucol preserves endothelial function by reduction of the endogenous nitric oxide synthase inhibitor level. Br J Pharmacol. 2002;135(5):1175–82.
201. Sener G, Ozer Sehirli A, et al. Taurine treatment protects against chronic nicotine-induced oxidative changes. Fundam Clin Pharmacol. 2005;19(2):155–64.
202. Wu QD, Wang JH, et al. Taurine prevents high-glucose-induced human vascular endothelial cell apoptosis. Am J Physiol. 1999;277(6 Pt 1):C1229–38.
203. Tan B, Jiang DJ, et al. Taurine protects against low-density lipoprotein-induced endothelial dysfunction by the DDAH/ADMA pathway. Vascul Pharmacol. 2007;46(5):338–45.

204. Xiao HB, Jun F, et al. Protective effects of kaempferol against endothelial damage by an improvement in nitric oxide production and a decrease in asymmetric dimethylarginine level. Eur J Pharmacol. 2009;616(1–3):213–22.
205. Maret W. Zinc coordination environments in proteins as redox sensors and signal transducers. Antioxid Redox Signal. 2006;8(9–10):1419–41.
206. Rios-Vazquez R, Marzoa-Rivas R, et al. Peroxisome proliferator-activated receptor-gamma agonists for management and prevention of vascular disease in patients with and without diabetes mellitus. Am J Cardiovasc Drugs. 2006;6(4):231–42.
207. Habib ZA, Havstad SL, et al. Thiazolidinedione use and the longitudinal risk of fractures in patients with type 2 diabetes mellitus. J Clin Endocrinol Metab. 2010;95(2):592–600.
208. Wang TD, Chen WJ, et al. Relation of improvement in endothelium-dependent flow-mediated vasodilation after rosiglitazone to changes in asymmetric dimethylarginine, endothelin-1, and C-reactive protein in nondiabetic patients with the metabolic syndrome. Am J Cardiol. 2006;98(8):1057–62.
209. Savoia C, Ebrahimian T, et al. Countervailing vascular effects of rosiglitazone in high cardiovascular risk mice: role of oxidative stress and PRMT-1. Clin Sci (Lond). 2010;118(9):583–92.
210. Kelly AS, Thelen AM, et al. Rosiglitazone improves endothelial function and inflammation but not asymmetric dimethylarginine or oxidative stress in patients with type 2 diabetes mellitus. Vasc Med. 2007;12(4):311–8.
211. Richir MC, Ellger B, et al. The effect of rosiglitazone on asymmetric dimethylarginine (ADMA) in critically ill patients. Pharmacol Res. 2009;60(6):519–24.
212. Mittermayer F, Schaller G, et al. Rosiglitazone prevents free fatty acid-induced vascular endothelial dysfunction. J Clin Endocrinol Metab. 2007;92(7):2574–80.
213. Khan NA, Wiernsperger N, et al. Characterization of metformin transport system in NIH 3T3 cells. J Cell Physiol. 1992;152(2):310–6.
214. Asagami T, Abbasi F, et al. Metformin treatment lowers asymmetric dimethylarginine concentrations in patients with type 2 diabetes. Metabolism. 2002;51(7):843–6.
215. Heutling D, Schulz H, et al. Asymmetrical dimethylarginine, inflammatory and metabolic parameters in women with polycystic ovary syndrome before and after metformin treatment. J Clin Endocrinol Metab. 2008;93(1):82–90.
216. Ozgurtas T, Oktenli C, et al. Metformin and oral contraceptive treatments reduced circulating asymmetric dimethylarginine (ADMA) levels in patients with polycystic ovary syndrome (PCOS). Atherosclerosis. 2008;200(2):336–44.
217. Marcovecchio ML, Widmer B, et al. Effect of acute variations of insulin and glucose on plasma concentrations of asymmetric dimethylarginine in young people with type 1 diabetes. Clin Sci (Lond). 2008;115(12):361–9.
218. Sydow K, Mondon CE, et al. Dimethylarginine dimethylaminohydrolase overexpression enhances insulin sensitivity. Arterioscler Thromb Vasc Biol. 2008;28(4):692–7.
219. Panza JA, Casino PR, et al. Effect of increased availability of endothelium-derived nitric oxide precursor on endothelium-dependent vascular relaxation in normal subjects and in patients with essential hypertension. Circulation. 1993;87(5):1475–81.
220. Nitenberg A, Paycha F, et al. Coronary artery responses to physiological stimuli are improved by deferoxamine but not by L-arginine in non-insulin-dependent diabetic patients with angiographically normal coronary arteries and no other risk factors. Circulation. 1998;97(8):736–43.
221. Creager MA, Gallagher SJ, et al. L-arginine improves endothelium-dependent vasodilation in hypercholesterolemic humans. J Clin Invest. 1992;90(4):1248–53.
222. Preli RB, Klein KP, et al. Vascular effects of dietary L-arginine supplementation. Atherosclerosis. 2002;162(1):1–15.
223. Blum A, Hathaway L, et al. Effects of oral L-arginine on endothelium-dependent vasodilation and markers of inflammation in healthy postmenopausal women. J Am Coll Cardiol. 2000;35(2):271–6.
224. Blum A, Hathaway L, et al. Oral L-arginine in patients with coronary artery disease on medical management. Circulation. 2000;101(18):2160–4.

225. Chin-Dusting JP, Alexander CT, et al. Effects of in vivo and in vitro L-arginine supplementation on healthy human vessels. J Cardiovasc Pharmacol. 1996;28(1):158–66.
226. Chin-Dusting JP, Kaye DM, et al. Dietary supplementation with L-arginine fails to restore endothelial function in forearm resistance arteries of patients with severe heart failure. J Am Coll Cardiol. 1996;27(5):1207–13.
227. Wilcken DE, Sim AS, et al. Asymmetric dimethylarginine (ADMA) in vascular, renal and hepatic disease and the regulatory role of L-arginine on its metabolism. Mol Genet Metab. 2007;91(4):309–17. discussion 308.
228. Arrigoni FI, Vallance P, et al. Metabolism of asymmetric dimethylarginines is regulated in the lung developmentally and with pulmonary hypertension induced by hypobaric hypoxia. Circulation. 2003;107(8):1195–201.
229. Celik T, Iyisoy A, et al. The beneficial effects of angiotensin-converting enzyme inhibitors on serum asymmetric dimethylarginine levels in the patients with cardiovascular disease. Int J Cardiol. 2010;142(1):107–9.
230. Hsueh WA, Bruemmer D. Peroxisome proliferator-activated receptor gamma: implications for cardiovascular disease. Hypertension. 2004;43(2):297–305.
231. Kielstein JT, Impraim B, et al. Cardiovascular effects of systemic nitric oxide synthase inhibition with asymmetrical dimethylarginine in humans. Circulation. 2004;109(2):172–7.
232. Mehta JL, Hu B, et al. Pioglitazone inhibits LOX-1 expression in human coronary artery endothelial cells by reducing intracellular superoxide radical generation. Arterioscler Thromb Vasc Biol. 2003;23(12):2203–8.
233. Ngo DT, Sverdlov AL, et al. Correlates of arterial stiffness in an ageing population: role of asymmetric dimethylarginine. Pharmacol Res. 2009;60(6):503–7.
234. Organisation, W.-W. H. (current). Cardiovascular Diseases. From http://www.who.int/topics/cardiovascular_diseases/en/.
235. Polikandriotis JA, Mazzella LJ, et al. Peroxisome proliferator-activated receptor gamma ligands stimulate endothelial nitric oxide production through distinct peroxisome proliferator-activated receptor gamma-dependent mechanisms. Arterioscler Thromb Vasc Biol. 2005;25(9):1810–6.
236. Pope AJ, Karrupiah K, et al. Role of dimethylarginine dimethylaminohydrolases in the regulation of endothelial nitric oxide production. J Biol Chem. 2009;284(51):35338–47.
237. Sydow K, Schwedhelm E, et al. ADMA and oxidative stress are responsible for endothelial dysfunction in hyperhomocyst(e)inemia: effects of L-arginine and B vitamins. Cardiovasc Res. 2003;57(1):244–52.

Potassium Channels Regulating the Electrical Activity of the Heart

6

Andrew Tinker and Stephen C. Harmer

6.1 Introduction

The potassium conductance in cardiac myocytes governs repolarization during the action potential, sets the resting membrane potential, and responds to hormonal and metabolic changes. Since the 1950s, the use of electrophysiological techniques has led to an appreciation of the large diversity of these currents and the reconstruction of their role in cardiac physiology using mathematical models [1, 2]. The $Na^+\backslash K^+$ ATPase is largely responsible for establishing the ionic gradients underlying excitability, but it is the temporally coordinated flux through sodium, calcium, and potassium ion channels that determines the trajectory and properties of the action potential in a myocyte. Ion channels are protein pores in the membrane that allow a high flux of ions down their electrochemical gradients and often show high selectivity between different ions. Cloning efforts revealed the molecular species underlying these proteins in the 1990s. During this time, it also became clear that genetic defects in these proteins were responsible for human cardiac disease. In this chapter, we are going to discuss this interface between human disease and basic potassium channel biology. In particular, we will focus on the molecular pathogenesis of diseases that have been associated with potassium channel defects, and the implications for therapeutics in both hereditary and the commoner non-hereditary cardiac pathology.

A. Tinker (✉) • S.C. Harmer
Department of Medicine, University College London,
London, UK
e-mail: a.tinker@ucl.ac.uk

D. Abraham et al. (eds.), *Translational Vascular Medicine*,
DOI 10.1007/978-0-85729-920-8_6,

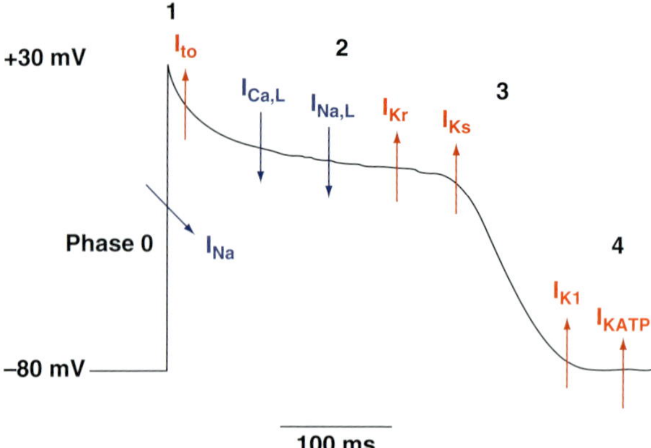

Fig. 6.1 A schematic of the ventricular cardiac action potential present in man. The fast upstroke seen in *phase 0* is generated by the activation of a sodium current (I_{Na}). The following downward deflection/notch, *phase 1*, is then formed by the activation and rapid inactivation of a K^+ current I_{to}. After *phase 1*, the action potential enters a plateau phase, early *phase 2*, which is maintained by the entry of Ca^{2+} ions through L-type Ca^{2+} channels ($I_{Ca,L}$) and a small amount of late sodium current ($I_{Na,L}$). Three K^+ currents, I_{Kr}, I_{Ks}, and I_{K1}, then act in a concerted fashion to repolarize the heart, *phases 2* and *3*. In *Phase 4*, I_{K1} and I_{KATP} act to set the membrane potential. *Blue arrows* indicate an inward flow of ions. *Red arrows* indicate an outward flow of ions

6.2 The Human Cardiac Action Potential and Three Repolarizing K^+ Currents

The starting point for discussion is the action potential in human ventricular cardiac myocytes. The initial depolarization is mediated by Na^+ entry via the sodium channel, and the subsequent plateau and repolarization is shaped by Ca^{2+} entry and outward K^+ currents as illustrated in Fig. 6.1. The action potential waveform actually varies within different regions of the heart and even between the endocardium and epicardium of the same chamber. For example, in the SA (sinoatrial) node, there are several unique currents such as the hyperpolarization-activated cation current responsible for pacemaker depolarization, the G-protein-gated inwardly rectifying K^+ current (GIRK/I_{KAch}), and the initial action potential depolarization is mediated by Ca^{2+} entry with little contribution from sodium currents [3]. In particular, we are going to focus on three K^+ currents responsible for the terminal repolarization of the ventricular cardiac action potential namely I_{Kr}, I_{Ks}, and I_{K1}. In the ventricular myocytes of large mammals, including man, there is a K^+ current that characteristically activates with a delay ("delayed rectifier") and this was originally assumed to be a single current ("I_K"). In contrast, in smaller mammals, such as the rat, terminal repolarization is determined by a transient outward K^+ current [4]. The use of E-4031 refined the picture of I_K leading to the pharmacological separation of two currents namely I_{Kr} and I_{Ks} [5]. I_{K1} is the classical strong inward rectifier first identified by Weidmann in sheep Purkinje fibers and subsequently in other species [6].

Potassium channels are oligomeric complexes consisting of pore-forming alpha subunits often in complex with beta subunits that can critically alter trafficking and function of the alpha subunit. The alpha subunits of voltage-gated channels (K_v) and inwardly rectifying channels (K_{ir}) are tetramers while twin pore channels (K_{2P}) are dimers [7]. The alpha subunits underlying I_{Kr} (HERG, Kv11.1) and I_{Ks} (KCNQ1, KvLQT1, Kv7.1) are members of the voltage-gated family of K^+ channels and have six transmembrane domains. In contrast, the pore-forming subunits of the I_{K1} are members of the inward rectifier family and have two transmembrane domains.

6.2.1 I_{Kr}

The use of E-4031 allowed the separation of I_K into two components. The drug-sensitive component was labeled I_{Kr} as it activated relatively rapidly compared to the drug-insensitive fraction (see below) [5]. The current is inwardly rectifying, and in contrast to the classical inward rectifiers, this arises from fast inactivation and not block by Mg^{2+} or polyamines [8, 9]. HERG was originally cloned from a brain cDNA library but its' significance in cardiac electrophysiology was not truly appreciated until it was linked with the long QT syndrome [10, 11]. The properties of HERG after heterologous expression are similar but not identical to the native I_{Kr} current [12, 13]. It has been proposed that HERG channels have a beta subunit, KCNE2, in a fashion similar to I_{Ks} (see below) and that defects in this protein can rarely lead to the long QT syndrome [14]. However, this has been disputed [15] and it is clear that KCNE2 can interact with a number of other ion channels [16–19].

6.2.2 I_{Ks}

The E-4031-insensitive current has exceptionally slow activation (and deactivation) kinetics and steady-state current amplitude is only achieved after seconds of depolarization [5]. These properties have led to the designation of $I_{K\text{"slow"}}$ abbreviated to I_{Ks}. However, this behavior is also physiologically important as it means that it is important late in repolarization and the current progressively accumulates during increases in heart rate as deactivation is incomplete.

I_{Ks} is composed of the pore-forming KCNQ1 and the auxiliary subunit KCNE1 [20, 21]. I_{Ks} is thought to be a complex of four alpha KCNQ1 subunits and probably two beta KCNE1 proteins [22]. In the absence of coexpression of KCNE1, KCNQ1 expression gives rise to smaller K^+ selective currents that also activate rapidly and inactivate upon prolonged depolarization. When the two subunits are expressed, currents are substantially enhanced compared to those occurring with expression of KCNQ1 alone. Furthermore, the activation and deactivation kinetics are markedly slowed, inactivation is lost, and the steady-state activation curve is shifted rightward to more depolarized potentials [20, 21]. The native current in cardiac myocytes is similar to that occurring after coexpression of the two subunits; however, it is also possible that there is some KCNQ1 that is not complexed with KCNE1. Secondly, it is known that during adrenergic beta receptor activation, the current is increased. This is important as increased Ca^{2+} entry through L-type Ca^{2+} channels would otherwise prolong the action potential duration during sympathetic activation. The underlying molecular mechanism by which this occurs is also unusual. The activation is mediated through protein kinase A but involves an anchoring protein known as yotiao [23]. This occurs both through direct channel phosphorylation by PKA which is dependent on the A-kinase anchoring abilities of yotiao but also seems to involve a direct effect of PKA phosphorylated yotiao on channel function [24].

6.2.3 I_{K1}

I_{K1} in ventricular myocytes has the biophysical properties of a classical strong inward rectifier namely that inward currents are more prominent than outward ones and the rectification properties are dependent on the membrane potential and potassium reversal potential. The isolation of the first member of the Kir2.0 family of inward rectifier (Kir2.1) was achieved using expression cloning [25]. Using homology approaches, the family now has six members [26, 27]. The exact isoforms and nature of the current in the heart are controversial. There is no dispute that Kir2.1 is of central importance in most species. For example, in the mouse, global genetic deletion of Kir2.1 leads to a complete loss of the current in ventricular myocytes [28]. Furthermore, genetic defects in the gene (KCNJ2) lead to Andersen–Tawil syndrome in man: a component of which is a prolonged QT interval (see below) [29–31]. However, a case has been made for a component of I_{K1} being constituted by Kir2.2, Kir2.3, Kir2.4, and heteromultimers of these isoforms with Kir2.1 [32, 33]. It is possible there are species differences and developmental changes.

6.3 K^+ Current Channelopathies Affecting the Heart

The main channelopathies affecting the heart involve the molecular counterparts underlying I_{Kr}, I_{Ks}, and I_{K1} and proteins that regulate these currents.

6.3.1 Long QT Syndrome

Long QT syndrome is characterized by prolongation of the rate-corrected QT interval on the ECG and this predisposes the individual to torsade-de-pointes (TdP) and subsequent sudden arrhythmic death due to ventricular fibrillation. The commonest correction is Bazett's ($QTc = QT/(R - R))^{0.5}$ but other corrections have been proposed. Probably the most accurate approach in a research setting is to examine the behavior of QT interval with heart rate on a beat-by-beat basis and compare this to normal individuals. It is also worth appreciating that it may not be the QT interval per se that is important. It is not solely the increase in action potential duration that is proarrhythmic and in fact all other things being equal this may well be antiarrhythmic

[34–36]. Instead the proarrhythmic potential depends on three other factors: (1) the spatial dispersion of the corrected QT interval, (2) that early repolarization is delayed leading to an action potential with a more triangular shape, and (3) that the action potential duration becomes unstable with varying heart rates predisposing to ventricular alternans [37]. Two clinical syndromes are distinguished in the hereditary disease. In the Romano–Ward syndrome (RWS), inheritance occurs in an autosomal-dominant pattern, while in the rarer autosomal-recessive Jervell and Lange-Nielsen syndrome (JLNS), there is profound hearing loss in addition to the prolonged QT interval and predisposition to sudden death [38, 39]. Numerous genetic studies have shown that approximately 90% of hereditary diseases are due to defects in K^+ channel alpha subunits in KCNQ1 (LQT1) and HERG (LQT2) [40].

A variety of mutations have been identified in KCNQ1 in LQT1. These are widely distributed throughout the protein with some evidence that mutations might cluster in the transmembrane and pore regions [40]. This is not surprising as these regions are responsible for the voltage sensor and the pore architecture. Much more rarely, the beta subunit, KCNE1, is affected in LQT5 [41, 42]. Mutations in KCNQ1 and KCNE1 can result in both RWS and JLNS. In an analogous fashion, mutations in HERG underlie LQT2 and the mutations are similarly widely distributed throughout the coding sequence. It has been proposed that mutations in KCNE2 underlie LQT6, and this subunit is a potential beta subunit of HERG. This interaction is however still controversial, and it is clear that the KCNE subunits are promiscuous in interacting with a number of K^+ and other channels (see above). Both missense and nonsense mutations can occur in the coding sequence and other genetic mechanisms can also be operative (see below). A mutation has also been identified in yotiao in one patient and the mutant A-kinase anchoring protein fails to mediate normal sympathetic modulation to the I_{Ks} current [43]. Andersen–Tawil syndrome (LQT7) is a rare syndrome characterized by periodic paralysis, cardiac arrhythmia, and dysmorphic features. It has an autosomal-dominant inheritance, and a number of mutations in Kir2.1 have been identified [29–31].

Clinically, the commonest cause of LQTS is the administration of drugs. A whole variety of pharmacophores developed for a wide range of diseases can prolong the QT interval and this has caused a major issue in drug development and post-marketing surveillance [44, 45]. There is considerable debate as to the most appropriate safety screening strategy (see [36]). Intriguingly the major molecular mechanism seems to be block of the HERG K^+ channel linking the acquired and hereditary causes of the disease (see below).

6.3.2 Short QT Syndrome

This is an intriguing syndrome in which the QT interval is dramatically shortened (QTc <320 ms) and there is a predisposition to sudden death. In addition, the ECG shows a virtual absence of the ST segment and tall peaked T-waves [46]. There seems to be some overlap with hereditary atrial fibrillation as these patients are also predisposed to this disease. Mutations that cause short QT syndrome have been identified in HERG, KCNQ1, and Kir2.1 [47–49].

6.3.3 Hereditary Atrial Fibrillation

Though rare as a cause of atrial fibrillation, hereditary disease is particularly interesting from a mechanistic point of view. In a number of families, a single gene mutation has been linked with the disease. With regard to the K^+ channels discussed here, mutations that cause atrial fibrillation have been described in KCNQ1, KCNE2, and Kir2.1 [50–52].

6.4 Disease Mechanisms

At the simplest level, a number of potential mechanisms might be operative. In the long QT syndrome, the ECG abnormality implies that the normal repolarizing K^+ currents are reduced, leading to an increase in the action potential duration (APD). This could occur because the protein is not transcribed and/or translated effectively, or alternatively a mutant protein is made that interferes with the normal cellular function of the protein. The short QT syndrome and hereditary atrial fibrillation represent the flip side of the coin. Under these circumstances, one would expect an increase in K^+ currents such that the QT interval is shorter and/or the atrial action potential duration and effective refractory period are reduced. In this case, one might envisage a gain-of-function effect in the K^+ channel proteins such that currents were increased under physiological conditions.

6.4.1 Genetic Issues

As discussed above, the protein may simply not be made but what type of genetic mechanisms underlies this? Aberrant splicing will interfere with the generation of a mature mRNA. Such mutations occur in about 5–7% of cases [40]. A second and less appreciated mechanism is nonsense-mediated decay [53]. This refers to degradation of mRNA containing a premature stop codon by cellular quality control mechanisms. Both frame-shift and nonsense mutations have the potential to do this and these occur in approximately 10–15% of cases. Recently, there has been a study describing such a mechanism in long QT with HERG mutations (W1001X and R1014X) [54]. It is also worth bearing in mind that compound mutations are relatively common in the long QT syndrome (~8% of probands). In addition, they lead to severe disease and are associated with a poor prognosis [55].

In hereditary LQTS, mutations show variable penetrance [56]. For example, Priori et al. studied nine families and estimated penetrance in these families to be 25% [57]. In view of this low penetrance, it has been suggested that sporadic cases of LQTS, for example, such as those induced by drugs, could in fact be a *forme fruste* of hereditary LQTS [58–60]. It is apparent that drug-induced long QT syndrome only occurs relatively rarely in patients given a particular pharmacophore and might only attract regulatory attention during post-marketing surveillance. In a study of 16 patients with acquired LQTS [61], only 1 patient had an identifiable mutation in the *HERG* gene. The remaining 15 patients showed no detectable mutations in *KCNQ1*, *KCNE1*, *KCNE2*, and *HERG* genes. Polymorphisms in KCNE2 (T8A) and the sodium channel SCN5A (S1102Y) in African Americans have been associated with a propensity to drug-induced LQTS [59, 60]. Finally, in a more recent study, 20 patients were tested with drug-induced disease and 40% were found to have mutations in long QT-related genes while the ascertainment rate in the hereditary disease was 52% [62]. Therefore, the prevalence of mutations leading to abnormal protein expression or function in the drug-induced syndrome remains an open question.

A related hypothesis is that individuals have a variable degree of cardiac ventricular repolarization reserve [63, 64]. In other words, it is possible that some individuals tolerate a diminution of repolarizing currents without physiological and clinical sequelae. The degree of such reserve may differ among individuals and contribute to the predisposition to drug-induced LQTS. Genome-wide association studies have been used to investigate heritability in long QT. Ten genetic loci were isolated, and some of these were predictable, for example, the K^+ channel and Na^+ channel genes. However, for others, such as the nitric oxide synthase 1 adapter protein, the association was unexpected. Subsequent functional studies revealed a role in action potential repolarization [65–67]. However, there were loci for which the link with cardiac excitability was opaque. For example, one SNP lay in the 3′ UTR of RNF207 a ring finger protein of unknown function, another in LITAF which is a DNA-binding protein and another group close to NRDG4/GINS3/CNOT1/SETD6 complex of genes.

6.4.2 Protein Function

If a mutant protein is translated, how might it generate pathophysiology? The majority of hereditary LQTS occurs as an autosomal-dominant syndrome (i.e., RWS). It is important to appreciate that in the cases occurring with K^+ channel alpha subunits, a dominant negative mechanism is likely to be operative [68]. As mentioned before, both KCNQ1 and HERG are tetrameric proteins and the dominant negative effect arises most prominently when the presence of a single mutant in a tetramer can inactivate or modify the function of the complex. In contrast, in the autosomal-recessive form of the disease (i.e., JLNS), the heterozygotes are asymptomatic. In JLNS, it is therefore unlikely that dominant negative mechanisms play a role, indicating that a simple loss of function is the predominant mechanism. Our own in vitro studies using heterologous expression largely bear out this generalization [63].

In LQTS, the mutations in the K^+ channel subunits most commonly lead to a loss of function [69, 70]. However, there are mutations that impair channel gating such that repolarizing currents are reduced. For example, in KCNQ1, the steady-state activation may be shifted due to slowing of voltage-dependent activation or an acceleration of deactivation at a given potential [63, 71–73]. Some of these mutations occur in RWS and it is important to consider how a single mutant subunit might influence the behavior of the wild-type subunit in a heteromultimeric complex. In contrast, in the short QT syndrome and hereditary atrial fibrillation, there is a predicted increase in repolarizing current. The V307L mutation in KCNQ1 in the short QT syndrome leads to a pronounced shift in the half maximal activation potential and an acceleration in the activation kinetics [48]. The S140G mutation described in a family with hereditary atrial fibrillation leads to an increase in current with instantaneous activation and deactivation and a linear current–voltage relationship [50].

In Andersen's syndrome, the disease mechanisms have not been as comprehensively investigated though many of the same principles apply [74]. One particularly interesting study correlated mutations in this channelopathy with known residues affecting the binding of phospatidylinositol bisphosphate (PIP_2) to the channel [75]. A number of common residues existed and the disease mutants were resistant to the activating actions of PIP_2 addition, potentially explaining the loss of channel function in cell membranes.

6.4.3 Cellular Mechanisms

The loss of function in hereditary LQTS was originally ascribed to the presence of non-functional channel complexes at the plasma membrane. However, it soon became apparent that other mechanisms could play a role in LQTS. Of these, the aberrant trafficking of channel complexes appears to play a major role in both LQT1 and LQT2 disease pathogenesis. The various trafficking checkpoints and cellular controls that may be important are illustrated in Fig. 6.2. Defects in the trafficking of HERG and KCNQ1 to the cell surface have been reported for a variety of LQT1 and 2 mutations [76–84]. In fact, in LQT2, it has been suggested that most mutations act to reduce I_{Kr} current density through defects in trafficking [85]. Whether this is also the case for LQT1 is less well established [76, 79, 80, 82, 83].

Of the mutations in HERG or KCNQ1 that cause defective trafficking the vast majority result in retention of the channel protein in endoplasmic reticulum (ER) [77, 79, 82]. In Fig. 6.3, we show a typical example of ER retention of a mutant KCNQ1. In general, it is thought that such mutations act to disrupt protein folding or complex assembly and can be found throughout both channels' structures. However, for HERG, it does appear that when mutations occur in regions that contain a highly ordered structure, e.g., α-helices or β-sheets, the dominant cause of loss of function is defective delivery to the plasma membrane [85]. This also appears to be the case for KCNQ1, although the majority of mutations appear to cluster in three regions, the S2–S3 linker, the pore, and the C-terminus [76, 79, 80, 82, 83, 86]. Although the location of mutations that affect trafficking is diverse, specific focus on mutations that occur in certain domains has helped to establish how these domains are involved in channel biogenesis in the secretory pathway. In HERG and KCNQ1, mutations that occur in the C-terminus have been extensively investigated [81, 83, 87–89]. In HERG, two nonsense C-terminal mutations, Q725X and R1014X, result in the formation of truncated proteins. Both proteins traffic abnormally, but only R1014X is able to form a tetrameric complex with wild-type HERG and suppress current in a dominant negative fashion [88]. In addition, it has been shown that the last 147 amino acids of the C-terminus of HERG act

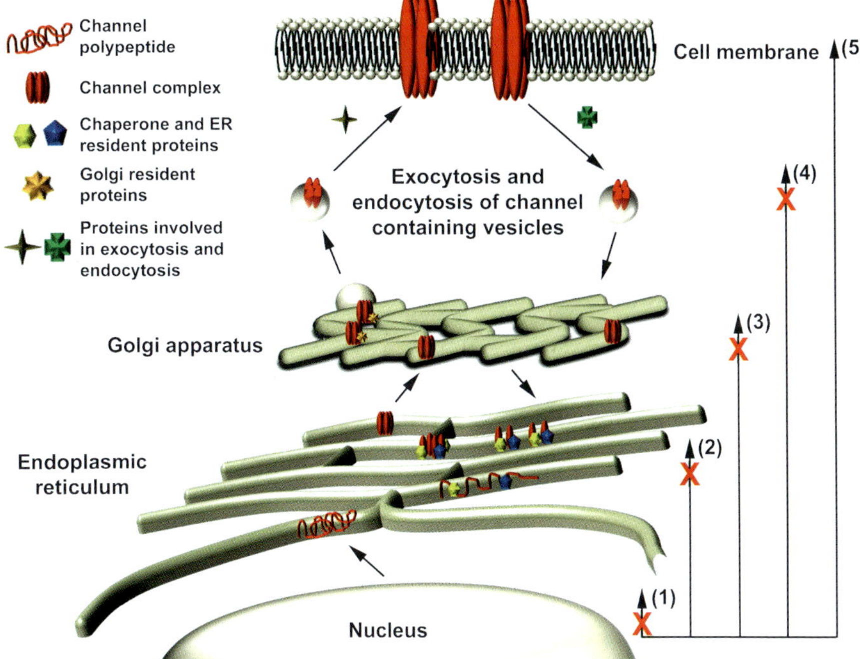

Fig. 6.2 Processing of normal and mutant K^+ channels in the secretory pathway. (*1*) Mutations lead to defects in transcription and translation and channel proteins are not synthesized. (*2*) Mutations lead to aberrant folding of the channel complex. Complexes that are incorrectly folded tend to be retained in the endoplasmic reticulum (*ER*) and are recognized as abnormal by chaperone and ER resident proteins that act to target these complexes to the proteasome for degradation. (*3*) If the mutant channels manage to pass cellular quality control in the ER, they can still be recognized as abnormal or fail to associate with golgi resident proteins, important for sorting and packaging, and be retranslocated back to the ER or targeted for degradation. (*4*) Mutations could affect the exocytosis and endocytosis of channel containing vesicles to and from the cell surface. (*5*) Normally processed channels. **X** = a block in the trafficking pathway

to mask an ER retention motif (RXR) located at position 1005–1007. Intriguingly, the surface delivery of this HERG deletion construct ($HERG_{\Delta 147}$) can be rescued by the coexpression of a 100-amino-acid peptide, as an ER targeted mini-gene, that spans the region containing the ER retention motif [89]. In KCNQ1, the C-terminal mutations, R518X, Q530X, T587M, and R594Q all cause significant ER retention [82, 83]. R518X and Q530X cannot act in a dominant negative manner and it is thought that the loss of the last ~150 amino acids impairs assembly through the removal of a tetramerization domain [63, 81, 82, 90]. A small C-terminal region of KCNQ1 has also been identified, residues 610–620, that is required for efficient trafficking to the cell surface. This region does not contain an ER retention motif, as is seen in HERG, but does provide a structural coiled coil domain that appears critical [91]. A putative ER retention motif (RXR) does however exist in the N-terminus and S2–S3 linker region of KCNQ1, and this motif is important for cell

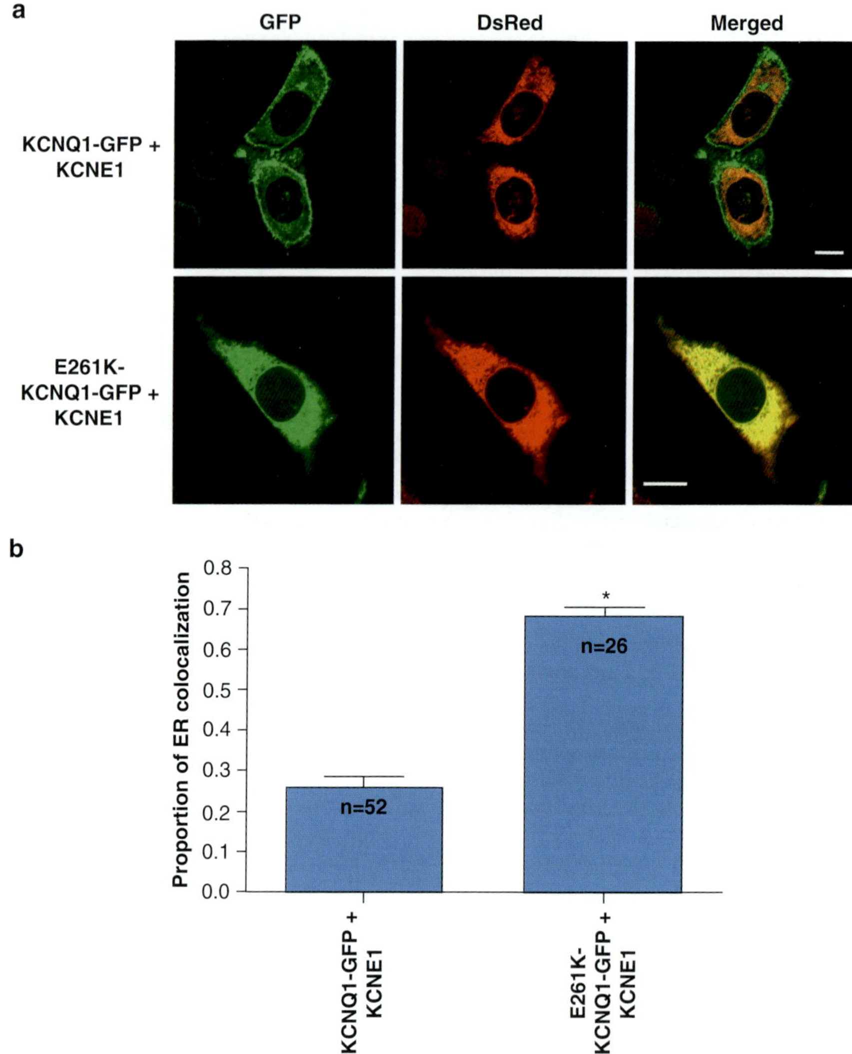

Fig. 6.3 An example of ER retention of a mutation in KCNQ1 causing LQT1. The LQT1 mutation E261K disrupts trafficking by acting to promote retention of the channel complex in the endoplasmic reticulum (*ER*). (**a**) Confocal images of Chinese hamster ovary (CHO)-K1 cells transfected with either wild-type KCNQ1-GFP or the LQT1 mutant E261K-KCNQ1-GFP in the presence of KCNE1 and the ER marker DsRed2-ER. Images are shown for GFP alone, DsRed2-ER alone, and the merged image. Colocalization between GFP and DsRed2-ER appears as yellow. Scale bar indicates 10 μm. (**b**) Mean data showing the proportion of wild-type and mutant channel ER colocalization. Data are presented as means ± SE. * $P<0.05$ compared with control (KCNQ1-GFP + KCNE1) (analysis made using a one-way ANOVA with Bonferroni post hoc test for multiple comparisons) [82]

surface delivery [86]. The LQT1 mutation, L191P, is located in the middle of this ER retention signal, and although it does not affect channel activation or deactivation kinetics, it does affect surface expression. In KCNQ1, there are two clusters of mutations that cause intracellular retention in the N-terminus, indicating that this region may also be important for trafficking [76]. Indeed, a structural motif has been identified, between residues 106 and 114, which based on structure prediction forms a short helix. Mutations in this region, Y111C and L114P, that may disrupt the structure of this short helix prevent normal trafficking of KCNQ1 and promote intracellular retention of the channel [76].

6.4.3.1 Beta Subunit Related Disease

Both KCNQ1 and perhaps HERG are thought to require the coexpression of β-subunits to reconstitute I_{Ks} and I_{Kr}: KCNE1 for KCNQ1 (MinK, IsK) and more controversially KCNE2 (MIRP1), for HERG [14, 20, 21]. Mutations in KCNE1 and KCNE2 account for LQT5 and LQT6 respectively. Mutations in KCNE1 act in general to cause a reduction in I_{Ks} current density [92]. Polymorphisms in KCNE2, T8A and Q9E, cause an increase in the sensitivity of HERG channels to drug-induced arrhythmia by LQTS-causing drugs [14, 59]. The mechanisms by which mutations/polymorphisms in KCNE2 cause disease are unclear. In comparison, several studies have investigated whether LQT5 mutations cause disease through defective trafficking. In general, these studies highlight that mutations in KCNE1 can promote defective trafficking of the I_{Ks} complex, for example, the mutations L51H, R98W, and T58P/L59P. However, in our opinion, the effects of defective trafficking in LQT5 do not appear to be as severe or as common as those seen for mutations in HERG or KCNQ1 [93–95]. The role of defective trafficking in LQT5 and LQT6 may also be complicated by the fact that KCNE1 and KCNE2 appear to exhibit promiscuity in alpha subunit interaction (as described above). For example, both subunits have been shown, in vitro, to modulate the biophysical properties of both HERG and KCNQ1 [96].

6.4.3.2 Mechanisms That Underlie the Defects in Trafficking

Mutations in HERG and KCNQ1 can result in trafficking defects but how do they act to reduce surface expression if they do not affect small peptide motifs? Before surface delivery occurs, proteins must first overcome cellular quality control. This occurs in the ER and golgi compartments, and here a large number of proteins can recognize if proteins are correctly folded and assembled (see [97]). Incorrectly folded proteins are targeted for degradation in an effort to prevent the passage of incorrectly folded complexes to the cell surface and/or the formation of toxic aggregates [97]. A number of studies have investigated whether mutant channels interact differentially with cellular quality control systems.

In the early stages of channel complex biogenesis, chaperones, which normally aid and promote the folding of proteins, have been shown to interact abnormally with mutant HERG and KCNQ1 channels. For example, the chaperones Hsp 70 and 90 interact with wild-type HERG and the specific Hsp90 inhibitor geldanamycin inhibits maturation and increases proteasomal degradation of wild-type HERG.

For the HERG mutations, R725W and G601S, the interaction with Hsp70 and Hsp90 is increased and these mutants remain tightly associated in the ER [98]. The Hsp40 type 1 chaperones, DJA1 and DJA2, also modulate HERG degradation and overexpression of both reduces HERG trafficking efficiency. The DJAs reduce HERG protein stability and the overexpression of DJA2 can reduce the partial rescue of trafficking seen for G601S when incubated at 26°C (discussed later) [99]. Another chaperone, FKBP38 (38-kDa FK506-binding protein), promotes HERG trafficking. FKBP38 immunoprecipitates and colocalizes with HERG in the ER and knockdown of FKBP38 causes a reduction of HERG trafficking. Interestingly, the overexpression of FKBP38 can partially rescue the mutant F805C [100].

Whether Hsp40, Hsp70, or Hsp90 plays a role in the trafficking of KCNQ1 or KCNQ1 mutants has not yet been investigated. However, the assembly and function of KCNQ1 is blocked by mutations that disrupt interaction with calmodulin (CaM). CaM is an obligate subunit for many ion channels and appears to act as a type of chaperone as it contributes to the control of channel assembly. CaM orchestrates the Ca^{2+}-controlled regulation of channel assembly. The formation of KCNQ1 tetramers requires CaM interaction with the C-terminus and mutations in IQ motifs; S373P (IQ1) and R518X (truncates the channel before IQ2) disrupt interaction of the channel with CaM [101, 102].

Mechanisms that regulate the rate of protein turnover are also altered by the presence of mutations in HERG or KCNQ1. A study by Gong et al. identified that degradation of the HERG mutant, Y611H, is enhanced in comparison to wild-type HERG and that this degradation is inhibited by the proteasomal inhibitors N-acetyl-L-leucyl-L-leucyl-L-norleucinal and lactacystin but not by the lysozyme inhibitor leupeptin. Inhibition of the proteasome also leads to the accumulation of polyubiquitinated HERG channels, indicating that the degradation of HERG is mediated by the cytosolic proteasome in a process that involves mannose trimming, polyubiquitination, and deglycosylation of mutant channels [103]. In a similar fashion, the LQT1 (N-terminal) mutants Y111C, L114P, and P117L are retained in the ER. All three exhibit reduced expression levels compared to wild-type KCNQ1 and radiolabeled pulse-chase experiments highlight that the reduced expression is not because of reduced rate of synthesis. Specifically, Y111C is in fact ubiquitinated and degraded in the proteasome more rapidly. The degradation of Y111C is also not dependent on Derlin 1, an ER resident protein implicated in the retrotranslocation of the cystic fibrosis transmembrane conductance regulator (CFTR) from the ER to the cytosol [104].

Mutations that cause defects in trafficking also appear able to disrupt/alter interactions with proteins that are involved in channel trafficking distal to the ER. For example, HERG normally interacts with the golgi resident protein, GM130, that plays a role in the packaging and sorting of specific vesicles. Trafficking-deficient mutations in the C-terminus of HERG, V822M, S818P, and R823W, located in the cyclic-nucleotide-binding domain, disrupt interactions with GM130 [105].

It is also possible that mutations in HERG or KCNQ1 may increase endocytosis (and degradation), reduce recycling, or decrease the rate of exocytosis. However, membrane levels of channel complexes are tightly regulated [106]. Indeed, the membrane density of KCNQ1 is regulated by Nedd4-2. Nedd4/Nedd4-like proteins bind to and ubiquitylate certain channels that contain a PY motif (L/PPxYxxΦ) in their intracellular C-terminus. Overexpression of a catalytically inactive form of Nedd4-2, that is able to antagonize the action of endogenous Nedd4-2, results in an increase of I_{Ks} current density in guinea pig cardiomyocytes. In HEK293 cells, the overexpression of Nedd4-2 increases ubiquitylation of KCNQ1 and reduces current density [107]. For HERG, channel density at the membrane appears to be strongly regulated by the concentration of extracellular K^+. Guo et al. have found that low extracellular K^+ promotes ubiquitination of the HERG channel–enhanced endocytosis and finally increased degradation of the channel [108, 109]. Additionally, the four and a half LIM domain protein 2 (FHL2) interacts with and regulates both I_{Kr} and I_{Ks} [110, 111]. Coexpression of FLH2 with HERG increases current density and results in a faster deactivation of the tail current [111]. FHL2 also appears to be able to regulate I_{Ks} and the expression of an antisense FHL2 construct reduces I_{Ks} current density [110]. It also appears that signals from the stress-axis can regulate I_{Ks} function. The expression of SGK1 (Serum and Glucocorticoid inducible Kinase 1) is regulated by cortisol and in vitro SGK1 stimulates I_{Ks}. SGK1 appears to increase I_{Ks} current density by phosphorylating PIKfyve which in turn promotes an increase in the exocytosis of KCNQ1/KCNE1 channels to the membrane via a Rab11-dependent pathway [112]. Intriguingly, a gain-of-function mutant in SGK1 is associated with shortening of the QT interval [113].

6.4.3.3 Trafficking and Acquired LQTS

Originally, it was thought that drugs that prolong the APD do so by acting as pore blockers, reducing HERG channel current density. However, a number of the drugs also inhibit the trafficking of HERG. In a thorough study, the ability of 100 compounds to inhibit HERG trafficking, 50 blockers and 50 non-blockers, was screened in a high-throughput system. This study identified that 40% of the HERG blockers studied were also able to inhibit trafficking [114]. Interestingly, some drugs not thought to block HERG directly, such as pentamidine or probucol, are able to affect trafficking without causing a direct block of I_{Kr} function [115, 116]. Importantly, these effects could be missed without screening for pharmacophores that reduce I_{Kr} through this mechanism. It is unclear as to how important these observations might be for drug development, but the implications are disconcerting.

It is surprising that the majority of cases of acquired LQT are due to drug interactions with I_{Kr} and not due to other repolarizing currents in the heart. The binding site for drugs that functionally block HERG has been identified by mutagenesis and has been modeled computationally. These studies have identified that the high-affinity drug-binding site compromises the amino acids G648, Y652, F656 in the S6 transmembrane domain and residues T623 and V625 of the pore helix [117, 118]. In particular, the antihistamine terfenadine, a drug removed from market due to proarrhythmia, interacts with residues Y652 and F656 [117]. The aromatic residues Y652 and F656 are unique to eag/erg K^+ channels, and this may explain why drugs that block HERG do not in general appear to affect the function of other channels that control APD [117].

6.4.3.4 Pharmacological Rescue of Trafficking Defects

Several methods have been developed for the rescue of trafficking defects in HERG. These methods, discussed in detail below, are varied and involve the use of reduced temperature incubation or nonspecific or specific pharmacological chaperones.

The concept that HERG trafficking could be rescued by incubation at reduced temperature came from the observation that the surface expression of the CFTR mutation ΔF508 could be increased by reducing temperature to 26°C [119]. At 37°C, the HERG mutations, N470D, R752W, and G601S, are retained in the ER, but when incubated at 26°C, all are able to fold correctly and traffic normally [78, 120, 121]. Interestingly, once at the membrane, these mutations appear to function normally. It is thought that a reduction in temperature provides more time for correct folding to occur which in turn reduces the level of targeted degradation [78, 120, 121]. We have tried to rescue the trafficking defect for mutations in KCNQ1 by reducing temperature. However, none of the trafficking defects for the mutations tested (R243H, E261D, L273F, R518X, Q530X, 1008delC, or R594Q) could be rescued [82]. A variety of low-molecular-weight compounds, that are believed to act as "nonspecific" chemical chaperones, such as dimethylsulfoxide (DMSO) and glycerol can also rescue trafficking for certain HERG mutants [78, 121]. Whether these agents can promote rescue of trafficking for KCNQ1 mutants has not been determined.

The use of specific pharmacological chaperones to aid/rescue trafficking has been particularly successful for a number of HERG mutants. Specific blockers of HERG channels, such as E-4031, are able to restore trafficking for N470D but not for all HERG mutations, for example, R725W is not rescued by E-4031 [78, 121]. In contrast, the specific I_{Ks} channel blockers and activators, Chromanol 293B and HMR-1556 respectively, are not able to rescue the trafficking defect seen for the KCNQ1 mutants R243H or E261D [82]. For HERG, the ability of pharmacological chaperones to rescue trafficking of the mutant G601S varies directly with their blocking potency. Ficker et al. have identified a binding site in the hydrophobic inner vestibule of HERG and have established that the ability for rescue was related to hydrophobicity and cationic charge [122]. In addition, they show that the mutants F805C and R823W could not be rescued. These data imply that rescue is domain limited and that mutations that occur in the pore are more readily rescued by pore-blocking pharmacological chaperones.

6.5 Therapeutic Considerations

For LQT1, drug therapy with beta-blockers is known to be effective; however, there have only been a few attempts to target drug therapy to the underlying channel mutation and/or disease mechanism [70]. The possibilities for mutation-specific therapy are actually quite broad. For LQT2, specific biophysical defects in HERG can be paradoxically overcome by increasing the extracellular K^+ concentration and there is evidence that K^+ supplementation is beneficial [123]. Nonsense mutations might be overcome by agents known to promote readthrough and one such agent is in use in clinical trials in other diseases [124]. Specific agents in LQT5, such as fenamates and stilbenes, might be efficacious [125]. Finally, gain-of-function mutations in Na^+ channels in LQT3 and in KCNQ1 and HERG in short QT and atrial fibrillation might be managed with agents known to block these channels [126, 127].

Whether the rescue of the function of mutants that are trafficked abnormally is feasible in the clinical situation remains a subject of debate [128]. There are two problems in a clinical setting. The first is that a lot of the techniques used, for example, low temperature, toxic chemicals, and channel-blocking agents cannot be used therapeutically. However, agents that rescue trafficking but do not cause channel block have been identified for HERG. Fexofenadine can rescue channel trafficking for N470D and G601S at concentrations that are ~350 fold lower than those that cause half maximal channel block [129]. In addition, thapsigargin, a sarcoplasmic/endoplasmic reticulum Ca^{2+} ATPase inhibitor, can also rescue the function of G601S without causing channel block [130]. A second practical point is that each mutation responds differently to a selected form of therapy and this means that clinical intervention would have to be mutation specific.

6.6 Conclusions

Potassium channels in the heart govern repolarization of the cardiac action potential. It has also become clear that they are involved in human disease and their abnormal function underlies hereditary and acquired disorders of cardiac rhythm. In one of these disorders, the long QT syndrome, abnormal trafficking of the KCNQ1 and HERG proteins, seems to be of major pathophysiological significance.

References

1. Noble D. The surprising heart: a review of recent progress in cardiac electrophysiology. J Physiol (Lond). 1984;353:1–50.
2. Luo CH, Rudy Y. A model of the ventricular cardiac action potential. Depolarization, repolarization, and their interaction. Circ Res. 1991;68(6):1501–26.
3. Boyett MR, Honjo H, Kodama I. The sinoatrial node, a heterogeneous pacemaker structure. Cardiovasc Res. 2000;47:658–87.
4. Josephson IR, Sanchez-Chapula J, Brown AM. Early outward current in rat single ventricular cells. Circ Res. 1984;54(2):157–62.
5. Sanguinetti MC, Jurkiewicz NK. Two components of cardiac delayed rectifier K+ current. Differential sensitivity to block by class III antiarrhythmic agents. J Gen Physiol. 1990;96(1):195–215.
6. Shah AK, Cohen IS, Datyner NB. Background K+ current in isolated canine cardiac Purkinje myocytes. Biophys J. 1987;52(4):519–25.
7. Tinker A. The assembly and targeting of potassium channels. In: Henley J, Moss SJ, editors. The assembly and targeting of ion channels. Oxford: Oxford University Press; 2002. p. 28–57.
8. Smith PL, Baukrowitz T, Yellen G. The inward rectification mechanism of the HERG cardiac potassium channel. Nature. 1996;379(6568):833–6.

9. Nichols CG, Lopatin AN. Inward rectifier potassium channels. Annu Rev Physiol. 1997;59: 171–91.
10. Warmke JW, Ganetzky B. A family of potassium channel genes related to eag in Drosophila and mammals. Proc Natl Acad Sci U S A. 1994;91(8):3438–42.
11. Curran ME, Splawski I, Timothy KW, Vincent GM, Green ED, Keating MT. A molecular basis for cardiac arrhythmia: HERG mutations cause long QT syndrome. Cell. 1995;80(5):795–803.
12. Sanguinetti MC, Jiang C, Curran ME, Keating MT. A mechanistic link between an inherited and an acquired cardiac arrhythmia: HERG encodes the IKr potassium channel. Cell. 1995;81(2):299–307.
13. Trudeau MC, Warmke JW, Ganetzky B, Robertson GA. HERG, a human inward rectifier in the voltage-gated potassium channel family. Science. 1995;269(5220):92–5.
14. Abbott GW, Sesti F, Splawski I, et al. MiRP1 forms IKr potassium channels with HERG and is associated with cardiac arrhythmia. Cell. 1999;97(2):175–87.
15. Weerapura M, Nattel S, Chartier D, Caballero R, Hebert TE. A comparison of currents carried by HERG, with and without coexpression of MiRP1, and the native rapid delayed rectifier current. Is MiRP1 the missing link? J Physiol. 2002;540(Pt 1):15–27.
16. Decher N, Bundis F, Vajna R, Steinmeyer K. KCNE2 modulates current amplitudes and activation kinetics of HCN4: influence of KCNE family members on HCN4 currents. Pflugers Arch. 2003;446(6):633–40.
17. Roepke TK, Kontogeorgis A, Ovanez C, et al. Targeted deletion of kcne2 impairs ventricular repolarization via disruption of I(K, slow1) and I(to, f). FASEB J. 2008;22(10):3648–60.
18. Brandt MC, Endres-Becker J, Zagidullin N, et al. Effects of KCNE2 on HCN isoforms: distinct modulation of membrane expression and single channel properties. Am J Physiol Heart Circ Physiol. 2009;297(1):H355–63.
19. Jiang M, Xu X, Wang Y, et al. Dynamic partnership between KCNQ1 and KCNE1 and influence on cardiac IKs current amplitude by KCNE2. J Biol Chem. 2009;284(24):16452–62.
20. Barhanin J, Lesage F, Guillemare E, Fink M, Lazdunski M, Romey G. K(V)LQT1 and IsK (minK) proteins associate to form the I(Ks) cardiac potassium current. Nature. 1996;384(6604): 78–80.
21. Sanguinetti MC, Curran ME, Zou A, et al. Coassembly of K(V)LQT1 and minK (IsK) proteins to form cardiac I(Ks) potassium channel. Nature. 1996;384(6604):80–3.
22. Chen H, Kim LA, Rajan S, Xu S, Goldstein SA. Charybdotoxin binding in the I(Ks) pore demonstrates two MinK subunits in each channel complex. Neuron. 2003;40(1):15–23.
23. Marx SO, Kurokawa J, Reiken S, et al. Requirement of a macromolecular signaling complex for beta adrenergic receptor modulation of the KCNQ1-KCNE1 potassium channel. Science. 2002;295(5554):496–9.
24. Chen L, Kurokawa J, Kass RS. Phosphorylation of the A-kinase-anchoring protein Yotiao contributes to protein kinase A regulation of a heart potassium channel. J Biol Chem. 2005;280(36):31347–52.
25. Kubo Y, Baldwin TJ, Jan YN, Jan LY. Primary structure and functional expression of a mouse inward rectifier potassium channel. Nature. 1993;362(6416):127–33.
26. Hibino H, Inanobe A, Furutani K, Murakami S, Findlay I, Kurachi Y. Inwardly rectifying potassium channels: their structure, function, and physiological roles. Physiol Rev. 2010;90(1):291–366.
27. Ryan DP, da Silva MR, Soong TW, et al. Mutations in potassium channel Kir2.6 cause susceptibility to thyrotoxic hypokalemic periodic paralysis. Cell. 2010;140(1):88–98.
28. Zaritsky JJ, Redell JB, Tempel BL, Schwarz TL. The consequences of disrupting cardiac inwardly rectifying K(+) current (I(K1)) as revealed by the targeted deletion of the murine Kir2.1 and Kir2.2 genes. J Physiol. 2001;533(Pt 3):697–710.
29. Sansone V, Griggs RC, Meola G, et al. Andersen's syndrome: a distinct periodic paralysis. Ann Neurol. 1997;42(3):305–12.
30. Plaster NM, Tawil R, Tristani-Firouzi M, et al. Mutations in Kir2.1 cause the developmental and episodic electrical phenotypes of Andersen's syndrome. Cell. 2001;105(4):511–9.
31. Tristani-Firouzi M, Jensen JL, Donaldson MR, et al. Functional and clinical characterization of KCNJ2 mutations associated with LQT7 (Andersen syndrome). J Clin Invest. 2002;110(3):381–8.

32. Liu GX, Derst C, Schlichthorl G, et al. Comparison of cloned Kir2 channels with native inward rectifier K+ channels from guinea-pig cardiomyocytes. J Physiol. 2001;532(Pt 1):115–26.
33. Schram G, Melnyk P, Pourrier M, Wang Z, Nattel S. Kir2.4 and Kir2.1 K(+) channel subunits co-assemble: a potential new contributor to inward rectifier current heterogeneity. J Physiol. 2002;544(Pt 2):337–49.
34. Akar FG, Yan GX, Antzelevitch C, Rosenbaum DS. Unique topographical distribution of M cells underlies reentrant mechanism of torsade de pointes in the long-QT syndrome. Circulation. 2002;105(10):1247–53.
35. Hondeghem LM, Carlsson L, Duker G. Instability and triangulation of the action potential predict serious proarrhythmia, but action potential duration prolongation is antiarrhythmic. Circulation. 2001;103(15):2004–13.
36. Hondeghem LM. Use and abuse of QT and TRIaD in cardiac safety research: importance of study design and conduct. Eur J Pharmacol. 2008;584(1):1–9.
37. Myles RC, Burton FL, Cobbe SM, Smith GL. The link between repolarisation alternans and ventricular arrhythmia: does the cellular phenomenon extend to the clinical problem? J Mol Cell Cardiol. 2008;45(1):1–10.
38. Ward OC. A new familial cardiac syndrome in children. J Ir Med Assoc. 1964;54:103–6.
39. Jervell A, Lange-Nielsen F. Congenital deaf-mutism, functional heart disease with prologation of Q-T interval and sudden death. Am Heart J. 1957;54:59–68.
40. Splawski I, Shen J, Timothy KW, et al. Spectrum of mutations in long-QT syndrome genes. KVLQT1, HERG, SCN5A, KCNE1, and KCNE2. Circulation. 2000;102(10):1178–85.
41. Schulze-Bahr E, Wang Q, Wedekind H, et al. KCNE1 mutations cause jervell and Lange-Nielsen syndrome. Nat Genet. 1997;17(3):267–8.
42. Tyson J, Tranebjaerg L, Bellman S, et al. IsK and KvLQT1: mutation in either of the two subunits of the slow component of the delayed rectifier potassium channel can cause Jervell and Lange-Nielsen syndrome. Hum Mol Genet. 1997;6(12):2179–85.
43. Chen L, Marquardt ML, Tester DJ, Sampson KJ, Ackerman MJ, Kass RS. Mutation of an A-kinase-anchoring protein causes long-QT syndrome. Proc Natl Acad Sci U S A. 2007; 104(52):20990–5.
44. Shah RR. The significance of QT interval in drug development. Br J Clin Pharmacol. 2002;54(2):188–202.
45. Fermini B, Fossa AA. The impact of drug-induced QT interval prolongation on drug discovery and development. Nat Rev Drug Discov. 2003;2(6):439–47.
46. Gaita F, Giustetto C, Bianchi F, et al. Short QT syndrome: a familial cause of sudden death. Circulation. 2003;108(8):965–70.
47. Brugada R, Hong K, Dumaine R, et al. Sudden death associated with short-QT syndrome linked to mutations in HERG. Circulation. 2004;109(1):30–5.
48. Bellocq C, van Ginneken AC, Bezzina CR, et al. Mutation in the KCNQ1 gene leading to the short QT-interval syndrome. Circulation. 2004;109(20):2394–7.
49. Priori SG, Pandit SV, Rivolta I, et al. A novel form of short QT syndrome (SQT3) is caused by a mutation in the KCNJ2 gene. Circ Res. 2005;96(7):800–7.
50. Chen YH, Xu SJ, Bendahhou S, et al. KCNQ1 gain-of-function mutation in familial atrial fibrillation. Science. 2003;299(5604):251–4.
51. Xia M, Jin Q, Bendahhou S, et al. A Kir2.1 gain-of-function mutation underlies familial atrial fibrillation. Biochem Biophys Res Commun. 2005;332(4):1012–9.
52. Yang Y, Xia M, Jin Q, et al. Identification of a KCNE2 gain-of-function mutation in patients with familial atrial fibrillation. Am J Hum Genet. 2004;75(5):899–905.
53. Frischmeyer PA, Vvan HA, O'Donnell K, Guerrerio AL, Parker R, Dietz HC. An mRNA surveillance mechanism that eliminates transcripts lacking termination codons. Science. 2002;295(5563):2258–61.
54. Gong Q, Zhang L, Vincent GM, Horne BD, Zhou Z. Nonsense mutations in hERG cause a decrease in mutant mRNA transcripts by nonsense-mediated mRNA decay in human long-QT syndrome. Circulation. 2007;116(1):17–24.
55. Westenskow P, Splawski I, Timothy KW, Keating MT, Sanguinetti MC. Compound mutations: a common cause of severe long-QT syndrome. Circulation. 2004;109(15):1834–41.

56. Roden DM, Lazzara R, Rosen M, Schwartz PJ, Towbin J, Vincent GM. Multiple mechanisms in the long-QT syndrome. Current knowledge, gaps, and future directions. The SADS Foundation Task Force on LQTS. Circulation. 1996;94(8):1996–2012.
57. Priori SG, Napolitano C, Schwartz PJ. Low penetrance in the long-QT syndrome: clinical impact. Circulation. 1999;99(4):529–33.
58. Napolitano C, Schwartz PJ, Brown AM, et al. Evidence for a cardiac ion channel mutation underlying drug-induced QT prolongation and life-threatening arrhythmias. J Cardiovasc Electrophysiol. 2000;11(6):691–6.
59. Sesti F, Abbott GW, Wei J, et al. A common polymorphism associated with antibiotic-induced cardiac arrhythmia. Proc Natl Acad Sci U S A. 2000;97(19):10613–8.
60. Splawski I, Timothy KW, Tateyama M, et al. Variant of SCN5A sodium channel implicated in risk of cardiac arrhythmia. Science. 2002;297(5585):1333–6.
61. Chevalier P, Rodriguez C, Bontemps L, et al. Non-invasive testing of acquired long QT syndrome: evidence for multiple arrhythmogenic substrates. Cardiovasc Res. 2001;50(2): 386–98.
62. Itoh H, Sakaguchi T, Ding WG, et al. Latent genetic backgrounds and molecular pathogenesis in drug-induced long-QT syndrome. Circ Arrhythm Electrophysiol. 2009;2(5):511–23.
63. Huang L, Bitner-Glindzicz M, Tranebjaerg L, Tinker A. A spectrum of functional effects for disease causing mutations in the Jervell and Lange-Nielsen syndrome. Cardiovasc Res. 2001;51(4):670–80.
64. Roden DM, George Jr AL. The cardiac ion channels: relevance to management of arrhythmias. Annu Rev Med. 1996;47:135–48.
65. Pfeufer A, Sanna S, Arking DE, et al. Common variants at ten loci modulate the QT interval duration in the QTSCD Study. Nat Genet. 2009;41(4):407–14.
66. Newton-Cheh C, Eijgelsheim M, Rice KM, et al. Common variants at ten loci influence QT interval duration in the QTGEN Study. Nat Genet. 2009;41(4):399–406.
67. Chang KC, Barth AS, Sasano T, et al. CAPON modulates cardiac repolarization via neuronal nitric oxide synthase signaling in the heart. Proc Natl Acad Sci U S A. 2008;105(11): 4477–82.
68. Herskowitz I. Functional inactivation of genes by dominant negative mutations. Nature. 1987;329(6136):219–22.
69. Moss AJ, Kass RS. Long QT syndrome: from channels to cardiac arrhythmias. J Clin Invest. 2005;115(8):2018–24.
70. Nerbonne JM, Kass RS. Molecular physiology of cardiac repolarization. Physiol Rev. 2005;85(4):1205–53.
71. Franqueza L, Lin M, Splawski I, Keating MT, Sanguinetti MC. Long QT syndrome-associated mutations in the S4-S5 linker of KvLQT1 potassium channels modify gating and interaction with minK subunits. J Biol Chem. 1999;274(30):21063–70.
72. Wang Z, Tristani Firouzi M, Xu Q, Lin M, Keating MT, Sanguinetti MC. Functional effects of mutations in KvLQT1 that cause long QT syndrome. J Cardiovasc Electrophysiol. 1999;10(6): 817–26.
73. Yang T, Chung SK, Zhang W, et al. Biophysical properties of 9 KCNQ1 mutations associated with long-QT syndrome. Circ Arrhythm Electrophysiol. 2009;2(4):417–26.
74. Bendahhou S, Fournier E, Sternberg D, et al. In vivo and in vitro functional characterization of Andersen's syndrome mutations. J Physiol. 2005;565(Pt 3):731–41.
75. Lopes CM, Zhang H, Rohacs T, Jin T, Yang J, Logothetis DE. Alterations in conserved Kir channel-PIP2 interactions underlie channelopathies. Neuron. 2002;34(6):933–44.
76. Dahimene S, Alcolea S, Naud P, et al. The N-terminal juxtamembranous domain of KCNQ1 is critical for channel surface expression – implications in the Romano-Ward LQT1 syndrome. Circ Res. 2006;99(10):1076–83.
77. Ficker E, Dennis AT, Obejero-Paz CA, Castaldo P, Taglialatela M, Brown AM. Retention in the endoplasmic reticulum as a mechanism of dominant-negative current suppression in human long QT syndrome. J Mol Cell Cardiol. 2000;32(12):2327–37.

78. Ficker E, Thomas D, Viswanathan PC, et al. Novel characteristics of a misprocessed mutant HERG channel linked to hereditary long QT syndrome. Am J Physiol Heart Circ Physiol. 2000;279(4):H1748–56.
79. Gouas L, Bellocq C, Berthet M, et al. New KCNQ1 mutations leading to haploinsufficiency in a general population – defective trafficking of a KvLQT1 mutant. Cardiovasc Res. 2004;63(1):60–8.
80. Sato A, Arimura T, Makita N, et al. Novel mechanisms of trafficking defect caused by KCNQ1 mutations found in long QT syndrome. J Biol Chem. 2009;284(50):35122–33.
81. Schmitt N, Schwarz M, Peretz A, Abitbol I, Attali B, Pongs O. A recessive C-terminal Jervell and Lange-Nielsen mutation of the KCNQ1 channel impairs subunit assembly. EMBO J. 2000;19(3):332–40.
82. Wilson AJ, Quinn KV, Graves FM, Bitner-Glindzicz M, Tinker A. Abnormal KCNQ1 trafficking influences disease pathogenesis in hereditary long QT syndromes (LQT1). Cardiovasc Res. 2005;67(3):476–86.
83. Yamashita F, Horie M, Kubota T, et al. Characterization and subcellular localization of KCNQ1 with a heterozygous mutation in the C terminus. J Mol Cell Cardiol. 2001;33(2):197–207.
84. Zhou Z, Gong Q, Epstein ML, January CT. HERG channel dysfunction in human long QT syndrome. Intracellular transport and functional defects. J Biol Chem. 1998;273(33):21061–6.
85. Anderson CL, Delisle BP, Anson BD, et al. Most LQT2 mutations reduce Kv11.1 (hERG) current by a class 2 (trafficking-deficient) mechanism. Circulation. 2006;113(3):365–73.
86. Pan N, Sun J, Lv CX, Li H, Ding JP. A hydrophobicity-dependent motif responsible for surface expression of cardiac potassium channel. Cell Signal. 2009;21(2):349–55.
87. Akhavan A, Atanasiu R, Shrier A. Identification of a COOH-terminal segment involved in maturation and stability of human ether-a-go-go-related gene potassium channels. J Biol Chem. 2003;278(41):40105–12.
88. Gong Q, Keeney DR, Robinson JC, Zhou Z. Defective assembly and trafficking of mutant HERG channels with C-terminal truncations in long QT syndrome. J Mol Cell Cardiol. 2004;37(6):1225–33.
89. Kupershmidt S, Yang T, Chanthaphaychith S, Wang Z, Towbin JA, Roden DM. Defective human Ether-à-go-go-related gene trafficking linked to an endoplasmic reticulum retention signal in the C terminus. J Biol Chem. 2002;277(30):27442–8.
90. Wiener R, Haitin Y, Shamgar L, et al. The KCNQ1 (Kv7.1) COOH terminus, a multitiered scaffold for subunit assembly and protein interaction. J Biol Chem. 2008;283(9):5815–30.
91. Kanki H, Kupershmidt S, Yang T, Wells S, Roden DM. A structural requirement for processing the cardiac K+ channel KCNQ1. J Biol Chem. 2004;279(32):33976–83.
92. Splawski I, TristaniFirouzi M, Lehmann MH, Sanguinetti MC, Keating MT. Mutations in the hminK gene cause long QT syndrome and suppress I-Ks function. Nat Genet. 1997;17(3):338–40.
93. Bianchi L, Shen Z, Dennis AT, et al. Cellular dysfunction of LQT5-minK mutants: abnormalities of IKs, IKr and trafficking in long QT syndrome. Hum Mol Genet. 1999;8(8):1499–507.
94. Harmer SC, Wilson AJ, Aldridge R, Tinker A. Mechanisms of disease pathogenesis in long QT syndrome type 5. Am J Physiol Cell Physiol. 2010;298(2):C263–73.
95. Krumerman A, Gao X, Bian JS, Melman YF, Kagan A, McDonald TV. An LQT mutant minK alters KvLQT1 trafficking. Am J Physiol Cell Physiol. 2004;286(6):C1453–63.
96. Abbott GW, Xu X, Roepke TK. Impact of ancillary subunits on ventricular repolarization. J Electrocardiol. 2007;40(6 Suppl):S42–6.
97. Aridor M. Visiting the ER: the endoplasmic reticulum as a target for therapeutics in traffic related diseases. Adv Drug Deliv Rev. 2007;59(8):759–81.
98. Ficker E, Dennis AT, Wang L, Brown AM. Role of the cytosolic chaperones Hsp70 and Hsp90 in maturation of the cardiac potassium channel HERG. Circ Res. 2003;92(12):e87–100.

99. Walker VE, Wong MJ, Atanasiu R, Hantouche C, Young JC, Shrier A. Hsp40 chaperones promote degradation of the HERG potassium channel. J Biol Chem. 2010;285(5):3319–29.
100. Walker VE, Atanasiu R, Lam H, Shrier A. Co-chaperone FKBP38 promotes HERG trafficking. J Biol Chem. 2007;282(32):23509–16.
101. Ghosh S, Nunziato DA, Pitt GS. KCNQ1 assembly and function is blocked by long-QT syndrome mutations that disrupt interaction with calmodulin. Circ Res. 2006;98(8): 1048–54.
102. Shamgar L, Ma L, Schmitt N, et al. Calmodulin is essential for cardiac IKS channel gating and assembly: impaired function in long-QT mutations. Circ Res. 2006;98(8):1055–63.
103. Gong Q, Keeney DR, Molinari M, Zhou Z. Degradation of trafficking-defective long QT syndrome type II mutant channels by the ubiquitin-proteasome pathway. J Biol Chem. 2005;280(19):19419–25.
104. Peroz D, Dahimene S, Baro I, Loussouarn G, Merot J. LQT1-associated mutations increase KCNQ1 proteasomal degradation independently of Derlin-1. J Biol Chem. 2009;284(8): 5250–6.
105. Roti EC, Myers CD, Ayers RA, et al. Interaction with GM130 during HERG ion channel trafficking. Disruption by type 2 congenital long QT syndrome mutations. Human Ether-à-go-go-Related Gene. J Biol Chem. 2002;277(49):47779–85.
106. Fortune ES, Chacron MJ. From molecules to behavior: organismal-level regulation of ion channel trafficking. PLoS Biol. 2009;7(9):e1000211.
107. Jespersen T, Membrez M, Nicolas CS, et al. The KCNQ1 potassium channel is down-regulated by ubiquitylating enzymes of the Nedd4/Nedd4-like family. Cardiovasc Res. 2007;74(1):64–74.
108. Guo J, Massaeli H, Xu J, et al. Extracellular K+ concentration controls cell surface density of IKr in rabbit hearts and of the HERG channel in human cell lines. J Clin Invest. 2009;119(9):2745–57.
109. Robertson GA. Endocytic control of ion channel density as a target for cardiovascular disease. J Clin Invest. 2009;119(9):2531–4.
110. Kupershmidt S, Yang IC, Sutherland M, et al. Cardiac-enriched LIM domain protein fhl2 is required to generate I(Ks) in a heterologous system. Cardiovasc Res. 2002;56(1):93–103.
111. Lin J, Lin S, Yu X, et al. The four and a half LIM domain protein 2 interacts with and regulates the HERG channel. FEBS J. 2008;275(18):4531–9.
112. Seebohm G, Strutz-Seebohm N, Birkin R, et al. Regulation of endocytic recycling of KCNQ1/ KCNE1 potassium channels. Circ Res. 2007;100(5):686–92.
113. Busjahn A, Seebohm G, Maier G, et al. Association of the serum and glucocorticoid regulated kinase (sgk1) gene with QT interval. Cell Physiol Biochem. 2004;14(3):135–42.
114. Wible BA, Hawryluk P, Ficker E, Kuryshev YA, Kirsch G, Brown AM. HERG-Lite: a novel comprehensive high-throughput screen for drug-induced hERG risk. J Pharmacol Toxicol Methods. 2005;52(1):136–45.
115. Guo J, Massaeli H, Li W, et al. Identification of IKr and its trafficking disruption induced by probucol in cultured neonatal rat cardiomyocytes. J Pharmacol Exp Ther. 2007;321(3): 911–20.
116. Kuryshev YA, Ficker E, Wang L, et al. Pentamidine-induced long QT syndrome and block of hERG trafficking. J Pharmacol Exp Ther. 2005;312(1):316–23.
117. Mitcheson JS, Chen J, Lin M, Culberson C, Sanguinetti MC. A structural basis for drug-induced long QT syndrome. Proc Natl Acad Sci U S A. 2000;97(22):12329–33.
118. Stansfeld PJ, Gedeck P, Gosling M, Cox B, Mitcheson JS, Sutcliffe MJ. Drug block of the hERG potassium channel: insight from modeling. Proteins. 2007;68(2):568–80.
119. Denning GM, Anderson MP, Amara JF, Marshall J, Smith AE, Welsh MJ. Processing of mutant cystic fibrosis transmembrane conductance regulator is temperature-sensitive. Nature. 1992;358(6389):761–4.
120. Furutani M, Trudeau MC, Hagiwara N, et al. Novel mechanism associated with an inherited cardiac arrhythmia: defective protein trafficking by the mutant HERG (G601S) potassium channel. Circulation. 1999;99(17):2290–4.

121. Zhou Z, Gong Q, January CT. Correction of defective protein trafficking of a mutant HERG potassium channel in human long QT syndrome. Pharmacological and temperature effects. J Biol Chem. 1999;274(44):31123–6.
122. Ficker E, Obejero-Paz CA, Zhao S, Brown AM. The binding site for channel blockers that rescue misprocessed human long QT syndrome type 2 ether-a-gogo-related gene (HERG) mutations. J Biol Chem. 2002;277(7):4989–98.
123. Compton SJ, Lux RL, Ramsey MR, et al. Genetically defined therapy of inherited long-QT syndrome. Correction of abnormal repolarization by potassium. Circulation. 1996;94(5):1018–22.
124. Kerem E, Hirawat S, Armoni S, et al. Effectiveness of PTC124 treatment of cystic fibrosis caused by nonsense mutations: a prospective phase II trial. Lancet. 2008;372(9640):719–27.
125. Abitbol I, Peretz A, Lerche C, Busch AE, Attali B. Stilbenes and fenamates rescue the loss of I(KS) channel function induced by an LQT5 mutation and other IsK mutants. EMBO J. 1999;18(15):4137–48.
126. Schwartz PJ, Priori SG, Locati EH, et al. Long QT syndrome patients with mutations of the SCN5A and HERG genes have differential responses to Na+ channel blockade and to increases in heart rate implications for gene-specific therapy. Circulation. 1995;92(12):3381–6.
127. Gaita F, Giustetto C, Bianchi F, et al. Short QT syndrome: pharmacological treatment. J Am Coll Cardiol. 2004;43(8):1494–9.
128. Kaufman ES, Ficker E. Is restoration of intracellular trafficking clinically feasible in the long QT syndrome?: The example of HERG mutations. J Cardiovasc Electrophysiol. 2003;14(3):320–2.
129. Rajamani S, Anderson CL, Anson BD, January CT. Pharmacological rescue of human K(+) channel long-QT2 mutations: human ether-a-go-go-related gene rescue without block. Circulation. 2002;105(24):2830–5.
130. Delisle BP, Anderson CL, Balijepalli RC, Anson BD, Kamp TJ, January CT. Thapsigargin selectively rescues the trafficking defective LQT2 channels G601S and F805C. J Biol Chem. 2003;278(37):35749–54.

Free Radicals, Oxidative Stress, and Cardiovascular Disease

7

K. Richard Bruckdorfer

7.1 Introduction

The evolution of cardiovascular disease occurs over many decades and has several phases associated with it. It is a process of two parts, one which involves the development of atherosclerotic plaque and the other the formation of thrombi most typically, but not exclusively, in the later stages of the disease. The association of free radicals and oxidative stress with some stages of this process is widely considered to be most relevant to the inflammatory elements of the disease. The principal phases leading to the formation of stable and unstable plaque are outlined in Table 7.1.

The purpose here is to examine the role of free radicals in cardiovascular disease and of antioxidants as prophylactic agents. The following questions will be considered:

Are free radicals always harmful?

What is the evidence that they play a role in the development of atherosclerosis?

Which radicals or reactive species are involved?

Do dietary antioxidants offer an effective means of therapy to prevent oxidation?

7.2 Free Radicals in Normal Physiology

There is a widespread assumption that free radicals and non-radical oxidizing species are always damaging to living cells and must always be suppressed. This is far from the truth, as is the common impression that they are of equivalent oxidizing power under all conditions and that they can be suppressed in the same way by all antioxidants.

K.R. Bruckdorfer
Structural and Molecular Biology, Faculty of Life Sciences,
University College London, London, UK
e-mail: k.bruckdorfer@ucl.ac.uk

D. Abraham et al. (eds.), *Translational Vascular Medicine*,

DOI 10.1007/978-0-85729-920-8_7,

Table 7.1 Sequence of events in the development of atherosclerotic plaque

Endothelial injury
Recruitment of monocyte/macrophages
Lipid deposition from lipoproteins into macrophages and their transformation into foam cells
Formation of a fibrotic plaque over the lipid layers by proliferation of smooth muscle cells in their fibroblastic phenotype
Further evolution to become vulnerable plaques linked to serious clinical events
Angiogenesis, calcification, fissuring, and thrombosis

Reactive oxygen species, a term which covers free radical and non-radical oxidants, are produced as by-products of normal metabolic processes. Cells have excellent endogenous antioxidant mechanisms which regulate their reactivity and these species have specific physiological purposes not least in primary defense mechanisms against microorganisms. It is only when their production becomes excessive, as is the case in inflammation, do they contribute to pathologies such as atherosclerosis and other inflammatory conditions. It was perhaps not until the discovery of the enzyme superoxide dismutase [1], and the realization that substantial amounts of free radicals are produced normally in the body, that this was fully understood.

In normal exercising muscle, free radicals are formed in tandem with increased mitochondrial respiratory activity. This evokes a protective response at gene level. Antioxidant enzymes such as hemoxygenase-1, other chaperone proteins, and antioxidant enzymes are induced to prevent damage to muscle protein [2]. Regular exercise promotes lasting antioxidant protection in response to the greater production of free radicals.

The main free radical produced through metabolic activity is superoxide anion, a by-product of mitochondrial oxidative phosphorylation, and the activity of NADPH oxidase. This has been estimated to be of the order of 1 kg per annum for a healthy individual [3]. Macromolecules can be also be radicalized and this may be an important step in the regulation of the activity of enzymes and other proteins where sulfydryl, histidyl, and tyrosyl residues are particularly vulnerable to radicalization. The main product of superoxide anion catabolism is the non-radical oxidant hydrogen peroxide, which is also formed during the metabolism of lysine. The peroxide is believed to have an important physiological role in the relaxation of resistance vessels which may be essential in exercise [4]. This effect mirrors the action of the free radical nitric oxide, biosynthesized from L-arginine, as an important vasodilator released from the endothelium of larger blood vessels [5]. Furthermore, free radicals have an important role in the normal life cycle of the cell by initiating apoptosis [6].

7.3 Discoveries Leading to an Understanding of the Role of LDL Oxidation in the Development of Atherosclerosis

Low-density lipoproteins (LDLs) are implicated in the development of cardiovascular disease, but also have an essential role in normal physiology. They transport much of the cholesterol from its site of biosynthesis in the liver or in the diet to peripheral tissues, where the sterol is required for the assembly of cell membranes and, in some cases, the formation of steroid hormones and bile salts. LDLs carry not only cholesterol, but dietary polyunsaturated fatty acids required for biosynthesis of longer chain fatty acids and membrane phospholipids and eicosanoids. Polyunsaturated fatty acids are found in the phospholipid and cholesterol ester fractions of LDL in amounts that are disproportionately large compared to the normal dietary intake of polyunsaturated fat. However, their double bonds are susceptible to oxidative attack by free radicals.

LDL transports a high proportion of dietary fat-soluble antioxidants, particularly tocopherols and carotenoids, which are delivered to the periphery. These are normally sufficient to protect the LDL from oxidation. Furthermore, water-soluble vitamin C accepts electrons from tocopheryl radicals formed during oxidative attack, thereby suppressing the propagation of fatty acid oxidation. LDLs are readily oxidized in the presence of transition metal ions through the formation of hydroxyl radicals generated in the presence of the traces of hydrogen peroxide (Eq. 7.1). These free radicals attack LDL polyunsaturated fatty acids peroxyl radicals, aldehydes, and other derivatives [7]. The formation of Schiff's bases with the ε-amino groups of lysine residues on apolipoprotein B100 impairs its recognition by LDL receptors. Incubation of LDL with cultured macrophages or endothelial cells leads to similar change in these lipoproteins [8] which become oxidized, more electronegative and are recognized by the scavenger receptors on macrophages SRA-1 and CD36. The macrophages arise in the artery by diapedesis of monocytes and transformation into the phagocytotic form subsequent to endothelial damage by hypercholesterolemia, smoking, or hypertension. The macrophages become foam cells as they engorge with lipid droplets of cholesterol esters derived from oxidatively modified LDL. These macrophage receptors also remove apoptotic cells and cell debris from the tissues as part of the normal senescence and repair process. In the presence of cytotoxic oxidized LDL, macrophage/foam cells cannot leave the atherosclerotic plaque and return the circulation as monocytes. Ultimately cell death ensues and cholesterol is released as cholesterol ester droplets or crystalline non-esterified cholesterol.

Formation of hydroxyl radicals:

$$Fe^{2+} + H_2O_2 \rightarrow {}^{\bullet}OH + OH^- + Fe^{3+} \tag{7.1}$$

7.4 Oxidants in Atherosclerotic Plaque

Attention has been given in recent years to the nature of the reactive species produced by macrophages. It is widely accepted that superoxide anions are not directly responsible for the oxidation of LDL. Hazen and colleagues have shown that myeloperoxidase in activated macrophages leads to the release of hypochlorous acid [9] which is a two-electron oxidant but not a free radical (Eq. 7.2). The enzyme contributes to the primary defense mechanisms against microorganisms. This oxidant chlorinates tyrosine residues on LDL and these are taken into the cells via a scavenger receptor mediated process.

The formation of hypochlorite catalyzed by myeloperoxidase:

$$H_2O_2 + Cl^- + H^+ \rightarrow HOCl + H_2O \tag{7.2}$$

Furthermore, another two-electron oxidant, peroxynitrite, is formed by the action of this enzyme using nitrite ions as the substrate (Eq. 7.3). The heme moiety of myeloperoxidase is essential for this conversion in its higher oxidation state and activates the formation of the nitrogen dioxide radical – a powerful oxidant.

Mechanisms leading to formation of nitrogen dioxide radicals from nitrite catalyzed by myeloperoxidase:

$$\begin{aligned} &\mathrm{MPO-heme\ (III)} + H_2O_2 \rightarrow \mathrm{MPO-heme\ (IV^*)} \\ &\mathrm{MPO-heme\ (IV^*)} + NO_2^- \rightarrow NO_{2.} + \mathrm{MPO-heme\ (IV)} \end{aligned} \tag{7.3}$$

The nitrogen dioxide radical nitrates susceptible tyrosyl residues in LDL and other proteins to 3-nitrotyrosine. Peroxynitrite may also be formed more directly by the reaction of nitric oxide and superoxide anion, both of which are produced in large amounts in activated macrophages as part of the primary defense mechanisms against infection (Eq. 7.4). Peroxynitrite readily decomposes at physiological pH to yield nitrogen dioxide radical and hydroxyl radical. The myeloperoxidase activity leaves a fingerprint of chlorinated and nitrated proteins in areas of plaque rich in macrophages. These findings have been extended to show that HDL is also capable of modification by hypochlorite and peroxynitrite [10]. This has profound effects on the ability of this lipoprotein to remove cholesterol from the macrophages through the ABC-A1 receptor and the enzyme phospholipid cholesterol acyltransferase (PCAT) which promotes cholesterol esterification in HDL. These proteins are integral to the reverse cholesterol transport system for the return of cholesterol to the liver and its excretion. Therefore, these oxidants inhibit the removal of cholesterol from plaque and enhance its deposition. Other relevant molecules may be modified by peroxynitrite. For example, nitration of fibrinogen may render this molecule more thrombotic by decreasing the stability of clots and increasing the risk of microthrombi [11]. A pro-thrombotic state is therefore induced by post-translational modification of fibrinogen. Similarly, plasmin activity is impaired by nitration [12].

The formation of peroxynitrous acid from nitric oxide:

$$\begin{aligned} &NO + O_{2^{\cdot-}} \rightarrow ONOO\text{-} + H^+ \rightarrow ONOOH \\ &ONOOH \rightarrow NO_{2.} + OH. \end{aligned} \tag{7.4}$$

7.5 Consequences of Lipoprotein Oxidation and Its Effects on Arterial Function

The products of lipoprotein oxidation are important to events than the formation of foam cells because of the formation of a wide range of oxidation products from polyunsaturated lipids. Many of these products are cytotoxic and genotoxic. Lysophosphatidylcholine is a potent detergent that damages the endothelium [13]. Lysophosphatidic acid is an activator of platelets [14] which, in concert with the diminished synthesis of endothelial nitric oxide a platelet inhibitor, is pro-thrombotic. Cholesterol oxides at low concentrations activate genes for chaperones and antioxidant enzymes which are protective for the endothelium, through the nuclear transcription factors LXR and RXR [15], thus providing some protection against the effects of oxidation. However, at higher concentrations, they are cytotoxic. Isoprostanes formed from the oxidation of polyunsaturated fatty acids are potent vasoconstrictors. The elevated plasma concentrations of isoprostanes are excellent markers for lipid oxidation in vivo [16].

7.6 Antioxidants and Cardiovascular Disease

Over the last 20 years, epidemiological studies pointed to a link between dietary antioxidants and cardiovascular disease. There was evidence of an independent inverse link between the consumption of fruit and vegetables and mortality from cardiovascular disease (see below). This allowed many to make the assumption that it resulted from the presence of antioxidants in these foods, although there are alternative explanations for such an association.

Esterbauer found that if α-tocopherol (but not β-carotene) was added exogenously to the LDL, or by oral doses to healthy subjects, these lipoproteins became much more resistant to oxidation [7]. The first endogenous material to become oxidized in the presence of cupric ions is α-tocopherol before the lipids oxidize. However, the susceptibility of the LDL to oxidation ex vivo was independent of the amount of *endogenous* α-tocopherol in LDL isolated from the cohort of donors. In the original experiments, the oral dosage of α-tocopherol was well in excess of the dietary norm. However, smaller amounts of the vitamin also increased resistance to oxidation [17]. Ascorbate also increases the resistance to oxidation: Electrons pass from the tocopherol in the LDL to ascorbate, limiting the oxidation of polyunsaturated fatty acids [18].

Experimental evidence in vitro showed that α-tocopherol has a number of inhibitory effects on processes which lead to the formation of atherosclerotic plaque, but it was not clear that these were due to its antioxidant function. Effects of α-tocopherol were even found in healthy individuals on platelet function which was attenuated by the vitamin [17] even at oral doses of 75 i.u. per day, much less than had been used in many trials, but still well above the intake from non-supplemented foods.

These findings suggested that antioxidants, at least α-tocopherol, may have an effect on thrombosis which is a key concomitant to myocardial infarction. These effects were minor compared with the actions of aspirin on platelet function which reduces the risk of infarction by about 20%.

The existence of antibodies to oxidized LDL in the plasma seemed to add supportive evidence for the oxidation hypothesis. The association of the concentration of these antibodies with extent of atherosclerosis in patients with cardiovascular disease was weak [19]. Indeed it has been suggested that the formation of these antibodies may be a protective response since dietary intervention to reduce oxidative stress increased the titer of circulation antibodies [20]. Indeed oral supplements of α-tocopherol increased the titer of circulating antibodies to oxidized LDL [21]. The autoimmune response to oxidized LDL is also associated with the contribution of bacterially derived antigens and the participation of Toll-like receptors in macrophages of atherosclerotic plaques [22]. Furthermore platelets express CD66 which links their activation to the presence of oxidized LDL in the circulation [23].

7.6.1 Epidemiological Studies on Antioxidants and Cardiovascular Disease

The experimental studies and the established inverse relationship between the consumption of fruit and vegetables and cardiovascular disease elicited a number of new studies on patients and populations which, for the most part, seemed to reinforce the central role of antioxidants as protective nutrients.

Coronary Heart Disease rates were known to be higher in areas where fruit and vegetable consumption was lowest [24]. In countries where consumption of fruit and vegetables was high, rates of CHD were lower [25]. Furthermore, vegetarians have lower rates of CHD [26]. The diets of over 75,000 nurses and nearly 39,000 male health professionals were compared showing a 31% reduction in the risk of stroke in the quintile eating the most fruit and vegetables compared with the quintile eating the least [27].

The Lyon Diet Heart Study on M.I. patients found that those who followed a "Mediterranean" diet had a significant reduction in the re-occurrence of myocardial infarction after 4 years [28]. Higher fruit and vegetable intakes lower the cardiovascular disease risk factors, blood cholesterol and blood pressure [29, 30]. These studies were the basis for further studies. More recent evaluations of the benefits of fruit and vegetables reinforce this view, but suggest that the evidence is not always strong [31] and that controlled nutritional prevention studies are scarce [32].

These and other findings led researchers to investigate whether the active factor in the fruit and vegetables could indeed be attributed to their antioxidant content. The MONICA study showed a north-south gradient in cardiovascular disease risk across Europe inverse to the gradient for plasma concentrations of vitamin E. These findings were supported by the results of case-control studies [33, 34]. Further studies demonstrated a lower risk of cardiovascular disease with a higher dietary intake of antioxidant nutrients [35, 36]. The European Prospective Investigation of Cancer (EPIC) study found a relationship between high levels of ascorbate and reduced risk of cardiovascular disease [37].

Prospective studies have investigated the contribution of vitamin supplements. In "The Nurses' Health Study" [35], where women in the highest (fifth) quintile of α-tocopherol consumption, many used supplements of vitamin E and had a 44% lower risk of CHD compared to those in the lowest quintile of intake of this vitamin. Those in the fourth quintile for dietary intake of vitamin E (mainly dietary vitamin rather than supplements) also had a lower risk (26% lower than the first quintile). In a cohort of men [33], the risk was also lower with higher α-tocopherol intakes. A total of 34,000 postmenopausal women in the Iowa Women's Health Study showed an inverse association between dietary vitamin E intake and deaths from CHD and stroke [38, 39]: Here vitamin E supplementation was not associated with protection from cardiovascular disease.

Intake of antioxidant flavonoids was shown to be inversely associated with CHD risk in several studies. The Iowa Women's Study found that increased intake of flavonoids was associated with a decreased risk of death from CHD [40]. A recent study from within the Iowa cohort showed a strong inverse association between CHD and intake of some types of catechins.

Selenium is important to the activity of certain antioxidant enzymes, particularly glutathione peroxidase. Patients with MI had low plasma selenium concentrations [41], but not all studies show an inverse relationship between plasma selenium concentrations and cardiovascular disease. Selenium is a micronutrient for which a significant proportion of the population of the UK and other countries has a marginal deficiency [42].

In other studies, the intake of fruit and vegetables has been associated with changes in markers for cardiovascular disease or in measurable physiological changes in arterial function. Many studies have looked at the relationship to these markers and the level of antioxidants in the diet or to measurements in the plasma. Plasma C-reactive protein is used as an index of inflammation, but also has been correlated strongly to the incidence and mortality from cardiovascular disease [43]. CRP plasma levels, in a prospective population study of 3,258 men aged 60–79, were also correlated inversely with plasma ascorbate concentrations and dietary vitamin C, even after adjustment for other confounding factors [44]. Another marker for endothelial dysfunction, the tissue plasminogen activator – 1, was also inversely correlated to these two parameters.

LDL and, to a greater extent and more permanently, oxidized LDL inhibit endothelium-dependent relaxation of arteries, a process mediated by the generation of nitric oxide. Organ bath studies with rabbit aortic rings showed that ascorbate reversed the actions of high concentrations of LDL but not that of oxidized LDL [45] In endothelial cell cultures, increasing concentrations of ascorbate within the physiological range could enhance the synthesis of nitric oxide [46], probably by increasing the biosynthesis of one of the co-factors for NO-synthase, tetrahydrobiopterin. Arterial dilatation, in response to acetylcholine in the human coronary artery, improved endothelial responses with pre-treatment with antioxidants using angiographic procedures. Relaxations of the brachial artery were measured using plethysmography or ultrasound techniques. Large oral doses of vitamin C (2 g) or direct infusion of the vitamin could improve these responses if they had been impaired by atherosclerosis or hypercholesterolemia [47].

7.7 Intervention Trials

The stage was set for intervention studies to determine whether antioxidants had an important therapeutic value alongside the statins.

In the last decade, a number of intervention studies have been completed: Many of them very expensive to run. In most cases, the antioxidants used were in quantities many fold greater than those found in any diet and at a level where their action would be considered pharmacological rather than fulfilling any nutritional requirement. There have been both primary and secondary intervention trials, some with a matrix design, so that the effects of single antioxidant, a combination of antioxidants with a statin, or a single antioxidant and fish oils could be tested [48, 49]. With a small number of exceptions where benefit was indicated, the majority of trials showed that antioxidant therapy did not decrease risk of cardiovascular disease. No benefit was demonstrated whether the antioxidants, mainly α-tocopherol, β-carotene, and ascorbic acid, were used individually or in combination. Indeed some, but not all of the trials, indicated negative effects especially among smokers [50–54]. The assessment of the major studies suggests that positive effects are only seen in patients who were experiencing oxidative stress [55]. Meta-analysis of these studies indicated that there is no real discernible benefit for antioxidant therapy [56].

One study found that α-tocopherol supplementation suppressed restenosis in surgically induced atherosclerosis [57]. The most recent reviews conclude that, though more work may be required, the future of antioxidants as therapeutic agents is bleak [58, 59]. Despite this, the sale of over-the-counter antioxidants and other dietary supplements remains a multi-billion dollar industry. In the food industry, they are frequently used for the preservation of foods. It is hard for consumers to avoid them.

The MRC/BHF Heart Protection Study was a large trial that examined the effects of a cocktail of antioxidant vitamins over 5 years (600 mg vitamin E, 250 mg vitamin C, and 20 mg β-carotene) or placebo in 20,536 UK adults (aged 40–80) with coronary disease, other occlusive arterial disease, or diabetes mellitus [48]. The supplements increased the blood levels of antioxidant vitamins, but without any significant reduction in mortality from vascular disease or cancer. The protection given by treatment with a cholesterol-lowering statin was evident and contrasted with the ineffectiveness of the antioxidant supplements. The GISSI-Prevenzione trial examined both the effects of vitamin E and dietary fish oils. The latter reduced the risk of death, non-fatal myocardial infarction, or stroke, but vitamin E supplementation (300 mg daily for 3.5 years) did not augment this effect [49].

7.8 What Could Be the Explanation?

An analysis of a large cohort of women aged between 60 and 79 years selected for the British Women's Heart and Health Study was performed in which plasma α-tocopherol and ascorbate concentrations were measured [60]. There was a strong associated with socioeconomic position indicators: the lower the social status, the lower the plasma concentration of these vitamins. Lower socioeconomic status is also strongly related to a higher incidence of cardiovascular disease. Tunstall-Pedoe demonstrated that social deprivation is a factor that is often neglected. [61]. These studies also exposed very low levels of plasma vitamin C in some individuals in the lowest socioeconomic groups [62]. Low levels of vitamin C are associated with smoking which may in turn affect food choice. These issues open the question whether the major intervention trials included only a small proportion of socially deprived individuals. This group may benefit from antioxidant therapy or nutritional advice to increase their plasma antioxidant levels. Low levels of ascorbate were also found in a study of the elderly and associated with an increased risk of cardiovascular disease and other diseases [63]. Genetic mutations which lead to impairment of the intestinal ascorbate co-transporter were found to have a small but significant lowering effect on blood ascorbate levels [64].

Antioxidants are not the sole nutritional factor in a diet rich in fruit and vegetables. An increased intake of these foods will lead to a beneficial decrease in other food components, particularly saturated fats. An increase in dietary fiber lowers blood cholesterol and improves glucose tolerance: Salt intakes are likely to be lower. There would also be an increase in the intake of polyunsaturated fatty acids with a concomitant decrease in LDL cholesterol.

The antioxidant content of the artery wall does not change significantly during the evolution of the atherosclerotic plaque. Only at the most advanced stage of the lesion is a reduction in the amount of α-tocopherol evident [65]. Plaque lipoproteins contained α-tocopherol alongside lipid oxidation products. Either the antioxidants do not function or they are radicalized in this environment. Antioxidants become pro-oxidants in the presence of metal ions as indicated above and are cyclically depleted and repleted through these radical forms.

The studies that showed hypochlorite and peroxynitrite were key oxidants in atherosclerosis were published after the main intervention studies had begun. The footprints of these oxidants are found in atherosclerotic plaque (Fig. 7.1) but also in other inflammatory diseases such as rheumatoid arthritis, Alzheimer's disease, and diabetes. The nitration and chlorination of proteins arises by the action of the enzyme myeloperoxidase in macrophages [9]. Serum myeloperoxidase levels have been associated with the future risk of coronary artery disease [66]. There is evidence that γ-tocopherol is a more effective antioxidant against peroxynitrite than α-tocopherol and that the flavonoid epi-gallocatechin or its gallate is yet more effective. These catechins are found in large amounts in green tea, chocolate, red wine, and fruit such as apples. Glutathione peroxidase, an antioxidant enzyme containing selenium, reduces peroxynitrite [67].

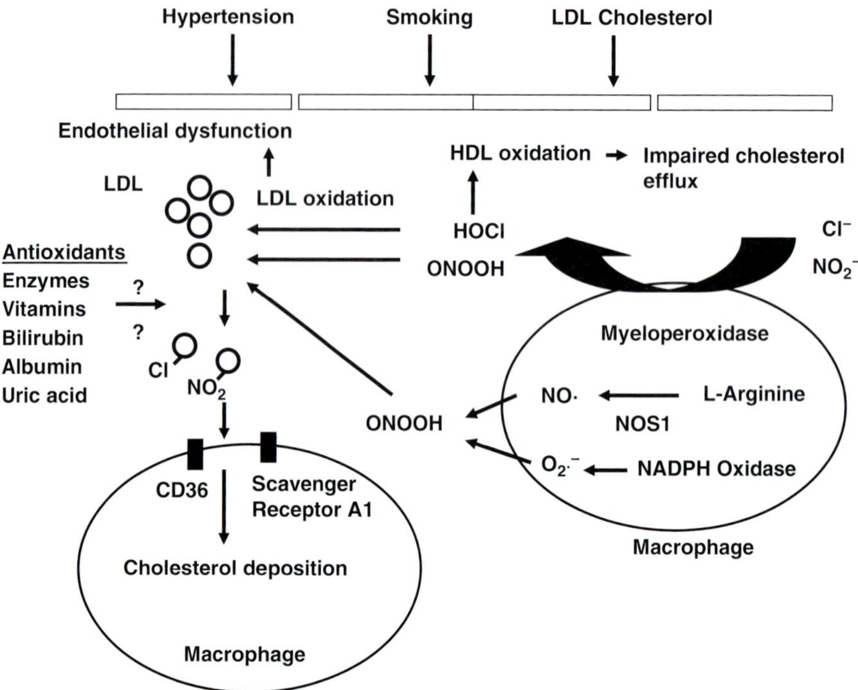

Fig. 7.1 The role of peroxynitrous acid and hypochlorous acid in the oxidation of low-density and high-density lipoproteins and its relevance to the development of atherosclerosis

Under experimental conditions, epi-gallocatechin gallate, at concentrations as low as 2 μM, inhibits protein nitration during the activation of blood platelets [68]. Although large amounts of these compounds may be consumed, levels in the plasma are low. The maximum concentration for this compound found in the plasma is 1 μM: The intracellular concentrations are unknown. Catechins are metabolized and eliminated in the form of glucuronides: Some of these products may have antioxidant activity. No major intervention studies have been attempted using these compounds, despite the current enthusiasm for them. The biological availability of potential antioxidants is important. A cocktail of these compounds and metabolites may have important collective antioxidant actions.

One of the key roles of the free radical nitric oxide released from the endothelium is to prevent the proliferation of smooth muscle cells and to maintain them in their contractile state, inhibiting differentiation into a fibroblastic phenotype. The fibroblastic phenotype biosynthesizes collagen and is abundant in fibrous plaque. Ascorbate is essential for the biosynthesis of collagen, specifically the hydroxylation of proline. NO has an inhibitory effect on collagen biosynthesis [69] whereas ascorbate opposes the inhibitory action of NO on collagen biosynthesis in skin fibroblasts [70]. These studies have been extended now to show that epi-gallocatechin gallate enhances the inhibitory action of NO. The role of catechins and other polyphenols in prevention of cardiovascular disease has not been established [71].

The upregulation of genes in response to oxidative stress leads to the increased synthesis of protective proteins. These include the enzyme hemoxygenase-1, which catalyzes the formation of the endogenous antioxidant bilirubin and is a chaperone. These actions are mediated through nuclear transcription factors, some of which are sensitive to the reduction/ oxidation status of the cell which can be changed by oxidation stress. Some of these proteins are regulated by NO through its interactions with sulfydryl groups on the transcription factors which are themselves proteins, e.g., NFκB, Nrf-2, and HIF-1. These factors operate through antioxidant response elements on the chromosome near regions where these antioxidant proteins are expressed. Siow and colleagues showed that polyphenols augment gene expression for antioxidant enzymes, chaperones, and increase NO biosynthesis through NFκB and Nrf-2 [72].

Jackson and co-workers [73] investigated the action of antioxidant vitamins on the expression of protective genes following muscular exercise. Changes were observed in the proteins of lymphocytes and skeletal muscle in untrained human subjects with and without supplementation with ascorbate (0.5 g/day for 8 weeks). There was an increase in lymphocyte superoxide dismutase, catalase activity, and the cellular content of HSP60 and HSP70 chaperone proteins in response to a low concentration of hydrogen peroxide, without ascorbate supplementation. After supplementation, the basal activity or content of the cellular proteins was slightly increased, but the cells gave an attenuated response to the peroxide. In muscle post exercise, there was a rise in HSP60 and HSP70 was diminished by supplementation with ascorbate, at least for HSP60.

7.9 Conclusions

A survey of the progress over 40 years of research into oxidative stress, free radicals, and the role of dietary antioxidants as therapeutic agents shows that it had its high points and distinctive low points. It is clear that there is more to do to understand the complex effects of dietary antioxidants and how they influence signaling mechanisms that respond to oxidative stress. Not all free radicals and reactive species are suppressed by the same dietary antioxidants. The simple "free radicals bad – antioxidants good" slogan is simply not adequate to comprehend how they may contribute to the lifelong process of atherosclerosis. It is clear that oxidative stress is important to its evolution. However, pharmacological doses of antioxidants do not prevent these pathological changes. It does seem that a diet rich in fruit and vegetables is beneficial and family resources may be better spent on them rather than expensive and ineffective supplements.

References

1. McCord JM, Fridovich I. Superoxide dismutase. An enzymic function for erythrocuprein (hemocuprein). J Biol Chem. 1969;244:6049–55.
2. Powers SK, Jackson MJ. Exercise-induced oxidative stress: cellular mechanisms and impact on muscle force production. Physiol Rev. 2008;88:1243–76.

3. Halliwell B, Gutteridge JMC. Free radicals in biology and medicine. Oxford: Oxford University Press; 1999.
4. Matoba T, Shimokawa H, Nakashima M, et al. Hydrogen peroxide is an endothelium-derived hyperpolarizing factor in mice. J Clin Invest. 2006;106:1521–30.
5. Bruckdorfer R. The basics about nitric oxide. Mol Aspects Med. 2005;26:3–31.
6. Warner HR. Apoptosis: a two-edged sword in aging. Ann N Y Acad Sci. 1999;887:1–11.
7. Esterbauer H, Dieber-Rotheneder M, Waeg G, Striegl G, Jurgens G. Biochemical, structural and functional properties of oxidised low-density lipoproteins. Chem Res Toxicol. 1990;3:77–91.
8. Henriksen T, Mahoney EM, Steinberg D. Enhanced macrophage degradation of biologically modified low density lipoprotein. Arteriosclerosis. 1983;3:149–59.
9. Nicholls SJ, Hazen SL. Myeloperoxidase and cardiovascular disease. Arterioscler Thromb Vasc Biol. 2005;25:1102–11.
10. Nicholls SJ, Zheng L, Hazen SL. Formation of dysfunctional high-density lipoprotein by myeloperoxidase. Trends Cardiovasc Med. 2005;15:212–9.
11. Vadseth C, Souza JM, Thomson L, et al. Pro-thrombotic state induced by post-translational modification of fibrinogen by reactive nitrogen species species. J Biol Chem. 2004;279:8820–6.
12. Nowak P, Kolodrziejczyk J, Wachowicz B. Peroxynitrite and fibrinolytic system: the effect of peroxynitrite on plasmin activity. Mol Cell Biochem. 2004;267:141–6.
13. Plane F, Bruckdorfer KR, Kerr P, Steuer A, Jacobs M. Oxidative modification of low-density lipoproteins and the inhibition of relaxations mediated by endothelium-derived nitric oxide in rabbit aorta. Br J Pharmacol. 1992;105:216–22.
14. Watson SP, McConnell RT, Lapetina EG. Decanoyl lysophosphatidic acid induces platelet aggregation through an extracellular action. Evidence against a second messenger role for lysophosphatidic acid. Biochem J. 1985;232:61–6.
15. Costet P, Luo Y, Wang N, Tall AR. Sterol-dependent transactivation of the ABC1 promoter by the liver X receptor/retinoid X receptor. J Biol Chem. 2000;275:28240–5.
16. Morrow JD, Roberts LJ. The isoprostanes: unique bioactive products of lipid oxidation. Prog Lipid Res. 1997;36:1–21.
17. Mabile L, Bruckdorfer KR, Rice-Evans C. Moderate supplementation with natural alpha-tocopherol decreases platelet aggregation and low-density lipoprotein oxidation. Atherosclerosis. 1999;147:177–85.
18. Jialal I, Vega GL, Grundy SM. Physiologic levels of ascorbate inhibit the oxidative modification of low density lipoprotein. Atherosclerosis. 1990;82:185–91.
19. Lehtimäki T, Lehtinen S, Solakivi T, et al. Autoantibodies against oxidized low density lipoprotein in patients with angiographically verified coronary artery disease. Arterioscler Thromb Vasc Biol. 1999;19:23–7.
20. Miller 3rd ER, Erlinger TP, Sacks FM, et al. A dietary pattern that lowers oxidative stress increases antibodies to oxidized LDL: results from a randomized controlled feeding study. Atherosclerosis. 2005;183:175–82.
21. Heitzer T, Ylä Herttuala S, Wild E, Luoma J, Drexler H. Effect of vitamin E on endothelial vasodilator function in patients with hypercholesterolemia, chronic smoking or both. J Am Coll Cardiol. 1999;33:499–505.
22. Vasdev S, Gill VD, Singal PK. Modulation of oxidative stress-induced changes in hypertension and atherosclerosis by antioxidants. Exp Clin Cardiol. 2006;11:206–16.
23. Podrez EA, Byzova TV, Febbraio M, et al. Platelet CD36 links hyperlipidemia, oxidant stress and a prothrombotic phenotype. Nat Med. 1997;13:1086–95.
24. Armstrong B, Doll R. Environmental factors and cancer incidence and mortality in different countries, with special reference to dietary practices. Int J Cancer. 1975;15:617–31.
25. Rimm EB, Ascherio A, Giovannucci E, et al. Vegetable, fruit and cereal fiber intake and risk of coronary heart disease among men. J Am Med Assoc. 1996;275:447–51.
26. Thorogood M. Do vegetables and fruit protect against coronary heart disease? Studies among vegetarians. In *Preventing coronary Heart Disease: The role of antioxidants, vegetables and fruit.* 1997 (Eds, Forum. NH). The Stationery Office, London.

27. Joshipura KJ, Ascherio A, Manson JE, et al. Fruit and vegetable intake in relation to risk of ishemic stroke. JAMA. 1999;282:1233–9.
28. De Lorgeril M, Salen P, Martin JL, et al. Mediterranean dietary pattern in a randomized trial: prolonged survival and possible reduced cancer rate. Arch Intern Med. 1998;158:1181–7.
29. Appel LJ, Moore TJ, Obarzanek E, et al. A clinical trial of the effects of dietary patterns on blood pressure. DASH Collaborative Research Group. N Engl J Med. 1997;336:1117–24.
30. Jenkins DJ, Popovich DG, Kendall CW, et al. Effect of a diet high in vegetables, fruit, and nuts on serum lipids. Metabolism. 1997;46:530–7.
31. Dauchet L, Amouyel P, Hercberg S, Dallongville J. Fruit and vegetable consumption and risk of coronary heart disease: a meta-analysis of cohort studies. J Nutr. 2007;136:2588–93.
32. Dauchet L, Amouyel P, Dollongeville P. Fruits vegetables and coronary heart disease. Nat Rev Cardiol. 2009;6:599–608.
33. Beaglehole R, Jackson R, Watkinson J, Scragg R, Yee RL. Decreased blood selenium and risk of myocardial infarction. Int J Epidemiol. 1990;19:918–22.
34. Riemersma RA, Oliver M, Elton RA, et al. Plasma antioxidants and coronary heart disease: vitamins C and E, and selenium. Eur J Clin Nutr. 1990;44:143–50.
35. Rimm EB, Stampfer MJ, Ascherio A, et al. Vitamin E consumption and the risk of coronary heart disease in men. N Engl J Med. 1993;328:1450–6.
36. Stampfer MJ, Hennekens CH, Manson JE, et al. Vitamin E consumption and the risk of coronary disease in women. N Engl J Med. 1993;328:1444–9.
37. Khaw KT, Bingham S, Welch A, et al. Relation between plasma ascorbic acid and mortality in men and women in EPIC-Norfolk prospective study: a prospective population study. European Prospective Investigation into Cancer and Nutrition. Lancet. 2001;357:657–63.
38. Kushi LH, Folsom AR, Prineas RJ, et al. Dietary antioxidant vitamins and death from coronary heart disease in postmenopausal women. N Engl J Med. 1996;334:1156–62.
39. Yochum LA, Folsom AR, Kushi LH. Intake of antioxidant vitamins and risk of death from stroke in postmenopausal women. Am J Clin Nutr. 2000;72:476–83.
40. Yochum L, Kushi LH, Meyer K, Folsom AR. Dietary flavonoid intake and risk of cardiovascular disease in postmenopausal women. Am J Epidemiol. 1999;149:943–9.
41. Navarro-Alarcon M, Lopez-Garcia de la Serrana H, Perez-Valero V, Lopez-Martinez C. Serum and urine selenium concentrations in patients with cardiovascular diseases and relationship to other nutritional indexes. Ann Nutr Metab. 1999;43:30–6.
42. Neve J. Selenium as a risk factor for cardiovascular diseases. J Cardiovasc Risk. 1996;3: 42–7.
43. Boekholdt SM, Hack SE, Sandhu MS, et al. C-reactive protein levels and coronary artery disease incidence and mortality in apparently healthy men and women: the EPIC-Norfolk prospective population study 1993–2003. Atherosclerosis. 2006;187:415–22.
44. Wannamethee SG, Lowe GD, Rumley A, Bruckdorfer KR, Whincup PH. Associations of vitamin C status, fruit and vegetable intakes, and markers of inflammation and hemostasis. Am J Clin Nutr. 2006;83:567–74.
45. Jacobs M, Plane F, Bruckdorfer KR. Native and oxidised LDL have differential inhibitory effects on endothelium derived relaxing factor in rabbit aorta. Br J Pharmacol. 1990;100: 21–6.
46. Heller R, Munscher-Paulig F, Grabner R, Till U. L-Ascorbic acid potentiates nitric oxide synthesis in endothelial cells. J Biol Chem. 1999;19:8254–60.
47. Ting HH, Timimi FK, Haley EA, Roddy MA, Ganz P, Creager MA. Vitamin C improves endothelium-dependent vasodilation in forearm resistance vessels of humans with hypercholesterolemia. Circulation. 1997;95:2617–22.
48. Heart Protection Study Collaborative Group. MRC/BHF Heart Protection Study of antioxidant vitamin supplementation in 20,536 high-risk individuals: a randomised placebo-controlled trial. Lancet. 2002;360:23–33.
49. GISSI-Prevenzione Investigators. Dietary supplementation with n-3 poly-unsaturated fatty acids and vitamin E after myocardial infarction: results from the GISSI-prevenzione trial. Lancet. 1999;354:447–55.

50. Alpha-Tocopherol Beta Carotene Cancer Prevention Study Group. The effect of vitamin E and beta carotene on the incidence of lung cancer and other cancers in male smokers. N Engl J Med. 1994;330:1029–35.
51. Tornwall ME, Virtamo J, Haukka JK, Albanes D, Huttunen JK. Alpha-tocopherol (vitamin E) and beta-carotene supplementation does not affect the risk for large abdominal aortic aneurysm in a controlled trial. Atherosclerosis. 2001;157:167–73.
52. Leppala JM, Virtamo J, Fogelholm R, et al. Controlled trial of alpha-tocopherol and beta-carotene supplements on stroke incidence and mortality in male smokers. Arterioscler Thromb Vasc Biol. 2000;20:230–5.
53. Omenn GS, Goodman GE, Thornquist MD, et al. Effects of a combination of beta-carotene and vitamin A on lung cancer and cardiovascular disease. N Engl J Med. 1996;334:1150–5.
54. Hennekens CH, Buring JE, Manson JE, et al. Lack of effect of long-term supplementation with beta carotene on the incidence of malignant neoplasms and cardiovascular disease. N Engl J Med. 1996;334:1145–9.
55. Willcox B, Curb JD, Rodriguez BL. Antioxidants in cardiovascular health and disease: key lessons from epidemiologic studies. Am J Cardiol. 2008;101(suppl):75D–86.
56. Clarke R, Armitage J. Antioxidant vitamins and risk of cardiovascular disease. Cardiovasc Drugs Ther. 2002;5:411–5.
57. Fang JC, Kinlay S, Beltrame J, et al. Effect of vitamins C and E on progression of transplant-associated arteriosclerosis: a randomised trial. Lancet. 2002;359:1108–13.
58. Katsiki N, Manes C. Is there a role for supplemented antioxidants in the prevention of atherosclerosis? Clin Nutr. 2009;28:3–9.
59. Leonard JA, Loscalzo J. Oxidative risk for atherothrombotic cardiovascular disease. Free Radic Biol Med. 2000;47:1673–706.
60. Lawlor DA, Davey Smith G, Kundu D, Bruckdorfer KR, Ebrahim S. Those confounded vitamins: what can we learn from the differences between observational versus randomised trial evidence? Lancet. 2004;363:1724–7.
61. Tunstall-Pedoe H, Woodward M, SIGN Group on Risk Estimation. By neglecting deprivation, cardiovascular risk scoring will exacerbate social gradients in disease. Heart. 2006;92: 307–10.
62. Wrieden WL, Hannah MK, Bolton-Smith C, Tavendale R, Morrison C, Tunstall-Pedoe H. Plasma vitamin C and food choice in the third Glasgow MONICA population survey. J Epidemiol Community Health. 2000;54:355–60.
63. Fletcher AE, Breeze E, Shetty PS. Antioxidant vitamins and mortality in older persons: findings from the nutrition add-on study to the Medical Research Council Trial of Assessment and Management of Older People in the Community. Am J Clin Nutr. 2003;78:999–1010.
64. Timpson NJ, Forouhi NG, Brion MJ, et al. Genetic variation at the SLC23A1 locus is associated with circulating concentrations of L-ascorbic acid (vitamin C): evidence from 5 independent studies with >15,000 participants. Am J Clin Nutr. 2010;92:375–82.
65. Upston JM, Terentis AC, Morris K, Keaney Jr JF, Stocker R. Oxidized lipid accumulates in the presence of alpha-tocopherol in atherosclerosis. Biochem J. 2002;363:753–60.
66. Meuwese MC, Stroes ESG, Hazen SL. Serum myeloperoxidase levels are associated with the future risk of coronary artery disease in apparently healthy individuals. J Am Coll Cardiol. 2007;50:159–65.
67. Sies H, Sharov VS, Klotz LO, Briviba K, Sies H, Sharov VS, et al. Glutathione peroxidase protects against peroxynitrite-mediated oxidations. A new function for selenoproteins as peroxynitrite reductase. J Biol Chem. 1997;272:27812–7.
68. Sabetkar M, Low SY, Naseem KM, Bruckdorfer KR. The nitration of proteins in platelets: significance in platelet function. Free Radic Biol Med. 2002;33:728–36.
69. Shukla A, Rasik AM, Shankar R. Nitric oxide inhibits wound collagen biosynthesis. Mol Cell Biochem. 1999;200:27–33.
70. Dooley A, Gao B, Shi-Wen X, Abraham DJ, Black CM, Jacobs M, et al. Effect of nitric oxide and peroxynitrite on type I collagen synthesis in normal and scleroderma dermal fibroblasts. Free Radic Biol Med. 2007;43:253–64.

71. Sies H. Polyphenols and health: update and perspectives. Arch Biochem Biophys. 2010;501:2–5.
72. Siow RC, Li FY, Rowlands DJ, de Winter P, Mann GE. Cardiovascular targets for estrogens and phytoestrogens: transcriptional regulation of nitric oxide synthase and antioxidant defense genes. Free Radic Biol Med. 2007;42:909–25.
73. Khassaf M, McArdle A, Esanu C, et al. Effect of vitamin C supplements on antioxidant defence and stress proteins in human lymphocytes and skeletal muscle. J Physiol. 2003;549:645–52.

Section III

Clinical Aspects of Cardiovascular Disease

The Takotsubo (Broken Heart Syndrome)

8

Lawrence S. Cohen

8.1 Introduction

An interesting syndrome was reported from Japan close to 20 years ago [1–5]. It had a varying nomenclature but the name takotsubo syndrome or apical ballooning syndrome was used frequently. More recently, the descriptive name, "broken heart" syndrome has been used. Also recently, the syndrome has been reported in patients from the United States and Europe [6–10].

The syndrome is found predominantly in postmenopausal women. It is sometimes mistaken for an acute myocardial infarction. The patient usually presents with chest pain or extreme weakness, has electrocardiographic changes mimicking an acute myocardial infarction, may have some elevation of cardiac biomarkers, and may present with hypotension or cardiogenic shock. Coronary arteriography reveals normal epicardial coronary arteries. Cardiac supportive therapy is usually successful in getting the patient through the acute event. Takotsubo cardiomyopathy is most often characterized by transient regional contractile dysfunction with hypokinesis or akinesis of the left ventricular apical segments and hyperkinesis of the basal segments. The term "takotsubo" was used by the original Japanese investigators as the left ventriculogram of a patient with the syndrome resembled a takotsubo, or Japanese octopus fishing pot (Fig. 8.1). In Japanese, "takotsubo" means "fishing pot for trapping octopus." These traps have a round bottom with a narrow neck. When

Dr. Cohen is the Ebenezer K. Hunt Professor Emeritus of Cardiology at Yale University School of Medicine, New Haven, CT 06510, USA

L.S. Cohen
Department of Cardiology, Yale University School of Medicine,
New Haven, CT, USA
e-mail: lawrence.s.cohen@yale.edu

D. Abraham et al. (eds.), *Translational Vascular Medicine*,
DOI 10.1007/978-0-85729-920-8_8, © Springer-Verlag London Limited 2012

Fig. 8.1 A takotsubo or "Japanese fishing pot." These traps have a round bottom with a narrow neck. When the octopus enters the takotsubo, it is trapped

the octopus enters the takotsubo, it is often trapped while the fisherman pulls the device to the surface. The other feature of the syndrome is that it most often is initiated by a severe emotional or psychological life event. These may include violent arguments, domestic abuse, death of a relative, learning of a catastrophic medical event in oneself or a close relative, or financial or gambling losses.

8.2 Case History

P.O. is a 68-year-old woman who had no cardiac history until May 2005 when she was 65 years of age. She had a history of surgical removal of uterine fibroids and the resection of a benign breast cyst. She was 10 years postmenopausal and had a lifelong history of Raynaud's phenomenon.

During a routine yearly checkup in April 2005, an electrocardiogram was totally normal. In early May 2005, she had a severely emotional and stressful afternoon at her mother's funeral. She had no chest pain but felt extremely weak and unwell. She was seen by her physician on May 19, 2005, where an electrocardiogram was quite abnormal. It revealed deep coved T waves in I, II, III, Avf, V3-V6. The QTc was 484 ms (Fig. 8.2).

On June 16, 2005, the T wave coving was less and the QTc was 450 ms (Fig. 8.3). On July 7, 2005, the abnormalities were starting to abate. The QTc was 445 ms (Fig. 8.4). By September 22, 2005, the ECG was virtually normal with no T wave abnormalities. The QTc was 412 ms (Fig. 8.5).

8.3 Epidemiology and Prevalence

Since first being described, the number of cases reported has increased. It is estimated that up to 2% of patients presenting with an acute coronary syndrome may actually be presenting with the takotsubo syndrome. This number of cases

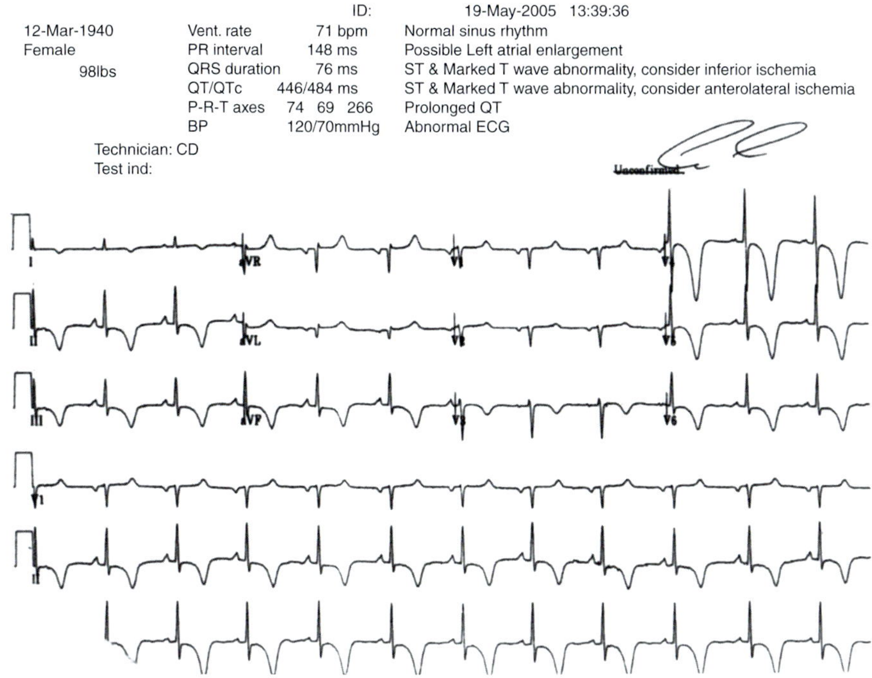

Fig. 8.2 ECG – May 19, 2005 – Deeply coved T waves. QTc 484 ms

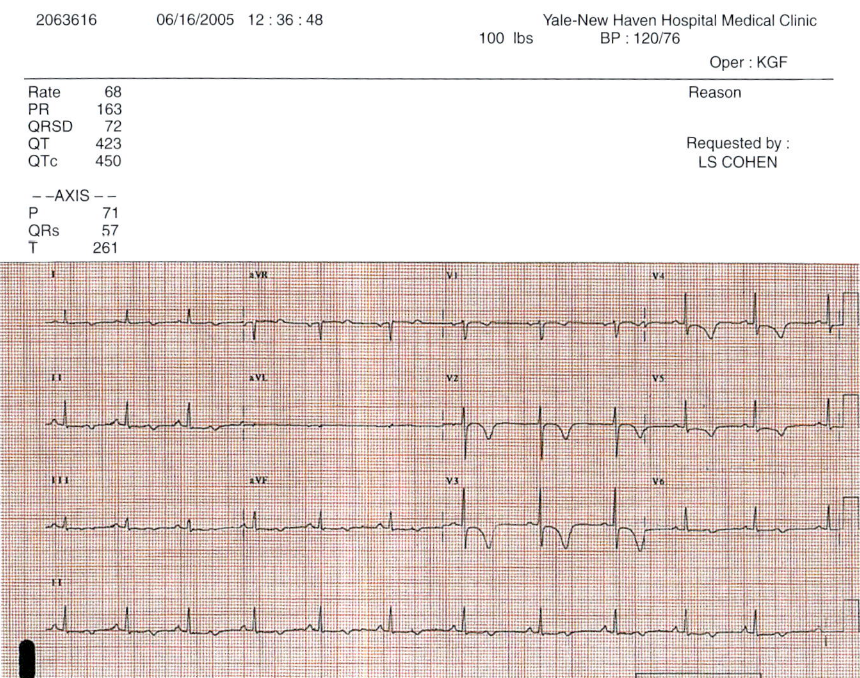

Fig. 8.3 ECG – June 16, 2005 – T wave coving improving. QTc 450 ms

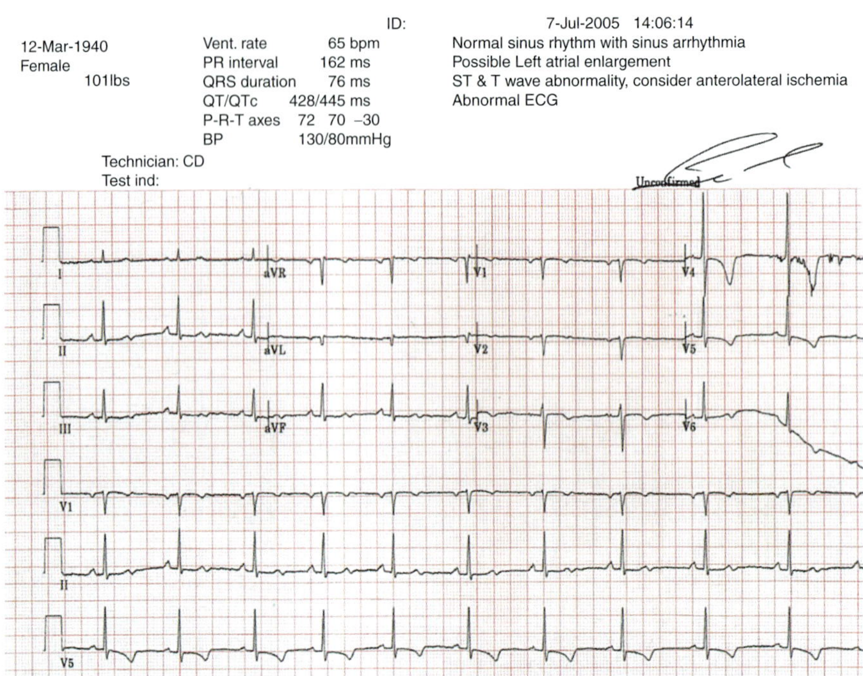

Fig. 8.4 ECG – July 7, 2005 – T wave coving improving. QTc 445 ms

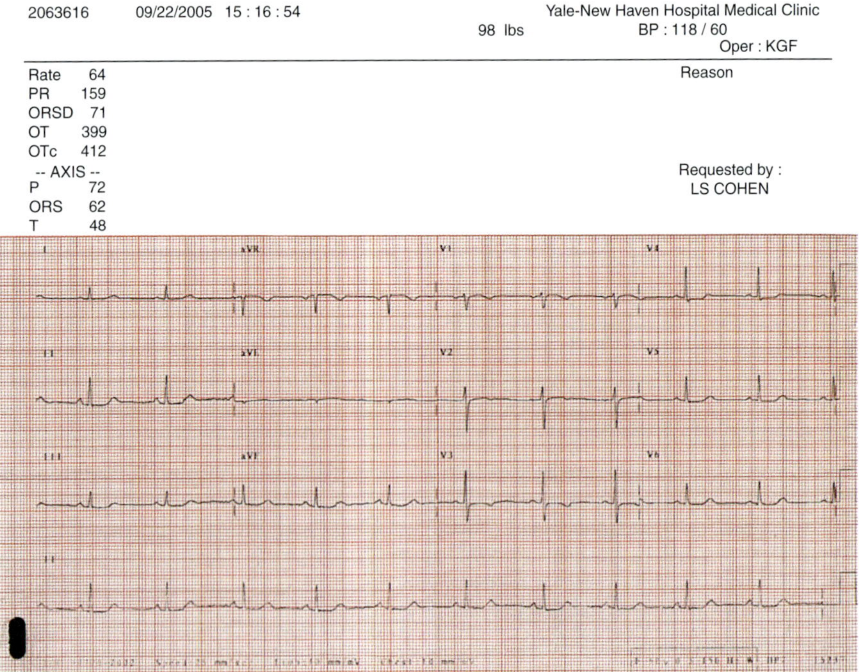

Fig. 8.5 ECG – September 22, 2005 – T waves normal. QTc 412 ms

has been reported both in the United States and in Europe. There is a strong predominance of postmenopausal women. The reasons for this are not clear but as will be discussed later, it is likely that sex hormones exert important influences on the sympathetic neurohumoral axis as well as on coronary vasoreactivity.

8.4 Clinical Presentation

Takotsubo cardiomyopathy is characterized by the acute onset of chest pain, dyspnea, and at times syncope. It is predominantly seen in postmenopausal women in their 50s or 60s. There is usually a severe emotional or physical event in the antecedent period leading up to the clinical presentation. News of an unexpected death or other such emotional trauma is common. The patient may be hypotensive and may require circulatory pharmacologic support. Similarly the patient may present with severe dyspnea and at times pulmonary edema. With appropriate hemodynamic support which at times may require an intra-aortic balloon pump, the immediate prognosis is favorable.

8.5 The Electrocardiogram

The electrocardiogram may mimic closely that of an acute anterior wall myocardial infarction. There is ST-segment elevation which may evolve into deeply coved T waves after the ST segment approaches baseline. There is usually a prolonged QT (QTc) interval which returns to normal somewhat more quickly then pathologic precordial Q waves if present.

8.6 Echocardiogram

The initial left ventricular ejection fraction is most often markedly depressed, at times as low as 20%. The typical contractile pattern demonstrates preserved basal function, moderate-to-severe dysfunction in the mid-ventricle, and apical akinesis or dyskinesis. Within a week's time, the left ventricular ejection fraction is usually improved and the mid-ventricular and apical segments are only mildly hypokinetic.

8.7 Cardiac Biomarkers

There is frequently a mild elevation of troponin T or troponin I levels. This elevation is by no means invariable and the biomarkers often remain normal. Similarly creatine kinase or creatine kinase MB levels may be normal or only minimally elevated.

8.8 Coronary Angiography and Ventriculography

Coronary angiography usually displays normal epicardial coronary arteries. At times there may be spasm recognized particularly in the left anterior descending coronary artery. The left ventriculogram usually displays typical apical ballooning and hypercontraction of the basal segments.

8.9 Pathogenesis

The etiology of the takotsubo syndrome is unclear. The most common explanation is that excessive catecholamine secretion and catecholamine cardiotoxicity are important. In support of this thesis, patients reported by Wittstein et al. had plasma levels of catecholamines (i.e., epinephrine, norepinephrine, and dopamine) on hospital day 1 or 2 two to three times higher than patients with Killip class IV myocardial infarction. Plasma levels of metanephrine and normetanephrine were also increased among patients with stress cardiomyopathy [10]. Further support for the thesis that a massive catecholamine discharge is relevant in this pathogenesis of the "broken heart" syndrome comes from the neurologic literature where patients with subarachnoid hemorrhage have been reported to develop profound electrocardiographic changes characterized by deep symmetrical T wave changes across the anterior precordium [11].

The mechanism underlying the association of catecholamine excess and electrocardiographic changes is not clear. One possibility is that catecholamine excess may lead to epicardial coronary artery spasm. Soufer and colleagues [12] have demonstrated that mental stress is a powerful initiator of a process that activates cerebral and adrenal pathways that lead to increased myocardial oxygen demand and simultaneously leads to coronary and peripheral vasoconstriction or spasm.

An alternative mechanism related to the above may be spasm of the myocardial microvasculature. The coronary microvasculature system is very responsive to sympathetic stimulation and may under periods of stress alter the delivery of oxygen to the myocardial muscle.

The idea of myocardial stunning was enunciated in an early paper addressing the effects of transient ischemia on myocardial contractility. Braunwald and Kloner [13] put forth the idea that transient coronary occlusion with reopening before necrosis occurs could lead to stunning or hibernation of the affected myocardium. With time, often up to a week or two, full function could be achieved. This phenomenon certainly mimics what is seen in women who develop the takotsubo syndrome. Lyon et al hypothesize that takotsubo cardiomyopathy is a form of myocardial stunning but with a different cellular effect than that secondary to myocardial ischemia. They believe that high levels of circulating epinephrine trigger a switch in intracellular signal trafficking in ventricular cardiomyocytes [14].

In addition, the fact that the overwhelming incidence of takotsubo cardiomyopathy occurs in women obviously raises the question of whether estrogen, or the lack thereof, plays a role in the pathogenesis of this syndrome. There is considerable experimental evidence that estrogen lack may also contribute. The fact that most patients with takotsubo cardiomyopathy are postmenopausal women is very suggestive. It has been shown that in postmenopausal women, estrogen supplementation enhances nitric oxide release and attenuates norepinephrine-induced vasoconstriction. The evidence is very strong that postmenopausal women who develop takotsubo cardiomyopathy are estrogen deficient. Further it is clear that the precipitating event causes an increase in norepinephrine and other catecholamines. The postmenopausal estrogen-deficient female may likely develop profound epicardial coronary artery vasoconstriction. The 17-beta-estradiol therapy lessens angina in postmenopausal women with normal coronary arteries [15]. In an experimental rat model, it has been shown that estrogen supplementation partially reversed the cardiac changes brought about by laboratory-induced stress [16].

Therefore patients with takotsubo cardiomyopathy clearly have exaggerated sympathetic activation. It is a hypothesis yet to be proved that the catecholamine excess associated with grief reacts on a coronary artery system primed for spasm and stunning due to estrogen lack.

The weight of evidence points to a multifactorial pathogenesis in patients who develop the takotsubo syndrome. It is akin to a physiologic perfect storm. An extreme emotional event unleashes a catecholamine surge. The postmenopausal estrogen-deficient female is particularly susceptible to the actions of the catecholamine surge. It is likely that a catecholamine-induced combination of vasoconstriction of the epicardial coronary arteries, constriction of the coronary microvasculature, and a direct effect on cardiomyocytes lead to a stunning effect on the left ventricular myocardium. With time, these changes resolve and the syndrome abates. Recognition of this syndrome leads to more rational and effective therapy.

References

1. Dote K, Sato H, Tateishi H, Uchida T, Ishihara M. Myocardial stunning due to simultaneous multivessel coronary spasms: a review of 5 cases. J Cardiol. 1991;21:203–14.
2. Kurisu S, Sato H, Kawagoe T, Ishihara M, Shimatani Y, Nishioka K, et al. Tako-tsubo-like left ventricular dysfunction with ST segment elevation: a novel cardiac syndrome mimicking acute myocardial infarction. Am Heart J. 2002;143:448–55.
3. Owa M, Aizawa K, Urasawa N, Urasawa N, Ichinose H, Yamamoto K, et al. Emotional stress-induced "ampulla cardiomyopathy" discrepancy between the metabolic and sympathetic innervation imaging performed during the recovery course. Jpn Circ J. 2001;65:349–52.
4. Abe Y, Kondo M, Matsuoka R, Araki M, Dohyama K, Tanio H. Assessment of clinical features in transient left ventricular apical ballooning. J Am Coll Cardiol. 2003;41:737–42.
5. Tsuchihashi K, Ueshima K, Uchida T, Ohemura N, Kimura K, et al. Transient left ventricular apical ballooning without coronary artery stenosis: a novel heart syndrome mimicking acute myocardial infarction. J Am Coll Cardiol. 2001;38:11–8.
6. Seth PS, Aurigemma GP, Krasnow JM, Tighe DA, Untereker WJ, Meyer TE. A syndrome of transient left ventricular apical wall motion in the absence of coronary disease: a perspective from the United States. Cardiology. 2003;100:61–6.
7. Sharkey SW, Lesser JR, Zenovich AG, Maron MS, Lindberg J, Longe TF, et al. Acute and reversible cardiomyopathy provoked by stress in women from the United States. Circulation. 2005;111:472–9.

8. Bybee KA, Prasad A, Barsness GW, Lerman A, Jaffe AS, Murphy JG, et al. Clinical characteristics and thrombolysis in myocardial infarction frame counts in women with transient left ventricular apical ballooning syndrome. Am J Cardiol. 2004;94:343–6.
9. Bybee KA, Kara T, Prasad A, Lerman A, Barsness GW, Wright RS, et al. Systematic review: transient left ventricular apical ballooning: a syndrome that mimics ST-segment elevation myocardial infarction. Ann Intern Med. 2004;141:858–65.
10. Wittstein IS, Theimann DR, Lima JA, Baughman KL, Schulman SP, et al. Neurohumoral features of myocardial stunning due to sudden emotional stress. N Engl J Med. 2005;352:539–48.
11. Sommargren CE, Zaroff JG, Banki N, Drew BJ. Electrocardiographic repolarization abnormalities in subarachnoid hemorrhage. J Electrocardiol. 2002;35:257–62.
12. Soufer R, Arrighi JA, Burg MM. Brain behavior, mental stress, and the neurocardiac interaction. J Nucl Cardiol. 2002;9:650–62.
13. Braunwald E, Kloner RA. The stunned myocardium: prolonged postischemic ventricular dysfunction. Circulation. 1982;66:1146–9.
14. Lyon AR, Rees PSC, Prasad S, Poole-Wilson PA. Stress (takotsubo) cardiomyopathy – a novel pathophysiological hypothesis to explain catecholamine-induced myocardial stunning. Nat Clin Pract Cardiovasc Med. 2008;5:22–9.
15. Rosano GMC, Peters NS, Leroy D, Lindsay DC, Sarrel PM, et al. 17-Beta Estradiol Therapy lessens angina in postmenopausal women with syndrome X. J Am Coll Cardiol. 1996;28:1500–5.
16. Ueyama T, Kasamatsu K, Jans T, Tsuruo Y, Ishikura F. Catecholamines and estrogen are involved in the pathogenesis of emotional stress-induced acute heart attack. Ann N Y Acad Sci. 2008;1148:479–85.

Lymphatic Vessels in Health and Disease

9

Elisabetta Weber, Francesca Sozio, Erica Gabbrielli, and Antonella Rossi

What made the long neglected lymphatic vessels an interesting aspect of vascular biology are two important discoveries: a lymphatic-specific growth factor, VEGF-C, and its receptor, VEGFR-3, and an excellent marker, D2-40.

9.1 The Discovery of VEGF-C and Its Receptor VEGFR-3

In 1995, the group of Alitalo in Helsinki found a specific receptor: Flt4, subsequently re-named vascular endothelial growth factor receptor-3 (VEGFR-3), which is initially expressed by blood and lymphatic developing vessels and later becomes restricted to lymphatic endothelium [1]. The year later the same group isolated and cloned from human prostatic carcinoma cells the ligand for VEGFR-3: vascular endothelial growth factor-C (VEGF-C), the first growth factor specific for lymphatic vessels [2]. Transgenic mice overexpressing VEGF-C have hyperplastic lymphatic vessels [3]. Defective VEGFR-3 signaling due to missense mutations has been reported in the congenital hereditary form of lymphedema: Milroy's disease [4, 5]. In this disease, lymphatic vessels are absent and lymphedema of the lower extremities is already present at birth and increases with age.

9.2 Lymphatic Markers

Research on lymphatic vessels has long been hampered by the difficulty to recognize them in common histological sections particularly when they are collapsed as they very often do. Lymphatic vessels are also easily confused with venules. A number of

E. Weber (✉) • F. Sozio • E. Gabbrielli • A. Rossi
Department of Neuroscience, Molecular Medicine Section,
University of Siena, Siena, Italy
e-mail: weber@unisi.it; sozio@unisi.it; gabbrielli2@unisi.it; rossianto@unisi.it

D. Abraham et al. (eds.), *Translational Vascular Medicine*,
DOI 10.1007/978-0-85729-920-8_9, © Springer-Verlag London Limited 2012

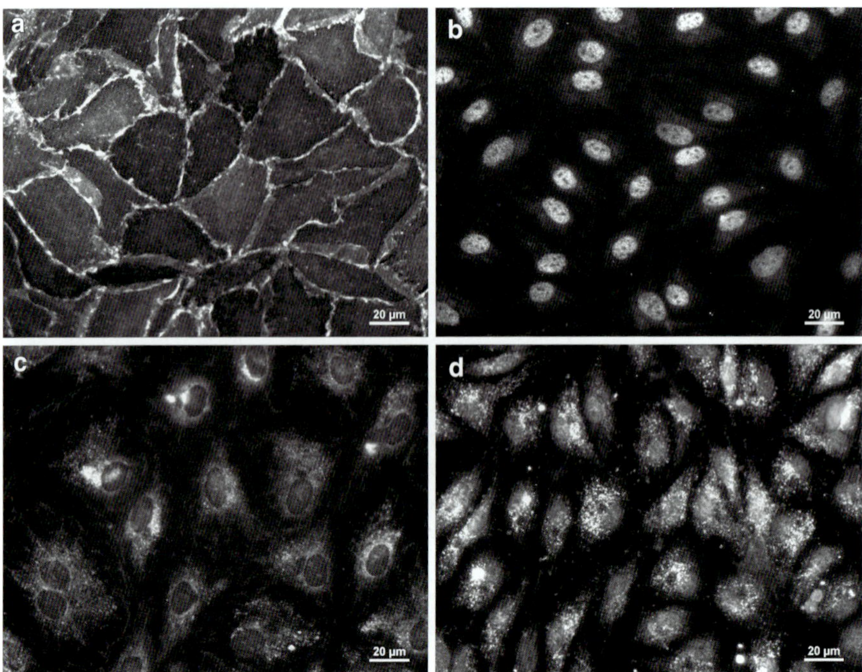

Fig. 9.1 Immunostaining of human dermal lymphatic microvascular endothelial cells in culture with lymphatic markers: (**a**) LYVE-1, (**b**) Prox-1, (**c**) VEGFR-3, (**d**) D240

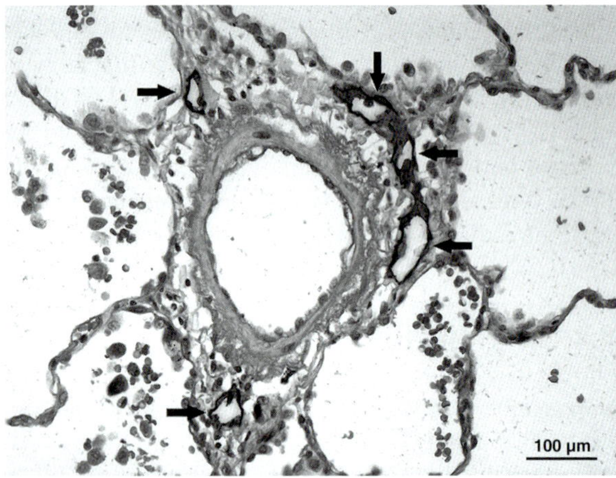

Fig. 9.2 Human lung: lymphatic vessels (*arrows*) stained in black by D240 around an artery

lymphatic markers have been proposed, but they had to be used in combination because a single marker often missed part of the lymphatic vessels present in a given tissue [6]. The most commonly used lymphatic markers besides VEGFR-3 are: Prox-1, the homologue of the Drosophila homeobox gene; Prospero, a master gene in specifying lymphatic fate [7, 8] – it is a nuclear marker; and LYVE-1 [9], the receptor for hyaluronan, homologue of CD44 for blood vessels. The immunostaining of cultured human lymphatic endothelial cells with these markers is illustrated in Fig. 9.1. But the ideal marker, reliable and strongly expressed in all lymphatic vessels, is D2-40 [10] (Fig. 9.2), a monoclonal antibody that recognizes podoplanin [11]. The role of podoplanin in lymphatic vessel biology is not well understood, but podoplanin knockout mice have defects in lymphatic vessels with congenital lymphedema and dilation of skin and intestinal lymphatic vessels [12].

9.3 Development of Lymphatic Vessels

How do lymphatic vessels develop has been the subject of a long debate. A very old theory by Sabin [13] said that lymphatic vessels arise from veins. Recent experimental evidence provided support to this theory: Primitive lymphatic vessels indeed bud from the cardinal vein at embryonic day 9 [14]. Some endothelial cells in the wall of the vein start expressing Prox-1 [7]; this gene determines lymphatic commitment. Prox1 null mice fail to develop any lymphatic vasculature. Prox1-expressing cells migrate and form primitive lymph sacs. Sprouting from primitive lymph sacs is made possible by VEGF-C stimulation. Homozygous deletion of VEGF-C in mouse embryos leads to the complete absence of the lymphatic vasculature, whereas heterozygous mice display severe hypoplasia [15]. Maturation of lymphatic vessels is controlled by several different factors including angiopoietins, FOXC2, Ephrin B2, Podoplanin. During maturation, the wall of lymphatic collecting vessels becomes provided with smooth muscle cells and valves are formed. In the absence of the forkhead transcription factor FOXC2, valves are inefficient and lymph flows back leading to a hereditary form of lymphedema with late onset, known as lymphedema-distichiasis because patients also have a double row of eyelashes [16].

9.4 Postnatal Lymphangiogenesis

Once lymphatic vessels are formed, several growth factors may promote postnatal lymphangiogenesis acting on receptors present in lymphatic endothelial cells. VEGFR-3 binds not only VEGF-C but also VEGF-D [15]. VEGF-D is dispensable during development but, when exogenously added, it rescues the impaired vascular sprouting in VEGF-C null mice. Lymphatic endothelial cells also have VEGFR-2, which binds VEGF-A. In adult lymphangiogenesis, VEGFR-2 and VEGFR-3 have cooperative and redundant roles of signaling [17]. This has important implications in therapy: Combined inhibition of both receptors may be more efficient in reducing tumor lymphangiogenesis than the inhibition of either receptor alone. Hepatocyte growth factor (HGF) is a novel potent lymphangiogenic factor that promotes lymphatic vessel formation and function independently from VEGFR-3 [18]. HGF receptor may be an interesting new target for inhibiting pathological lymphangiogenesis.

In the cornea, which normally is avascular, during inflammation, new lymphatic vessels are formed [19]. These newly formed lymphatics do not originate from the preexisting ones of the limbus but they rather arise in the center of the cornea due to the transdifferentiation of CD11b-positive macrophages that express lymphatic markers, Prox1, podoplanin, and LYVE-1. Macrophages have also been shown to be in vitro able to form lymphatic capillaries in matrigel.

9.5 Lymphatic Vessels and Tumors

The role of lymphatic vessels in tumor spreading has been extensively studied and is beyond the objectives of this chapter. Of particular interest is however the recent report that VEGF-A binding to VEGFR-2 in tumors not only induces angiogenesis but also tumor and sentinel lymph node lymphangiogenesis, promoting lymphatic metastasis. Non-metastatic sentinel lymph nodes have been shown to contain increased numbers of enlarged LYVE-1-positive sinusoids [20], confirming the old seed and soil hypothesis: Tumor cells prepare the soil (the lymph node) where they are going to be seeded during metastatic diffusion.

9.6 Lymphatic System Organization

Initial lymphatic vessels arise bluntly in the interstitium where they drain fluids and macromolecules escaped from blood capillaries and venules. Initial lymphatic vessels, improperly known as capillaries, are larger than blood capillaries, with a characteristically tortuous, irregular profile. Their wall is extremely thin, made only by endothelial cells without pericytes. They may contain valves. ECs of initial lymphatic vessels are large, oak-leaf shaped, with overlapping flaps sealed by "buttons" that contain, like the "zippers" of collecting vessels, VE-cadherin and tight junction–associated proteins [21]. Buttons may open and close without disrupting junctional integrity to allow fluid entrance.

From initial lymphatic vessels, lymph flows into larger vessels, provided with valves, precollectors, whose wall has an alternation of thinner tracts made solely by endothelial and thicker tracts in which the endothelium is irregularly surrounded by smooth muscle cells [22, 23]. Precollectors drain into collecting vessels, characterized by larger dimensions and a continuous wrapping of smooth muscle cells. Their course is interrupted by lymph nodes. The largest lymphatic vessels, the thoracic duct and the right lymphatic duct, eventually convey lymph into the large veins at the base of the neck.

Under transmission electron microscopy, lymphatic vessels are characterized by a discontinuous basement membrane, which may be for long tracts absent, and anchoring filaments (Fig. 9.3) which connect the abluminal membrane of endothelial cells with the surrounding extracellular matrix [24].

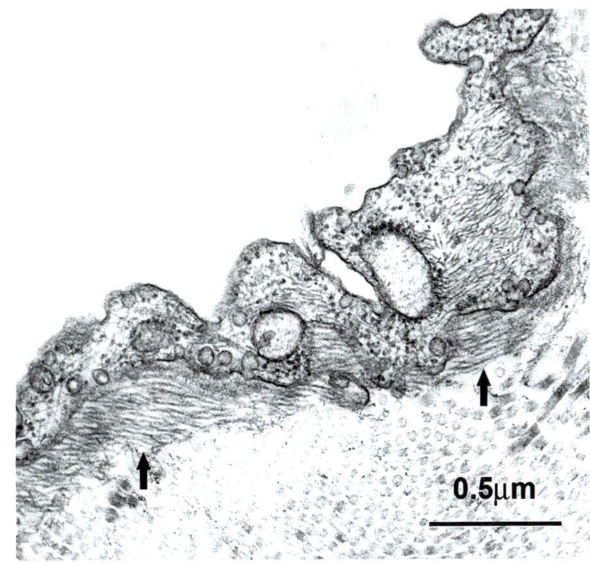

Fig. 9.3 Transmission electron micrograph of a precollector. Anchoring filaments (*arrows*) are clearly visible beneath the endothelium [23])

9.7 Anchoring Filaments

Anchoring filaments have long been postulated to favor interstitial fluid drainage by pulling apart interendothelial junctions in edema [25]. They are made of fibrillin [26], a large (approximately 350 kDa) and ancient molecule [27], present even in jellyfish. In those animals that have a circulatory system, fibrillin during development forms a track for the deposition of elastin, the protein that confers elasticity to blood vessels. This is called the "structural" role of fibrillin. Fibrillin also has an "instructive" role due to its capacity to sequester transforming growth factor-β (TGF-β) and bone morphogenetic protein complexes in the extracellular matrix [28].

Around skin initial lymphatic vessels, fibrillin microfibrils, establish a connection with elastic fibers forming a fibrillo-elastic apparatus [29] that, under the mechanical solicitations of the surrounding connective tissue, dilates lymphatics favoring lymph formation and then allows the lymphatic to resume the original dimensions.

Fibrillin is produced by several types of cells; it was first found in the cell culture medium of fibroblasts [30], but it is also deposited in the extracellular matrix. We found that cultured bovine lymphatic endothelial cells obtained from the largest lymphatic vessel, the thoracic duct, also produce fibrillin (Fig. 9.4) [31] and the related protein microfibril-associated glycoprotein-1 or MAGP-1 [32].

Based on literature and personal data, we have recently proposed [33] that the role of fibrillin-containing anchoring filaments in lymphatic vessels might be much more sophisticated than previously thought. A schematic diagram is illustrated in

Fig. 9.4 An irregular web of fibrillin microfibrils deposited by cultured lymphatic endothelial cells in the underlying matrix

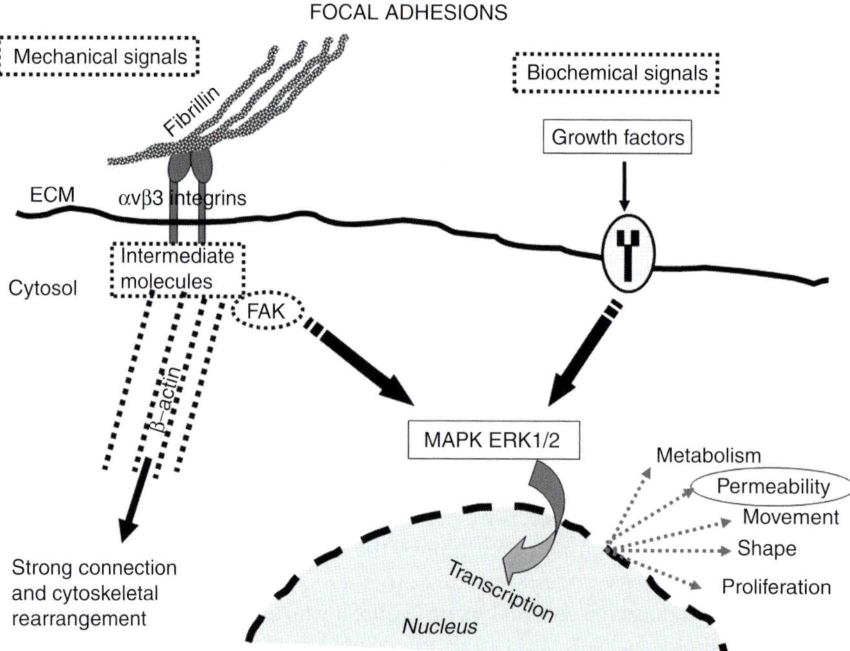

Fig. 9.5 Schematic representation of signal transduction at focal adhesions

Fig. 9.5. Briefly, fibrillin contains an RGD (arginine-glycine-aspartic acid) motif capable of binding to integrins at focal adhesions [28]. Focal adhesions are the molecular devices responsible for the transduction of mechanical signals from the extracellular matrix into biochemical signals inside the cytoplasm [34]. They are formed by clusterings of integrins. Since integrins have no enzymatic activity, many of the signaling functions of focal adhesions rely on the phosphorylation on tyrosine of an associated molecule: focal adhesion kinase (FAK). FAK phosphorylation triggers a cascade of phosphorylations that causes actin and cytoskeletal rearrangement so that cells may strongly connect with the matrix and modify their shape [35]. Molecular cascades triggered by FAK are also directed toward the nucleus. The short duration signals of tyrosine phosphorylation are converted into long-lasting serine-threonine phosphorylations by mitogen-activated protein kinases (MAPK). On MAPK converge not only mechanical stimuli acting on focal adhesions but also biochemical signals acting on receptors, for instance, growth factors contained in serum or in cell culture medium. MAPK isoforms ERK1-(44 kDa) and ERK2-(42 kDa) are phosphorylated, leave the cytoplasm and enter the nucleus [36], where they act on the promoter of genes for transcriptional modulation. Thus, a number of cell activities, including metabolism, proliferation, and permeability, may be modulated [37, 38].

We applied static stretching to bovine thoracic duct segments and lymphatic endothelial cells cultured on elastic membranes and evaluated the expression of ERK1/2 by Western blotting [33]. The stretching of isolated thoracic duct segments and lymphatic endothelial cells cultured on elastic membranes activated the expression of MAPK ERK1/2. ERK1/2 activation occurs also in cells deprived of growth factors and grown with only 0.1% serum. The cells exposed to 20% serum with endothelial cell growth supplement (ECGS) express ERK1/2 independently from mechanical stimulation via the receptorial route of activation of ERK1/2. Signal transduction may thus occur in lymphatic endothelial cells in response to mechanical stimulation of focal adhesions or via receptor activation by growth factors. Lymphatic endothelial cells would respond to these stimuli modifying their permeability. Lymph formation would so be precisely and continuously adapted to functional requirements.

9.8 Lymphatic Vessels in SSc Skin

Vascular involvement is frequent in scleroderma, but the role of the lymphatic vasculature is poorly known. Interestingly, systemic sclerosis (SSc) patients have no clinical evidence of lymphedema in spite of the profound alterations of their skin, which might potentially affect lymphatic circulation. In the skin of SSc patients, angiogenesis is insufficient despite severe hypoxia which is a major pro-angiogenic stimulus. VEGF-A is strongly overexpressed in the skin and serum of SSc patients [39], and serum levels of VEGF correlate with the development of fingertip ulcers [40] Prolonged exposure to VEG-A leads however to formation of a chaotic vessel network with megacapillaries and reduced blood flow, resembling the disturbed vessel morphology of SSc patients [41, 42]. Circulating levels of VEGF-C and local expression of its lymphatic receptor VEGFR-3 in the skin have been reported to be also increased in patients with scleroderma [43].

The only report on lymphatic vessels in the skin of SSc patients is a fluorescence microlymphography study by A.J.Leu et al. [44] showing that in SSc, the clinically

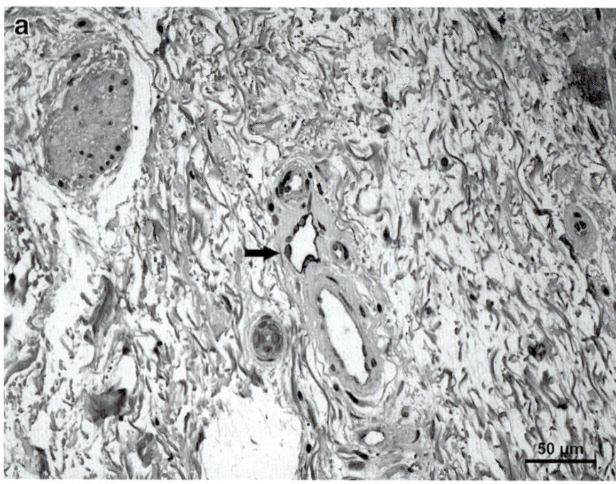

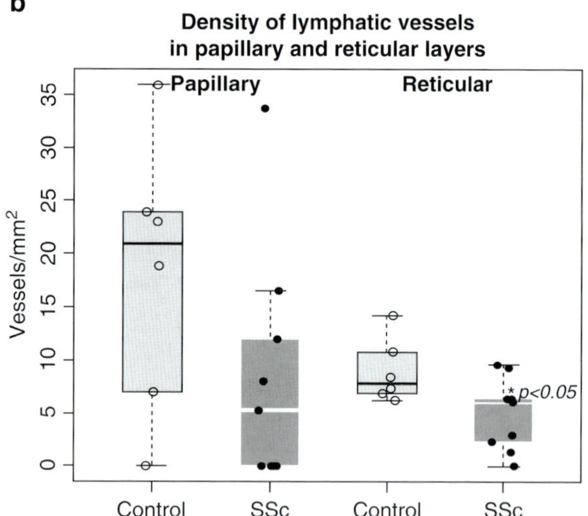

Fig. 9.6 (**a**) Only one lymphatic vessel (*arrow*) is present in this micrograph of the reticular dermis of a patient affected by SSc. (**b**) The density of lymphatic vessels in the reticular dermis of patients affected by SSc is significantly lower than in controls ($P<0.05$)

affected areas have a pattern of lymphatic microangiopathy, characterized by increased length of the visualized lymphatic capillaries and cutaneous backflow or even the complete absence of stained microlymphatics.

We sought to determine whether lymphatic vessels are affected in SSc [45] postulating that they might be decreased in number as in other fibrotic diseases due to inhibition of lymphangiogenesis by overexpression of TGF-β1 [46, 47] or dilated as in conditions of chronic lymphostasis [48] and in other autoimmune diseases [49, 50]. Forearm skin biopsies of SSc patients (4 with the diffuse and 5 with the limited form) and healthy volunteers were fixed in formalin and embedded in paraffin. Double immunolabeling was performed with the lymphatic marker D2-40 followed by a panendothelial antibody to von Willebrand factor (vWF). Lymphatic and blood vessels were so easily recognized by their different staining, brown and red, respectively. Both in controls and SSc biopsies, the density of lymphatic and blood vessels

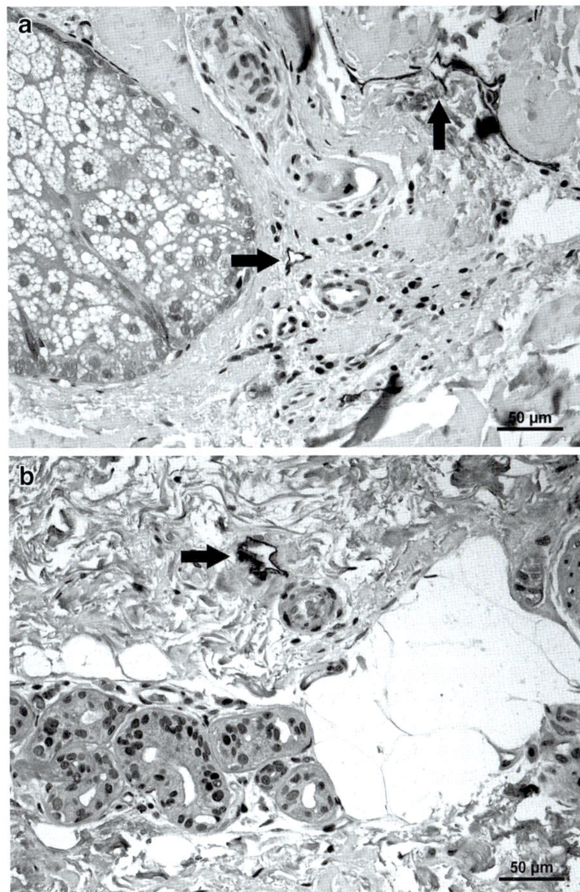

Fig. 9.7 Periglandular lymphatics are spared in SSc. Lymphatic vessels (*arrows*) around a sebaceous (**a**) and sweat (**b**) gland

in the papillary dermis resulted markedly greater, and their mean area conversely smaller, than in the reticular dermis. In SSc, in the reticular dermis, the density of lymphatic vessels was significantly lower than in controls, and a similar trend (although not reaching statistical significance) was observed in the papillary layer (Fig. 9.6).

Interestingly, periglandular lymphatic vessels were preserved in scleroderma (Fig. 9.7). To assess whether this could be due to local production of lymphangiogenic factors, we stained some sections with a polyclonal antibody to VEGF-C and we found that the epithelial cells of glands were strongly immunoreactive for VEGF-C.

Although the mean outer area was similar in the two groups, in the reticular dermis, the percentage of inner luminal area (Fig. 9.8), which can be considered a sign of dilation, was significantly greater in SSc with respect to controls ($p<0.05$). This difference was mainly due to the dilation of periglandular lymphatics. Lymphatics not associated with glands were similar in the two groups.

In conclusion, in SSc lymphatic vessels decrease in number due to diminution of the lymphatic vessels of the reticular dermis not associated with glands. Periglandular

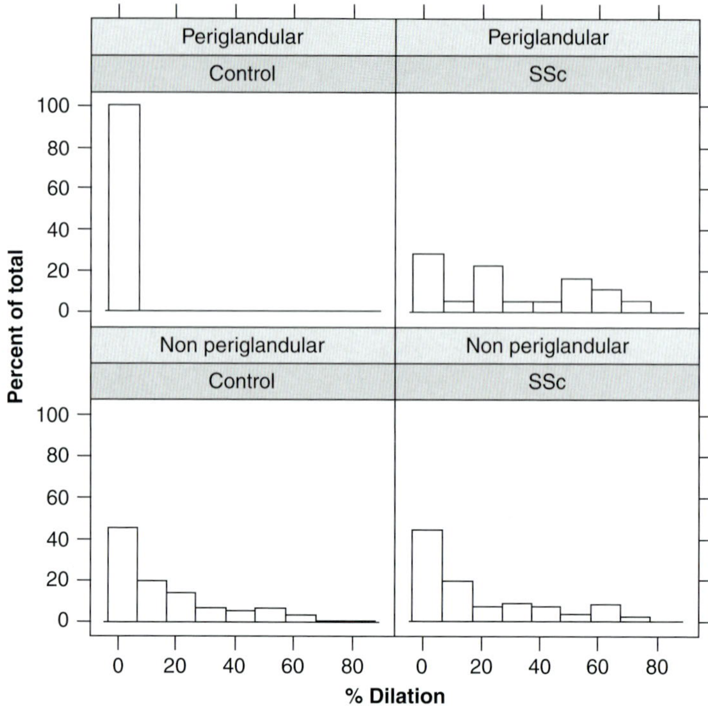

Fig. 9.8 Percentage of dilation in periglandular and not periglandular lymphatics in SSc and in controls. Periglandular lymphatic vessels only are dilated in SSc

lymphatics are in fact spared in SSc, possibly due to VEGFC produced by the epithelium of glands and dilated, interpretatively as a compensatory mechanism.

9.9 Perspectives for the Future

Experimental evidence suggests possible perspectives in therapy: congenital lymphedema might be treated by manipulation of VEGFR-3 signaling [51] or alternatively other lymphangiogenic factors like HGF. Inhibition of VEGFR-3 might be exploited to prevent lymphatic spreading of tumors. Inhibition of VEGFR-2 signaling by a well-known anti-angiogenic drug, Avastin, may also be useful to prevent lymphangiogenesis induced by VEGF-A through VEGFR-2 in regional lymph nodes [17, 20]. Also the opposite is true: tumoral angiogenesis is stimulated also via VEGFR-3. Blockade of this receptor, which is normally restricted to lymphatic endothelium, but is upregulated in tumors, has been shown to suppress angiogenic sprouting in a mouse model [52]. Targeting VEGFR-3 may thus provide additional efficacy for anti-angiogenic therapies in cancer.

As to secondary lymphedema, which is most often caused by surgical ablation of lymph nodes particularly in breast cancer, a new approach has been recently proposed:

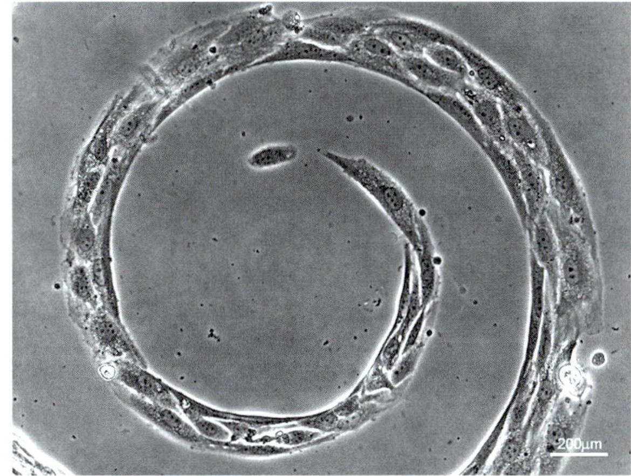

Fig. 9.9 Lymphatic endothelial cells cultured on a negative spiral pattern obtained by photoimmobilization of Hyal on aminosilanized glass grow on the glass domains, avoiding Hyal, and align along the spiral

Axillary Reverse Mapping [53]. A blue dye is injected dermally or subcutaneously in the arm and reaches the axillary nodes. The resulting blue lymph nodes that drain the upper arm can, in most cases, be preserved except when too close to or coincident with the sentinel lymph node. It has been shown that even when most axillary nodes are metastatic, the blue ones are not. This simple technique has proved safe and effective and, if one considers the burden of a life-long invalidating condition as lymphedema, it seems reasonable that axillary reverse mapping should enter standard surgical procedures as the sentinel lymph node one. Since also collectors are colored in blue, this technique also facilitates performing lymphatico-venous anastomoses [54].

9.10 Tissue Engineering of Lymphatic Vessels

Tissue-engineered blood vessels have been successfully implanted in humans, particularly in children with congenital vascular malformations [55]. The tissue-engineered vessel grew with the child with no need of re-intervention. Tissue-engineered vessels can be made with autologous endothelial cells taken from a peripheral vein, expanded in culture, and seeded onto a reabsorbable polymer or with autologous bone marrow cells [56].

Research on tissue engineering of lymphatic vessels is still in its infancy. Due to the fragility of their wall, it is unfeasible that the same approaches that have led to pioneer successful implants of tissue-engineered blood vessels in man may be applied to lymphatics. Basic research is needed to understand the strategies that can be useful for lymphatic vessel tissue engineering. Under this respect, micropatterned surfaces with different geometries based on the alternation of hyaluronan domains that prevents cell adhesion and aminosilanized glass ones that promote it have been proved effective in orienting lymphatic endothelial cell growth [32, 57, 58]. Cells may be induced to align along the desired direction and also actin cytoskeleton is accordingly oriented (Fig. 9.9). Fibrillin deposition is also influenced by the geometry of the substrate. Being able to condition cell growth, orientation, and metabolic activities may help in designing tissue-engineered vessels capable of adapting to the functional requirements of the environment.

Acknowledgments We wish to thank D.J. Abraham, C.P. Denton, K. Khan (UCL Medical School, London), E.A. Renzoni (Royal Brompton Hospital, London), and P. Sestini (University of Siena) for their collaboration in the study of lymphatic vessels in scleroderma skin.

References

1. Kaipainen A, Korhonen J, Mustonen T, et al. Expression of the fms-like tyrosine kinase 4 gene becomes restricted to lymphatic endothelium during development. Proc Natl Acad Sci U S A. 1995;92:3566–70.
2. Joukov V, Pajusola K, Kaipainen A, et al. A novel vascular endothelial growth factor, VEGF-C, is a ligand for the Flt4 (VEGFR-3) and KDR (VEGFR-2) receptor tyrosine kinases. Embo J. 1996;15:290–8.
3. Jeltsch M, Kaipainen A, Joukov V, et al. Hyperplasia of lymphatic vessels in VEGF-C transgenic mice. Science. 1997;276:1423–5.
4. Irrthum A, Karkkainen MJ, Devriendt K, et al. Congenital hereditary lymphedema caused by a mutation that inactivates VEGFR3 tyrosine kinase. Am J Hum Genet. 2000;67:295–301.
5. Karkkainen MJ, Ferrell RE, Lawrence EC, et al. Missense mutations interfere with VEGFR-3 signalling in primary lymphoedema. Nat Genet. 2000;25:153–9.
6. Sleeman JP, Krishnan J, Kirkin V, et al. Markers for the lymphatic endothelium: in search of the holy grail? Microsc Res Tech. 2001;55:61–9.
7. Wigle JT, Oliver G. Prox1 function is required for the development of the murine lymphatic system. Cell. 1999;98:769–78.
8. Hong YK, Harvey N, Noh YH, et al. Prox1 is a master control gene in the program specifying lymphatic endothelial cell fate. Dev Dyn. 2002;225:351–7.
9. Banerji S, Ni J, Wang SX, et al. LYVE-1, a new homologue of the CD44 glycoprotein, is a lymph-specific receptor for hyaluronan. J Cell Biol. 1999;144:789–801.
10. Kahn HJ, Marks A. A new monoclonal antibody, D2-40, for detection of lymphatic invasion in primary tumors. Lab Invest. 2002;82:1255–7.
11. Schacht V, Dadras SS, Johnson LA, et al. Up-regulation of the lymphatic marker podoplanin, a mucin-type transmembrane glycoprotein, in human squamous cell carcinomas and germ cell tumors. Am J Pathol. 2005;166:913–21.
12. Schacht V, Ramirez MI, Hong YK, et al. T1alpha/podoplanin deficiency disrupts normal lymphatic vasculature formation and causes lymphedema. Embo J. 2003;22:3546–56.
13. Sabin FR. On the origin of the lymphatic system from the veins and the development of the lymph hearts and thoracic duct in the pig. Am J Anat. 1902;1:367–91.
14. Detmar M, Hirakawa S. The formation of lymphatic vessels and its importance in the setting of malignancy. J Exp Med. 2002;196:713–8.
15. Alitalo K, Tammela T, Petrova TV. Lymphangiogenesis in development and human disease. Nature. 2005;438:946–53.
16. Petrova TV, Karpanen T, Norrmén C, et al. Defective valves and abnormal mural cell recruitment underlie lymphatic vascular failure in lymphedema distichiasis. Nat Med. 2004;10:974–81.
17. Goldman J, Rutkowski JM, Shields JD, et al. Cooperative and redundant roles of VEGFR-2 and VEGFR-3 signaling in adult lymphangiogenesis. Faseb J. 2007;21:1003–12.
18. Kajiya K, Hirakawa S, Ma B, et al. Hepatocyte growth factor promotes lymphatic vessel formation and function. Embo J. 2005;24:2885–95.
19. Maruyama K, Ii M, Cursiefen C, et al. Inflammation-induced lymphangiogenesis in the cornea arises from CD11b-positive macrophages. J Clin Invest. 2005;115:2363–72.

20. Hirakawa S, Kodama S, Kunstfeld R, et al. VEGF-A induces tumor and sentinel lymph node lymphangiogenesis and promotes lymphatic metastasis. J Exp Med. 2005;201:1089–99.
21. Baluk P, Fuxe J, Hashizume H, et al. Functionally specialized junctions between endothelial cells of lymphatic vessels. J Exp Med. 2007;204:2349–62.
22. Sacchi G, Weber E, Aglianò M, et al. The structure of superficial lymphatics in the human thigh: precollectors. Anat Rec. 1997;247:53–62.
23. Sacchi G, Weber E, Aglianò M, et al. Lymphatic vessels of the human heart: precollectors and collecting vessels. A morpho-structural study. J Submicrosc Cytol Pathol. 1999;31:515–25.
24. Leak LV, Burke JF. Fine structure of the lymphatic capillary and the adjoining connective tissue area. Am J Anat. 1966;118:785–809.
25. Casley Smith JR. Are the initial lymphatics normally pulled open by anchoring filaments? Lymphology. 1980;13:120–9.
26. Gerli R, Ibba L, Fruschelli C. Ultrastructural cytochemistry of anchoring filaments of human lymphatic capillaries and their relation to elastic fibers. Lymphology. 1991;24:105–12.
27. Kielty CM, Wess TJ, Haston L, et al. Fibrillin-rich microfibrils: elastic biopolymers of the extracellular matrix. J Muscle Res Cell Motil. 2002;23:581–96.
28. Ramirez F, Sakai LY. Biogenesis and function of fibrillin assemblies. Cell Tissue Res. 2010;339:71–82.
29. Gerli R, Ibba L, Fruschelli C. A fibrillar elastic apparatus around human lymph capillaries. Anat Embryol. 1990;181:281–6.
30. Sakai LY, Keene DR, Engvall E. Fibrillin, a new 350-kD glycoprotein, is a component of extracellular microfibrils. J Cell Biol. 1986;103:2499–509.
31. Weber E, Rossi A, Solito R, et al. Focal adhesion molecules expression and fibrillin deposition by lymphatic and blood vessel endothelial cells in culture. Microvasc Res. 2002;64:47–55.
32. Weber E, Rossi A, Gerli R, et al. Micropatterned hyaluronan surfaces promote lymphatic endothelial cell alignment and orient their growth. Lymphology. 2004;37:15–21.
33. Rossi A, Weber E, Sacchi G, et al. Mechanotransduction in lymphatic endothelial cells. Lymphology. 2007;40:102–13.
34. Burridge K, Turner CE, Romer LH. Tyrosine phosphorylation of paxillin and pp 125FAK accompanies cell adhesion to extracellular matrix: a role in cytoskeletal assembly. J Cell Biol. 1992;119:893–903.
35. Schaller MD, Borgman CA, Cobb BS, et al. pp 125FAK a structurally distinctive protein-tyrosine kinase associated with focal adhesions. Proc Natl Acad Sci U S A. 1992;89:5192–6.
36. Rubinfeld H, Seger R. The ERK cascade: a prototype of MAPK signaling. Mol Biotechnol. 2005;31:151–74.
37. Sun HW, Li CJ, Chen HQ, et al. Involvement of integrins, MAPK, and NF-kappaB in regulation of the shear stress-induced MMP-9 expression in endothelial cells. Biochem Biophys Res Commun. 2007;353:152–8.
38. Coulthard LR, White DE, Jones DL, et al. p38(MAPK): stress responses from molecular mechanisms to therapeutics. Trends Mol Med. 2009;15:369–79.
39. Choi JJ, Min DJ, Cho ML, et al. Elevated vascular endothelial growth factor in systemic sclerosis. J Rheumatol. 2003;30:1529–33.
40. Distler O, Del Rosso A, Giacomelli R, et al. Angiogenic and angiostatic factors in systemic sclerosis: increased levels of vascular endothelial growth factor are a feature of the earliest disease stages and are associated with the absence of fingertip ulcers. Arthritis Res. 2002;4:R11.
41. Distler O, Distler JH, Scheid A, et al. Uncontrolled expression of vascular endothelial growth factor and its receptors leads to insufficient skin angiogenesis in patients with systemic sclerosis. Circ Res. 2004;95:109–16.
42. Cutolo M, Pizzorni C, Tuccio M, et al. Nailfold videocapillaroscopic patterns and serum autoantibodies in systemic sclerosis. Rheumatology. 2004;43:719–26.
43. Chitale S, Al-Mowallad AF, Wang Q, et al. High circulating levels of VEGF-C suggest abnormal lymphangiogenesis in systemic sclerosis. Rheumatology (Oxford). 2008;47:1727–8.
44. Leu AJ, Gretener SB, Enderlin S, et al. Lymphatic microangiopathy of the skin in systemic sclerosis. Rheumatology (Oxford). 1999;38:221–7.

45. Rossi A, Sozio F, Sestini P, et al. Lymphatic and blood vessels in scleroderma skin, a morphometric analysis. Hum Pathol. 2010;40:366–74.
46. Oka M, Iwata C, Suzuki HI, et al. Inhibition of endogenous TGF-beta signaling enhances lymphangiogenesis. Blood. 2008;111:4571–9.
47. Clavin NW, Avraham T, Fernandez J, et al. TGF-beta1 is a negative regulator of lymphatic regeneration during wound repair. Am J Physiol Heart Circ Physiol. 2008;295:H2113–27.
48. Tanaka T, Damião AO, Gabriel Júnior A, et al. Protein-losing enteropathy in systemic lupus erythematosus. Rev Hosp Clin Fac Med Sao Paulo. 1991;46:34–7.
49. Pruim B, Strutton G, Congdon S, et al. Cutaneous histiocytic lymphangitis: an unusual manifestation of rheumatoid arthritis. Australas J Dermatol. 2000;41:101–5.
50. Takiwaki H, Adachi A, Kohno H, et al. Intravascular or intralymphatic histiocytosis associated with rheumatoid arthritis: a report of 4 cases. J Am Acad Dermatol. 2004;50:585–90.
51. Jurisic G, Detmar M. Lymphatic endothelium in health and disease. Cell Tissue Res. 2009;335:97–108.
52. Tammela T, Zarkada G, Wallgard E, et al. Blocking VEGFR-3 suppresses angiogenic sprouting and vascular network formation. Nature. 2008;454:656–60.
53. Thompson M, Korourian S, Henry-Tillman R, et al. Axillary reverse mapping (ARM): a new concept to identify and enhance lymphatic preservation. Ann Surg Oncol. 2007;14:1890–5.
54. Casabona F, Bogliolo S, Ferrero S, et al. Axillary reverse mapping in breast cancer: a new microsurgical lymphatic-venous procedure in the prevention of arm lymphedema. Ann Surg Oncol. 2008;15:3318–9.
55. Shin'oka T, Imai Y, Ikada Y. Transplantation of a tissue-engineered pulmonary artery. N Engl J Med. 2001;344:532–3.
56. Matsumura G, Hibino N, Ikada Y, et al. Successful application of tissue engineered vascular autografts: clinical experience. Biomaterials. 2003;24:2303–8.
57. Pasqui D, Rossi A, Barbucci R, et al. Hyaluronan and sulphated hyaluronan micropatterns: effect of chemical and topographic cues on lymphatic endothelial cell alignment and proliferation. Lymphology. 2005;38:50–65.
58. Rossi A, Pasqui D, Barbucci R, et al. The topography of microstructured surfaces differently affects fibrillin deposition by blood and lymphatic endothelial cells in culture. Tissue Eng Part A. 2009;15:525–33.

10 Importance of Subtype Selectivity for Endothelin Receptor Antagonists in the Human Vasculature

Janet J. Maguire and Anthony P. Davenport

10.1 Endothelin Pathway in the Human Vasculature

The endothelins (ETs) are a family of three endogenous peptides: ET-1, ET-2, and ET-3 that are structurally similar in being comprised of 21 amino acids [1, 2]. In man, ET-2 differs from ET-1 by only two amino acids and both isoforms mediate their action via two G-protein-coupled receptors, ET_A [3, 4] and ET_B [5]. In contrast, ET-3 differs by six amino acids, representing more substantial changes, and is the only isoform that can distinguish between the two receptor subtypes, having a similar potency at the ET_A receptor as ET-1 and ET-2 but much lower affinity for the ET_B subtype [6, 7]. The deleterious actions of ET are mainly mediated by the ET_A receptor, whereas ET_B activation results in many of the beneficial effects of the peptide, frequently acting as a regulatory counterbalance [7].

Two distinct therapeutic strategies have emerged to block the unwanted action of ET in pathophysiological conditions: receptor antagonists [8] and inhibitors of the endothelin-converting enzymes (ECE-1 [9] and ECE-2 [10]), the major synthetic pathway in the human vasculature [11]. Bosentan (Tracleer) was the first ET antagonist to be introduced into the clinic for the treatment of pulmonary arterial hypertension (PAH [12]) and is a mixed ET_A/ET_B antagonist blocking both receptors. This was followed by ambrisentan (Letairis, Volibris) in 2007, reported to display modest ET_A selectivity [13], and the more ET_A-selective antagonist sitaxentan (Thelin)

J.J. Maguire
University of Cambridge, Centre for Clinical Investigation,
Addenbrooke's Hospital,
Cambridge, UK
e-mail: jjm1003@medschl.cam.ac.uk

A.P. Davenport (✉)
Clinical Pharmacology Unit, University of Cambridge,
Centre for Clinical Investigation, Addenbrooke's Hospital,
Cambridge, UK
e-mail: apd10@medschl.cam.ac.uk

D. Abraham et al. (eds.), *Translational Vascular Medicine*,
DOI 10.1007/978-0-85729-920-8_10, © Springer-Verlag London Limited 2012

[14, 15]. While mixed ET_A/ET_B and ET_A-selective antagonists have become established as having therapeutic benefit in PAH [16, 17], the relative merits of the two classes continue to be debated [18–20]. To date both ET_A/ET_B or modest ET_A-selective antagonists are thought to have little or no efficacy in chronic heart failure and further trials with more ET_A-selective antagonists are unlikely [21]. More promising clinical uses are in chronic kidney disease where the ET system is increasingly recognized as an important pathway [22–27] and where efficacy has been demonstrated with experimental ET_A-selective antagonists in acute trials [28]. ET receptors are also emerging as new therapeutic targets in autoimmune disorders of the vasculature such as scleroderma [29] and remarkably in cancer [30–32], particularly the treatment of refractory cancer of the prostrate by the ET_A-selective antagonist ZD4054 [32]. This is notably the first G-protein-coupled receptor in Family A to be targeted for the treatment of cancer. Inhibitors of ECE are represented by SLV-306 (Daglutril). This compound is an orally active mixed enzyme inhibitor of both ECE and neutral endopeptidase (NEP) and a Phase II trial has been completed in 2010 by Solvay for the treatment of essential hypertension and congestive heart failure [33–35]. It is not yet clear whether lowering levels of endogenous ET changes the ratio of ET-1:ET-3 which could then impact on the relative activation of the two receptor subtypes. Significantly, ET antagonists represent a spectrum of selectivity that has the potential to be exploited for extending the therapeutic targets for this class of compound. The objective of this review is to consider the importance of subtype selectivity for ET receptor antagonists in the human vasculature.

10.1.1 ET-1

ET-1 is the most abundant isoform in the human cardiovascular system and is one of the most powerful constrictors of human vessels discovered [7]. ET-1 plays a major physiological role in regulating vascular function in most, if not all, organs systems including heart, kidney, lungs, and liver. Overproduction in pathophysiological conditions may lead to vasospasm, particularly where there is endothelial cell dysfunction and associated loss of opposing vasodilators such as nitric oxide, prostacyclin, and endothelium-derived hyperpolarizing factor. The peptide is thought to stimulate proliferation in multiple cell types, including vascular smooth muscle cells, as well as contributing to fibrosis and inflammation – processes associated with vascular remodeling.

10.1.2 Dual Synthetic Pathway in Endothelial Cells and Interaction with ET_A and ET_B Receptors

The primary source of ET-1 within vessels is the endothelial cells although other cell types that synthesize the peptide could also modulate vascular reactivity. These include perivascular neurons in the periphery, perivascular astrocytes in the CNS, and, under pathophysiological conditions such as atherosclerosis, macrophages and monocytes. ET is synthesized in a three-step process. Initially pre-pro-ET-1 is cleaved by a signal peptidase to proET-1, which is in turn cleaved by a furin enzyme to an inactive precursor big ET-1 which is subsequently transformed to the mature, biologically active peptide by the action of the pathway-specific ECE-1. ECE-1 is present within the small secretory vesicles of the constitutive pathway from where ET-1 is continuously released to maintain normal vascular tone. A second enzyme, ECE-2, is also present within the vesicles and functions at an acidic pH [6] that may occur under pathophysiological conditions associated with ischemia. Unusually for vasoactive peptides, ET-1 is also synthesized by ECE-1 and stored in specialized Weibel-Palade bodies within endothelial cells until released following an external physiological or pathophysiological stimulus (the regulated pathway) to produce further vasoconstriction [11, 36].

10.1.3 ET_A Receptors and Vasoconstriction

In the human vasculature, ET-1, released by these two distinct exocytotic pathways, can potentially interact with the ET_A receptors that predominate on the underlying smooth muscle. ET_A receptors are widely expressed on vascular smooth muscle cells throughout the human cardiovascular system and mediate vasoconstriction. Under pathophysiological conditions, ET_A activation may contribute to proliferation, apoptosis, and fibrosis within the vessel wall. In some, but not all, human vessels, a small population of ET_B receptors (usually <15%) are present and these may also mediate constriction [37, 38]. Haynes and Webb [39] were the first to report that infusion of an ET_A-selective peptide antagonist, BQ-123, into healthy volunteers via the brachial artery using venous occlusion plethysmography caused progressive vasodilatation. This is consistent with ET-1 being continuously released by the constitutive pathway to cause vasoconstriction and is unusual as antagonists of other vasoconstrictors, such as angiotensin II, do not alter blood flow in normotensive individuals.

10.1.4 ET_B Receptors and Vasodilatation

ET-1 also interacts with endothelial cell ET_B receptors. Although representing about 1% of the weight of the vessel wall, endothelial cells line the vasculature of every organ and tissue in the body that receives blood supply and have a combined mass comparable to some endocrine glands. Infusion of an ET_B selective antagonist, BQ788, caused systemic vasoconstriction in healthy volunteers, showing that the main consequence of activation of endothelial ET_B receptors by tonically secreted ET-1 was the physiological basal release of nitric oxide [40]. The interaction of ET-1 feeding back onto endothelial receptors to release nitric oxide not only limits ET_A-mediated vasoconstriction by stimulation of vascular cyclic GMP, but also limits further ET-1 release, emphasizing the importance of ET_B receptors as a counter-regulatory pathway.

In agreement, and importantly, where different concentrations of ET-1 have been compared, infusions of low doses of exogenous ET-1 into the brachial artery caused vasodilatation, but this was followed by sustained vasoconstriction of the forearm vascular bed at higher doses [41]. Initially, it was surprising to find in studies, knocking out the ET-1 gene, that ET-1+/− heterozygous mice (which produced lower levels of ET-1 in plasma and lung tissue than wild-type) developed *elevated* blood pressure and mild hypertension, rather than the fall in blood pressure that might have been expected [42]. These results suggest that ET-1 has an essential physiological role in cardiovascular homeostasis. Low levels promote vasodilatation, whereas higher and pathophysiological concentrations of ET-1 increase blood pressure and total peripheral vascular resistance. Interestingly, renal and pulmonary circulations are particularly sensitive to the vasoconstrictor effects of ET-1. Thus, in the vasculature, nitric oxide and other dilators are crucial in balancing the ET system, but these may be reduced and absent in pathophysiological conditions. Furthermore, alternative pathways for ET-1 synthesis from big ET-1 by vascular smooth muscle (see Sect. 2.1) result in ET-1 binding immediately to ET_A receptors without activation of the endothelial ET_B feedback pathway to oppose vasoconstriction.

10.1.5 ET_B Clearing Receptors, Diuresis, and Natriuresis

In addition to releasing vasodilators, the ET_B receptor also functions as a "clearing receptor," to internalize the ligand–receptor complex and remove ET-1 from the circulation [43–45]. As a result, the plasma half-life of ET-1 is comparatively short. In the human heart, when the ratio of ET_A:ET_B receptors is measured, ET_A receptors are more abundant (>60%). In marked contrast, while autoradiography reveals ET_A receptors also predominate on the smooth muscle of the vasculature in human lung, kidney, and liver, these organs are characterized by particularly high densities of the ET_B subtype, reflecting that they are rich in endothelial cells [46]. For example, the lungs have one of the highest densities of ET receptors (~9,600 fmol/g protein) compared with other peripheral tissues and even higher than the brain (~5,000 fmol/g protein). In lung, ET_B receptors are present on airway smooth muscle (and mediate bronchoconstriction), epithelial cells, and vascular smooth muscle cells, but the majority are present on the endothelium. Similarly, in human kidney, ET_B receptors comprise 70% of the ET receptors in both the cortex and medulla. ET_B receptors localize to endothelial cells throughout the renal vasculature consistent with their roles in endothelium-dependent vasodilatation and as clearing receptors, removing ET-1 from circulation [47, 48]. ET_B receptors are also present on epithelial cells throughout the tubular epithelium, particularly the inner medullary collecting duct cells where the major action of ET-1 is to promote beneficial diuresis and natriuresis [47, 48]. As a result, evidence is emerging that ET_A-selective antagonists might be superior to mixed blockade, as antagonism of ET_B receptors may be undesirable.

ET-1 is also a very potent and sustained vasoconstrictor of the hepatic vasculature [49], and preclinical in vivo studies have suggested that ET antagonists could be new therapeutic agents in the treatment of portal hypertension [50]. Interestingly, the isolated perfused liver avidly extracts proportionately more ET-1 than the lungs, with 80% uptake in a single pass. This is hypothesized to occur mostly through binding to ET_B receptors on hepatic stellate cells and is reduced in conditions such as cirrhosis [51]. Portal hypertension remains a major cause of morbidity and mortality in patients with cirrhosis of the liver, but only about a third of patients respond to current therapies and new treatments are urgently needed. In human cirrhosis, plasma levels of ET-1 are enhanced and elevated concentrations in the liver are thought to be a consequence of both increased synthesis and decreased clearance [52]. Bosentan has been tested in a single patient and shown to beneficially reduce hepatic venous gradient over time [53]. The cellular expression of ET subtypes has not been studied in detail in human liver, and the precise identity of cells expressing ET_B receptors is unclear. A small number of animal studies have addressed whether ET_A receptor-selective antagonists provide an advantage over nonselective agents in ameliorating the effects of portal hypertension; the majority of these data indicate that selective antagonists may be sufficient [54, 55]. Thus, animal studies and a single clinical observation support a role for ET-1 in portal hypertension, but there are as yet insufficient human data to draw conclusions regarding optimum receptor selectivity for therapeutic ET receptor blockade in this condition.

The rapid clearance of ET labeled with the positron-emitting isotope [^{18}]F from the circulation can be visualized in vivo using positron emission tomography in animal models [56]. In these studies, the distribution of ET into all major organs can be measured and confirms that the major sites for clearance of circulating [^{18}F]-ET-1 are the lungs, the liver, and the kidney, with little uptake by other tissues. Binding could not be displaced with BQ788 administered *after* infusing the radioligand, in agreement with the proposed internalization of ET-1 by ET_B receptors and degradation in the lysosome. In contrast, infusion of BQ788 prior to injecting [^{18}F]-ET-1 significantly reduced clearing in lung and kidney by 85%, although importantly the amount of [^{18}F]-ET-1 significantly increased in the liver as the label was no longer cleared by ET_B receptors and now bound to the ET_A subtype. Surprisingly, binding of [^{18}F]-ET-1 could not be visualized to receptors within the heart in the control animal, but binding was detected in this organ when ET_B receptors were blocked by the antagonist. These results show that clearance of ET-1 was mediated by the ET_B receptor in the lung, kidney, and to a certain extent by the liver, and crucially, this prevents binding of ET-1 to the heart. This mechanism is important in limiting the detrimental vasoconstrictor effect caused by upregulation of ET-1 in the vascular system associated with disease.

ET_B receptors are expressed by a number of cell types in addition to endothelial cells including epithelial and smooth muscle cells. Currently, there are no antagonists that distinguish between these receptors, but endothelial ET_B receptors have been selectively deleted in mice [57, 58]. This did not alter the remaining ET_B (or ET_A) receptor expression which was confirmed by radioligand binding and autoradiography. As expected, clearance of an intravenous bolus of labeled ET-1 was impaired in these knockout animals compared with controls. An ET_B antagonist reduced clearance in controls but not in the knockout mouse providing clear evidence that endothelial ET_B receptors are mainly, if not exclusively, responsible for ET clearance from the circulation.

10.2 Alternative Pathways for ET Synthesis

10.2.1 Tissue-Specific Conversion of Big ET-1 by Non-endothelial Cell ECE and Effect of ECE/NEP Inhibitors

A key question will be to determine what effect (if any) the lowering of ET levels by inhibiting synthesis has on ET receptors (Fig. 10.1). Some big ET-1 circulates in plasma but does not bind to vascular ET receptors until cleaved to ET-1 by converting enzymes present on smooth muscle [59]. Interestingly, ECE activity is increased in endothelium-denuded human vessels with atherosclerosis [60] suggesting that

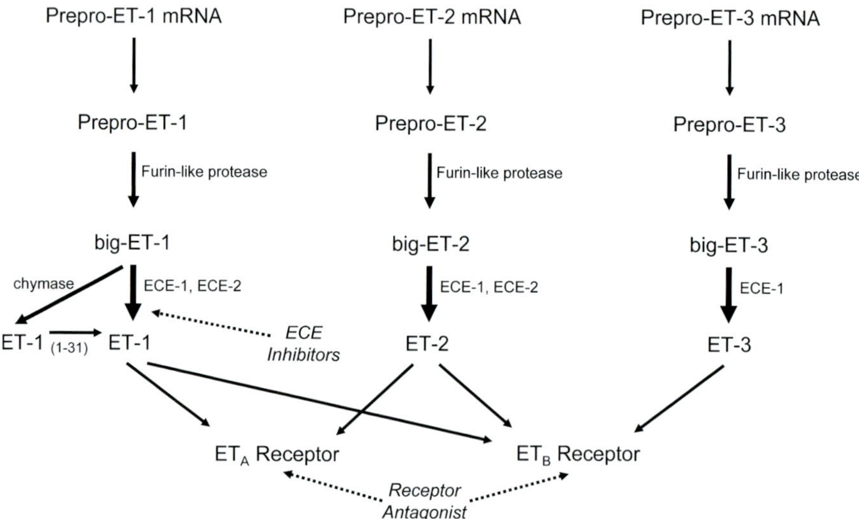

Fig. 10.1 ET pathway in the human cardiovascular system. All three ET isoforms are synthesized by a three-step process. For ET-1 and ET-2, this consists of an initial proteolytic cleavage of the signal peptidase of preproET-1, a second cleavage of proET-1, to big ET-1-by a furin-like enzyme. Transformation to the mature, biologically active peptides is mainly by the action of ECE-1 but also by ECE-2 within endothelial cells. Further processing may occur by smooth muscle ECE or via alternative pathways catalyzed by chymase for ET-1. ET-3 is synthesized by a similar pathway but not by the endothelium. Following release from endothelial cells, ET-1 interacts predominantly with ET_A receptors on the underlying smooth muscle. In some, but not all, human vessels, a small population of ET_B receptors can also mediate constriction. Some ET-1 may also interact with endothelial ET_B receptors to act as a feedback mechanism to limit the constrictor response by the release of vasodilators such as nitric oxide. Low levels of ET-1 can also be detected in the plasma thought to be the result of overspill from the endothelium. ET_B receptors present in organs that are rich in this subtype, the kidney and lungs, remove ET-1 from the circulation by internalization followed by lysosomal degradation. Targets for therapeutic intervention are currently ECE and by blocking the ETA or both subtypes by antagonists

nonendothelial ECE may contribute to increased plasma/tissue ET levels in disease. To date, orally active dual inhibitors of both NEP and ECE have been developed, rather than purely ECE selective [61, 62]. These have the potential advantage over selective ECE inhibitors of reducing plasma ET and increasing plasma concentrations of the atrial and brain natriuretic peptides, both beneficial vasodilators. The first study has been carried out on the acute effect of single oral doses of the NEP/ECE inhibitor SLV 306 [63]. This measured, in 15 normotensive volunteers, the blood pressure response to infused big ET-1 at doses, determined in pilot studies, likely to lead to a rise in mean arterial pressure of approximately 20 mmHg. SLV 306 dose dependently attenuated the rise in blood pressure after big ET-1 infusion. This was accompanied by a corresponding increase in the big ET-1/ET-1 ratio in a concentration-dependent manner consistent with systemic ECE inhibition, preventing metabolism of the enzyme substrate, big ET-1, to its active metabolite, ET-1. Plasma atrial natriuretic peptide levels also increased as predicted.

This process of big ET-1 conversion can be imaged in the living animal by infusion of [^{18}F]-big ET-1 to quantify tissue-specific conversion to [^{18}F]-ET-1 which immediately binds to ET_A receptors on the vascular smooth muscle [64]. Infused [^{18}F]-big ET-1 was rapidly cleared from the circulation with a half-life (t½) of less than 3 min. Whole body images showed highest uptake of radioactivity in two major organs, the liver and lungs, which could be significantly reduced using phosphoramidon, an inhibitor of ECE and NEP, consistent with inhibition of enzyme conversion and subsequent reduction of [^{18}F]-ET-1 receptor binding. The ET_A antagonist, FR139317, did not alter half-life of [^{18}F]-big ET-1 (t½ = 2.5 min) in the plasma, but radioactivity uptake was reduced in all tissues consistent with binding of the cleavage product [^{18}F]-ET-1 to this subtype rather than to ET_B receptors. Plasma levels of big ET-1 are also elevated in pathophysiological conditions such as PAH. It is significant that the lungs were an important site for big ET-1 conversion, suggesting that overexpression of big ET-1, with subsequent cleavage to ET-1 and binding to ET_A receptors, is an additional source of peptide in PAH. Plasma levels of ET-1 are also elevated in renal failure. Interestingly in the kidney, in marked contrast to liver and lungs, there was no binding to renal ET receptors reflecting excretion of [^{18}F]-big ET-1 unchanged without conversion to ET-1. In agreement with animal studies, big ET-1 can be detected in urine of normal human subjects [65] and levels are increased in patients with acute myocardial infarction, chronic renal failure, essential hypertension, and vasospastic angina pectoris. These results suggest that excretion of unmetabolized big ET-1 by the kidney may be an important mechanism for removal of the precursor and, although not yet tested, may be a site of removal of increased plasma big ET-1 in volunteers treated with NEP/ECE inhibitors.

10.2.2 Non-ECE Pathways: The Serine Protease Chymase

One of the unexpected findings of Yangisawa and colleagues [66] was the presence of significant amounts of ET-1/ET-2 in the ECE-1/ECE-2 double knockout mouse embryos, suggesting other proteases must be significantly involved in the tissue production of mature ET-1 and ET-2. This study has important implications for the action of NEP/ECE inhibition on the ET pathway and has led to the search for alternative synthetic pathways to ECE.

The serine protease chymase, which is present in mast cells, can mediate an additional conversion pathway by cleaving the Tyr^{31}–Gly^{32} peptide bond of big ET-1 to generate ET-1(1–31), which is in turn converted to the mature peptide by cleaving the Trp^{21}–Val^{22} bond [67, 68]. Importantly, ET-1(1–31) was equipotent compared with big ET-1 in causing vasoconstriction in human isolated vessels, including coronary arteries, and this was associated with the appearance of measurable levels of ET-1 in the bathing medium, consistent with conversion to the mature peptide. ET-1(1–31) competed for specific [^{125}I]-ET-1 binding to ET_A and ET_B receptors in human heart with a single affinity, indicating little or no selectivity for the subtypes. Vasoconstriction was fully blocked by ET_A-selective antagonists, reflecting the predominance of the ET_A receptor on vascular smooth muscle [69]. The precise physiological role of mast cells within the human blood vessels is unclear, but following degranulation, which may occur under pathophysiological conditions, the mast cell chymase is associated with interstitial spaces with the potential to convert circulating big ET-1 and provide a further source of ET-1. Mast cell expression is increased in cardiovascular disease, for example, in atherosclerotic lesions. In pathophysiological conditions, it is possible that the contribution of this pathway within the vasculature, leading to overexpression of ET-1, may be underestimated particularly in conditions of endothelial malfunction where opposing levels of endogenous vasodilators may be reduced. It is unclear whether under conditions of NEP/ECE inhibition, the rising levels of big ET-1 would favor increased conversion by the serine protease pathway, thus increasing the pressor effect via ET_A receptors or whether excretion of unmetabolized big ET-1 by the kidney would be sufficient to remove the elevated levels of precursors.

10.3 ET-2: The Forgotten Isoform

ET-2 remains the least studied of the endothelin isopeptides and much less is known of its function and location than for ET-1 and ET-3. Messenger RNA encoding ET-2 [70] together with the peptide [71] is present in the human cardiovascular system including failing hearts. Both mRNA [71] and the precursor big ET-2 are detected in the cytoplasm of endothelial cells [72] and ET-2 may also be released from these cells in addition to ET-1. Intriguingly, big ET-2 levels are higher in normal human plasma than big ET-1 [73] and plasma levels of ET-2 are detectable, with an average value in 40 volunteers of 0.9 ± 0.03 pmol/l. ET-2 differs from ET-1 by only two amino acids and binds with a similar affinity as ET-1 to both receptor subtypes [74] and it is as potent a vasoconstrictor of isolated vessels as ET-1 ([37].

Recently, a global knockout of ET-2 revealed a distinct phenotype exhibiting growth retardation and changes in energy homeostasis. Importantly, given the current therapeutic targets of ET antagonists, changes in lung morphology and function were also observed [75]. While the importance of the ET-2 signaling pathway is not yet clear, big ET-2-like immunoreactivity has been detected in human lungs [76] and some of the alternatively spliced variants for ET-2 mRNA contain sites for the posttranscriptional processing of preproET-2 into mature ET-2 and posttranscriptional processing may be disrupted or altered in these variants [70]. A detailed investigation into the specific contribution, if any, of ET-2 to human diseases such as PAH has not yet been carried out.

10.4 ET-3: The Receptor Subtype Selective Isoform

ET-3 and its precursor big ET-3 circulate in the human plasma although at lower concentrations than ET-1 [73] and ET-3 is present in other tissues including the heart [71]. Endothelial cells do not synthesize ET-3, but alternative sources may be from the adrenal gland [77] with conversion of big ET-3 to the mature peptide within the vasculature [78]. ET-3 is the only endogenous isoform that distinguishes between the two subtypes with at least 100-fold lower affinity at the ET_A receptor.

10.5 Is There a Shift Toward ET_B-Mediated Vasoconstriction in Human Disease?

In human coronary artery disease, there is no functional evidence for an upregulation in ET_B receptors. Variable responses to the ET_B agonist sarafotoxin S6c were obtained in control vessels ($n=15$) and diseased coronary arteries containing atherosclerotic lesions ($n=16$) with 40% and 50% of arteries not responding to S6c, respectively. While S6c contracted the remaining vessels, there was no significant difference in the maximum response to S6c observed between the two groups. In agreement, there was no significant alteration in medial ET_B subtype density observed in diseased vessels compared to control, with ET_A receptors still comprising over 90% of the total ET receptor population in both diseased and control arteries. These results imply that there is no increase in the importance of the constrictor ET_B receptor in human coronary artery disease [79].

These results from in vitro experiments are supported by a substantial clinical study on the effect of ET antagonists in 39 patients with coronary atherosclerosis, or risk factors for coronary artery disease, undergoing cardiac catheterization. In agreement with forearm blood flow studies in healthy volunteers, selective ET_B receptor antagonism in this group caused coronary microvascular constriction, without affecting epicardial coronary tone or endothelial function. Treatment with combined ET_A and ET_B blockade dilated coronary conduit and resistance vessels and improved endothelial dysfunction of the epicardial coronary arteries. This evidence therefore suggests that ET-1 acting predominantly via ET_A receptors contributes to basal constrictor tone and in disease to endothelial dysfunction. ET_B activation mediated beneficial coronary vasodilatation in these patients indicating that selective ET_A blockade may have greater therapeutic potential than nonselective agents, particularly for treatment of endothelial dysfunction in atherosclerosis [80].

In human pulmonary resistance arteries with an internal diameter of 150–200 μm, ET-1 caused the expected sustained vasoconstriction, but the responses to low concentrations of peptide could be blocked by ET_B antagonists. In contrast, higher concentrations above 1 nM were blocked by an ET_A but not an ET_B antagonist, suggesting that at levels of ET-1 in the pathophysiological range, ET_A receptors will be activated [81]. Davie and colleagues [82] carried out an extensive study on the distribution of ET receptors in pulmonary arteries with an internal diameter of about 500–1,000 μm from pulmonary hypertensive patients versus control subjects, using in vitro autoradiography, so that the ratio of the two subtypes could be quantified in the small arteries. ET receptor density in distal arteries and lung parenchyma was twofold greater in these patients compared with controls. Although distal arteries possessed a higher proportion of medial smooth muscle ET_B receptors than proximal arteries, there was no change in any vessel in the ratio of the two subtypes in patients compared with controls and therefore no shift toward increased ET_B expression. These results are consistent with ET_B-mediated constrictor responses at low ET concentrations, but in the absence of an upregulation in receptor density, it is unlikely there would be increased ET_B constrictor response in this patient group.

10.6 Endothelin Antagonists and Receptor Selectivity

10.6.1 Rationale for ET-1 Receptor Blockade: How Do We Define Selectivity?

The definition of selectivity depends on the measurement of the affinity (the equilibrium dissociation constant or K_D) of each compound at the two receptor subtypes and the comparison of these affinities to give a ratio of selectivity [6, 7]. This classification of antagonists will crucially depend on how affinities were measured and this varies between investigators. Many of the reported affinities for endothelin receptor antagonists are based on assays using cloned receptor subtypes each expressed in separate cell lines. Artificially expressed receptors may not reflect and correspond to the affinities measured in native tissues for a number of reasons, such as differences in posttranslational modifications and expression at much higher densities than is encountered in native human tissue. Affinities for antagonists should ideally be measured in competition binding assays from their ability to compete for the binding of radiolabeled ET-1 since this is the predominant endogenous ligand that needs to be blocked in the clinic. However, in some cases, selectivity is calculated using radiolabeled ET-1 to identify ET_A receptors but radiolabeled ET-3 to identify ET_B.

ET receptor antagonists are classified as either selective for one subtype or alternatively as mixed antagonists that block both receptors. The classification is usually made by the manufacturer (Fig. 10.2) but there is no agreed definition and there are anomalies. We have proposed that antagonists that are ET_A-selective should display more than 100-fold selectivity for the ET_A subtype and those that block both ET_A and ET_B (mixed antagonists) should demonstrate less than 100-fold ET_A selectivity. The rationale for this is shown in Fig. 10.3. The Langmuir isotherm for the theoretical occupancy of ET receptor subtypes is shown for an ET_A-selective compound that has an affinity of 1 nM for the ET_A receptor but 100 nM for the ET_B, that is 100-fold ET_A selectivity. Occupancy is calculated using the formula $L^*/(K_D+L^*)$, where

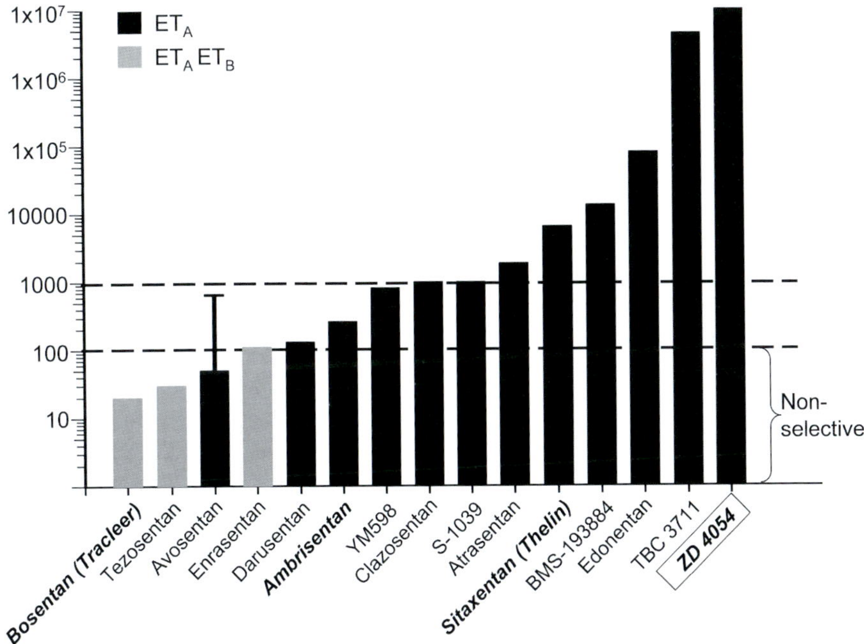

Fig. 10.2 Reported degree of selectivity of ET receptor antagonists for ET_A receptors versus classification by manufactures as either mixed or ETA-selective

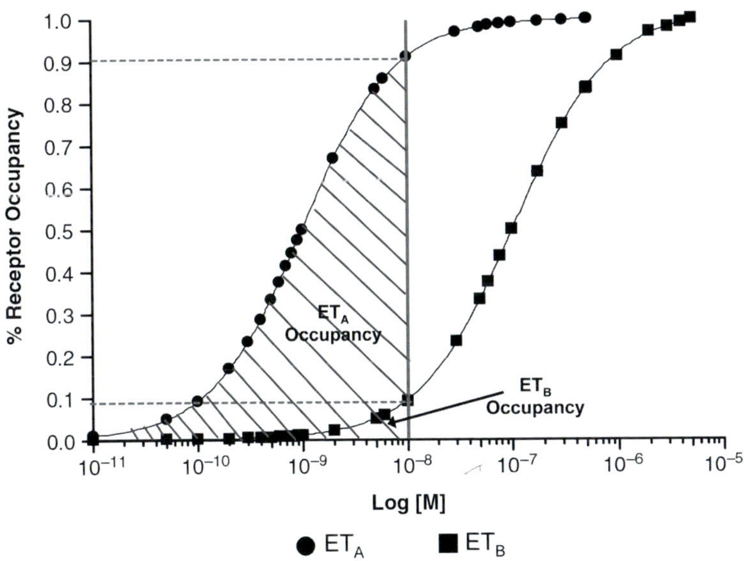

Fig. 10.3 Langmuir isotherm for a compound with 100-fold selectivity for ET_A over ET_B receptors

L^* = free ligand concentration (M) and K_D is the affinity constant (M). At a concentration of 10 nM, 90% of ET_A receptors are predicted to be blocked but less than 10% of the ET_B.

Compounds displaying 100-fold selectivity are therefore useful, at least in vitro where an ET_A-selective concentration can be accurately achieved. However, 100-fold selectivity is likely to represent the minimum that can be used in vivo to achieve selective ET_A receptor blockade. An increase in the concentration of such an antagonist by only one log unit results in a significant (50%) occupancy of ET_B receptors (Fig. 10.3). In practice, if selective ET_A blockade is desired, then compounds of higher selectivity are needed for testing in vivo in animal models or in clinical trials and experimental medicine to be certain that there is no significant activation of ET_B receptors. Ideally, compounds of greater than a 1,000-fold selectivity are needed for in vivo studies to ensure ET_A selectivity. Fortunately, the most widely used compounds for preclinical as well as clinical studies, that are also commercially available, are the very highly selective peptide antagonists, FR139317, BQ-123 (both ET_A-selective), and BQ-788 (ET_B). In addition, subtype selective radiolabeled ligands for the pharmacological characterization of receptors, $[^{125}I]$-PD151242 or $[^3H]$-BQ123 (ET_A) and $[^{125}I]$-BQ3020 or $[^{125}I]$-IRL1620 (ET_B), also display a high degree of selectivity. If these pharmacological tools have been used in studies, the results can be interpreted with confidence that the expected subtype is blocked without affecting the other [6, 7].

The situation is more complex for nonpeptide antagonists. In Fig. 10.2, examples of the selectivity of compounds used in clinical trials have been calculated from data published by the manufacturer when characterizing the compounds for the first time and using the manufacturer's own classification as to whether they considered the compound mixed or ET_A selective. The figure includes the three antagonists currently available for the treatment of PAH: bosentan, sitaxentan, and ambrisentan. The selectivity spectrum ranges from bosentan, the first mixed antagonist to be introduced clinically, to the markedly ET_A-selective sitaxentan to ZD4054, that is in Phase III clinical trial for refractory prostate cancer, that remarkably is reported to have no affinity for ET_B receptors. For compounds that display marginal selectivity, such as ambrisentan, the interpretation of results is less clear as it is difficult to be certain whether, in particular studies, the antagonist has been used at a concentration that blocked both receptors or just the ET_A receptor. Additional confusion arises as some compounds that have similar marginal ET_A selectivity such as enrasenatan and darusentan were reported by their manufactures to be a mixed antagonist and an ET_A-selective antagonist respectively.

10.6.2 Measuring Selectivity

One approach toward measuring selectivity is to compare the affinity constants for a particular antagonist measured from its ability to compete for the binding of $[^{125}I]$-ET-1 in the same assay to both native receptors using human tissue, the therapeutic

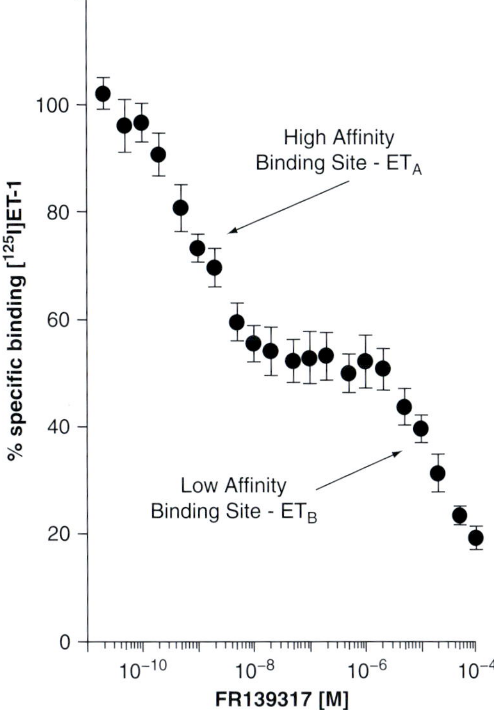

Fig. 10.4 An example of a competition binding curve for the inhibition of a fixed concentration of [^{125}I]-ET-1 (0.1 nM) binding to ET receptors by increasing concentrations of unlabeled antagonist (FR139317) in human heart. Over the concentration range tested, the antagonist competed in a biphasic manner and a two-site fit was preferred to a one-site or three-site model using LIGAND. The high affinity site corresponded to the ET_A receptor (K_D = 7 nM), the low affinity site to the ET_B K_D = 104 μM, giving ~15,000 fold ET_A selectivity

target, rather than using artificially expressed human receptors or animal tissues. ET receptor subtypes are present in left ventricle of the human heart in a ratio of about 60% ET_A to 40% ET_B which is ideal for accurately measuring affinity constants for antagonists against both receptors in the same tissue. Having determined the K_D of radiolabeled ET-1 for the target receptors in a saturation assay, this information is used to determine the ability of unlabeled antagonists, tested over a much wider concentration range (typically 10 pM–100 μM), to compete for the binding of a fixed concentration of [^{125}I]-ET-1 in human left ventricle. An example of a competition curve is shown in Fig. 10.4, visualized by plotting the amount of [^{125}I] ET-1 bound as a percentage of total specific [^{125}I] ET-1 binding (specific binding in the absence of competitor) against the $\log_{10}$ of the molar concentration of the competing ligand. A steep competition curve is usually indicative of binding to a single population of receptors. However, increasing concentrations of unlabeled FR139317 inhibited the binding of [^{125}I]-ET-1 biphasically. Computer-based programs such as LIGAND are used to mathematically model the curve and to measure whether a two-site fit is statistically a better fit than a one-site model. In this case, a two-site fit was preferred, consistent with FR139317 binding with high affinity to the ET_A site but with low, micromolar affinity to the ET_B receptors, giving >200,000 fold selectivity for the ET_A receptor ([83], Table 10.1).

10.6.3 Comparison of the Selectivity of ET Antagonists Determined in Human Tissues Versus Cloned Human or Animal Receptors

A crucial question is therefore how do values of antagonist affinity obtained from either cloned human receptors or animal tissues relate to those obtained in human tissues that express both receptor subtypes and which are the intended target for endothelin antagonists clinically, specifically the heart, lungs, kidney, and vasculature? We have identified a number of antagonists, belonging to different structural classes (Fig. 10.5) that have a spectrum of reported ET receptor affinity and selectivity, and have pharmacologically characterized their activity at ET receptors in these human tissues. Table 10.1 shows the literature reported selectivity of compounds determined for human cloned receptors or in cells/animal tissues that endogenously express only (or predominantly) one or other receptor subtype. Those antagonists that we have investigated belonging to the peptide, sulfonamide, or

Table 10.1 Receptor subtype binding affinity and selectivity of endothelin antagonists determined in model systems and human left ventricle

	Cloned receptors[a]	Human left ventricle		
Antagonist	Reported ET_A selectivity	K_D ET_A	K_D ET_B	ET_A selectivity
Peptides				
BQ123	653 [85]	0.73±0.22 nM [86]	24.3±2.0 μM [86]	33,288 [86]
FR139317	7,300 [87]	1.20±0.28 nM [88]	287±93 μM [88]	239,167 [88]
PD151242	ND	7.21±2.80 nM [89]	104±23 μM [89]	14,424 [89]
PD142893	1.7 [90]	0.30±0.03 μM[b]	1.17±0.14 μM[b]	4[b]
Sulfonamides				
Ro-462005	1.5 [91]	One site fit 50.3±9.5 μM [88]		Non-selective [88]
Bosentan	20 [92]	One site fit 77.6±7.9 nM [88]		Non-selective [88]
Sitaxentan	7,000 [93]	1.65±0.80 nM[b]	327±134 μM[b]	198,182[b]
BMS 182874	1,042 [94]	590±100 nM[b]	Not detectable[b]	>10,000[b]
Carboxylic acids				
SB209670	34 [95]	One site fit 0.67±0.14 nM [96]		Non-selective [96]
PD156707	2,600 [97]	0.92±0.38 nM [98]	13.3±2.1 μM [98]	14,457 [98]
L-749329	65 [99]	One site fit 303.5±34.3 nM[b]		Non-selective[b]
Myceric acids				
50235	>1,000 [100]	162±61 nM [88]	171±42 μM [88]	1,056 [88]
S97-139	1,000 [101]	45.3±25 nM[b]	47.6±9.9 μM[b]	1051[b]

ND Not determined

[a]Or cells/animal tissues that endogenously express one receptor subtype exclusively or predominantly

[b]Unpublished data

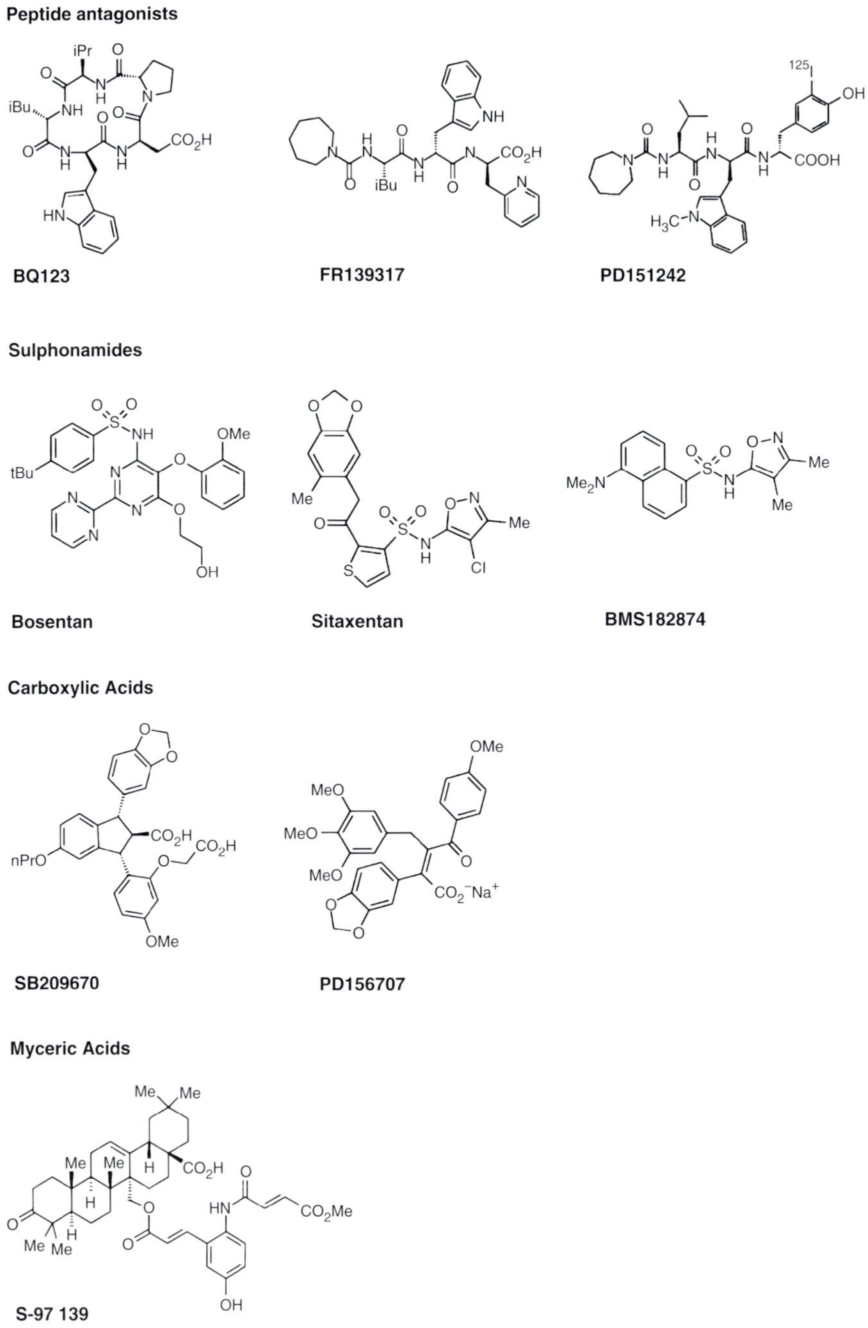

Peptide antagonists

BQ123 **FR139317** **PD151242**

Sulphonamides

Bosentan **Sitaxentan** **BMS182874**

Carboxylic Acids

SB209670 **PD156707**

Myceric Acids

S-97 139

Fig. 10.5 Structures of the most widely used ET peptide antagonists and examples of structures from the families of nonpeptide antagonists

carboxylic acid groups comprise both reported ET_A-selective and mixed antagonists. From our human data, it is apparent that for these groups of compounds, the degree of selectivity for the ET_A-selective compounds is markedly increased from 800 to 8,000 fold in model cell/tissue systems to more than 10,000 fold in the human left ventricle binding assay. As expected, those that are reported to be mixed antagonists do not distinguish between the two receptors in human heart and a one-site fit is statistically preferred by data analysis. Interestingly, the myceric acid derivatives have about 1,000-fold ET_A selectivity in both the model receptor assays and in human cardiac tissue.

The binding experiments in human heart are carried out in the presence of bovine serum albumin (BSA) to more closely reflect in vivo conditions, as some of these compounds are known to show appreciable binding to plasma proteins [84]. Indeed, in the absence of BSA, the potency of, for example, sitaxentan for both the ET_A and ET_B receptor is increased and ET_A selectivity is maintained (unpublished data ET_A K_D 0.06 nM; ET_B K_D 13 μM; ET_A selectivity >230,000 fold). However, there are differences in the binding assays between the model systems and the human tissues that may account for some of the observed discrepancies in selectivity. Where animal tissues or cells have been employed, species differences in the ET receptors may need to be considered when comparing reported antagonist data to data obtained in human tissues. In some cases, the different radioligands, [^{125}I]-ET-1 and [^{125}I]-ET-3, are used to label ET_A and ET_B receptors respectively in both cloned/animal tissue experiments, whereas [^{125}I]-ET-1 is used to label both populations of receptor subtypes in the human left ventricle.

If selective blockade of the vascular ET_A receptor is clinically desirable, how predictable of functional potency is the affinity of an antagonist determined in a receptor binding assay? To address this question we have carried out an additional study to measure how well the ET_A affinity of selective and mixed antagonists, determined in binding experiments, reflects their potency as functional antagonists at the ET_A receptor in vitro. We carried out Schild analysis of data obtained from the antagonism of ET-1-induced vasoconstriction in human isolated coronary artery and/or saphenous vein (an ET_A response [37, 38]) for representative antagonists from each structural group and compared the Schild-derived affinity values (ET_A K_B) to the ET_A affinity (K_D) determined in binding experiments in the same vascular tissue. The resulting data were expressed as a K_B/K_D ratio (Table 10.2) and it can be seen that for most of the antagonists tested, their ability to block ET-1 vasoconstriction was 10–1,000 fold less than predicted by their binding affinity determined in the same tissue. The degree of K_D to K_B discrepancy did not appear to relate to structural class, although the peptide antagonists were particularly less effective as functional antagonists than predicted by their binding affinity. To what extent these data can be extrapolated to the clinical setting is unclear, but it may be that for some antagonists, such as bosentan, the concentration required to achieve sufficient receptor occupancy in vivo for clinical efficacy may be much greater than predicted by in vitro binding assays. While this is not necessarily a problem for compounds that are either nonselective or have a very marked ET_A selectivity, it may mean that those compounds that have a more marginal ET_A selectivity may have to be administered at doses at which ET_B occupancy will become apparent and so these compounds will not behave as selective ET_A antagonists at clinically effective doses.

Table 10.2 Comparison of ET_A receptor affinity for endothelin antagonists determined in binding and functional assays in human coronary artery and saphenous vein

		Binding	Functional	
Antagonist	Vascular preparation	ET_A K_D	ET_A K_B [a]	K_B/K_D
Peptides				
BQ123	Saphenous vein	0.55±0.17 nM [102]	141 nM [37]	256
	Coronary artery	0.85±0.03 nM [102]	91 nM [37]	107
FR139317	Saphenous vein	0.56±0.01 nM[b]	87 nM [37]	156
	Coronary artery	0.41±0.13 nM [74]	126 nM [37]	307
PD151242	Coronary artery	0.51±0.07 nM [103]	1.1 μM [103]	2,157
Sulfonamides				
Ro-462005	Saphenous vein	0.15±0.01 μM[b]	1.4 μM[b]	9
	Coronary artery	0.19±0.04 μM [74]	2.4 μM[b]	12
Bosentan	Saphenous vein	32.2±3.2 nM[b]	1.6 μM[b]	50
	Coronary artery	2.94±0.95 nM [74]	2.9 μM[b]	967
BMS 182874	Saphenous vein	580±40 nM[b]	219 nM[b]	0.4
Carboxylic acids				
SB209670	Saphenous vein	11.2±1.4 nM[b]	12 nM[b]	1.1
PD156707	Saphenous vein	0.5±0.13 nM [98]	2 nM [98]	4
	Coronary artery	0.15±0.06 nM [98]	8 nM [98]	40
L-749329	Saphenous vein	66.7±7.5 nM[b]	6.5 nM[b]	0.1
Myceric acids				
50235	Coronary artery	6.8±2.9 nM [104]	1.1 μM [104]	157

[a]K_B derived from Schild data with slope constrained to one or from Gaddum-Schild equation
[b]Unpublished data

10.7 Conclusions

The differential distribution and function of ET receptor subtypes provides the rationale for using two distinct pharmacological strategies, mixed or ET_A-selective antagonism. To exploit this difference for selective compounds, it is essential to be able to achieve concentrations where ET_A receptors are blocked, but there is no significant ET_B receptor occupancy. While this can be achieved in vitro with >100-fold selectivity, in vivo antagonists such as sitaxentan which display at least 1,000-fold selectivity may be the minimum to resolve this hypothesis.

Acknowledgments We thank Dr. Neil Davie for discussion, Mrs. Rhoda Kuc for critically reading the manuscript, and the British Heart Foundation (grant numbers PG/09/050/27734 and RG/10/077/28300) and an investigator initiated grant from Pfizer for support.

References

1. Yanagisawa M, Kurihara H, Kimura S, et al. A novel potent vasoconstrictor peptide produced by vascular endothelial cells. Nature. 1988;332:411–5.
2. Inoue A, Yanagisawa M, Kimura S, et al. The human endothelin family: three structurally and pharmacologically distinct isopeptides predicted by three separate genes. Proc Natl Acad Sci USA. 1989;86:2863–7.

3. Hosoda K, Nakao K, Tamura N, et al. Organization, structure, chromosomal assignment, and expression of the gene encoding the human endothelin-A receptor. J Biol Chem. 1992;267: 18797–804.
4. Arai H, Hori S, Aramori I, Ohkubo H, Nakanishi S. Cloning and expression of a cDNA encoding an endothelin receptor. Nature. 1990;348:730–2.
5. Sakurai T, Yanagisawa M, Takuwa Y, et al. Cloning of a cDNA encoding a non-isopeptide-selective subtype of the endothelin receptor. Nature. 1990;348:732–5.
6. Davenport AP. International Union of Pharmacology. XXIX. Update on endothelin receptor nomenclature. Pharmacol Rev. 2002;54:219–26.
7. Davenport AP, Maguire JJ. Endothelin. Handb Exp Pharmacol. 2006;176(Pt 1):295–329.
8. Palmer MJ. Endothelin receptor antagonists: status and learning 20 years on. Prog Med Chem. 2009;47:203–37.
9. Xu D, Emoto N, Giaid A, et al. ECE-1: a membrane-bound metalloprotease that catalyzes the proteolytic activation of big endothelin-1. Cell. 1994;78:473–85.
10. Emoto N, Yanagisawa M. Endothelin-converting enzyme-2 is a membrane-bound, phosphoramidon-sensitive metalloprotease with acidic pH optimum. J Biol Chem. 1995;270:15262–8.
11. Russell FD, Davenport AP. Secretory pathways in endothelin synthesis. Br J Pharmacol. 1999;126:391–8.
12. Rubin LJ, Badesch DB, Barst RJ, et al. Bosentan therapy for pulmonary arterial hypertension. N Engl J Med. 2002;346:896–903.
13. Vatter H, Seifert V. Ambrisentan, a non-peptide endothelin receptor antagonist. Cardiovasc Drug Rev. 2006;24:63–76.
14. Galie N, Naeije R, Burgess G, Dilleen M. 3-year survival of patients treated with sitaxentan sodium (Thelin®) for pulmonary arterial hypertension. European Society of Cardiology, Barcelona, Spain; August 29–September 2, 2009.
15. Benza RL, Mehta S, Keogh A, et al. Sitaxsentan treatment for patients with pulmonary arterial hypertension discontinuing bosentan. J Heart Lung Transplant. 2007;26:63–6.
16. Liu C, Chen J, Gao Y, Deng B, Liu K. Endothelin receptor antagonists for pulmonary arterial hypertension. Cochrane Database Syst Rev. 2009;3:CD004434.
17. McLaughlin VV, Archer SL, Badesch DB, et al. ACCF/AHA 2009 expert consensus document on pulmonary hypertension: a report of the American College of Cardiology Foundation Task Force on Expert Consensus Documents and the American Heart Association: developed in collaboration with the American College of Chest Physicians, American Thoracic Society, Inc., and the Pulmonary Hypertension Association. Circulation. 2009;119:2250–94.
18. Dupuis J, Hoeper MM. Endothelin receptor antagonists in pulmonary arterial hypertension. Eur Respir J. 2008;31:407–15.
19. Vachiery J-L, Davenport AP. The endothelin system in pulmonary and renal asculopathy – “Les liaisons dangereuses”. Eur Respir Rev. 2009;18:260–1.
20. Davie NJ, Schermuly RT, Weissmann N, Grimminger F, Ghofrani HA. The science of endothelin-1 and endothelin receptor antagonists in the management of pulmonary arterial hypertension: current understanding and future studies. Eur J Clin Invest. 2009;39 Suppl 2:38–49.
21. Dhaun N, Pollock DM, Goddard J, Webb DJ. Selective and mixed endothelin receptor antagonism in cardiovascular disease. Trends Pharmacol Sci. 2007;28:573–9.
22. Kohan DE. Endothelins in the normal and diseased kidney. Am J Kidney Dis. 1997;29:2–26.
23. Karet FE, Davenport AP. Localization of endothelin peptides in human kidney. Kidney Int. 1996;49:382–7.
24. Karet FE, Kuc RE, Davenport AP. Novel ligands BQ123 and BQ3020 characterize endothelin receptor subtypes ETA and ETB in human kidney. Kidney Int. 1993;44:36–42.
25. Kohan DE. Endothelin, hypertension and chronic kidney disease: new insights. Curr Opin Nephrol Hypertens. 2010;19:134–9.
26. Longaretti L, Benigni A. Endothelin receptor selectivity in chronic renal failure. Eur J Clin Invest. 2009;39 Suppl 2:32–7.
27. Neuhofer W, Pittrow D. Endothelin receptor selectivity in chronic kidney disease: rationale and review of recent evidence. Eur J Clin Invest. 2009;39 Suppl 2:50–67.

28. Goddard J, Johnston NR, Hand MF, et al. Endothelin-A receptor antagonism reduces blood pressure and increases renal blood flow in hypertensive patients with chronic renal failure: a comparison of selective and combined endothelin receptor blockade. Circulation. 2004;109:1186–93.
29. Abraham D, Dashwood M. Endothelin-role in vascular disease. Rheumatology (Oxford). 2008;47 Suppl 5:23–4.
30. Bhalla A, Haque S, Taylor I, Winslet M, Loizidou M. Endothelin receptor antagonism and cancer. Eur J Clin Invest. 2009;39 Suppl 2:74–7.
31. Bagnato A, Spinella F, Rosano L. The endothelin axis in cancer: the promise and the challenges of molecularly targeted therapy. Can J Physiol Pharmacol. 2008;86:473–84.
32. Growcott JW. Preclinical anticancer activity of the specific endothelin A receptor antagonist ZD4054. Anticancer Drugs. 2009;20:83–8.
33. Dickstein K, De Voogd HJ, Miric MP, et al. Effect of single doses of SLV306, an inhibitor of both neutral endopeptidase and endothelin-converting enzyme, on pulmonary pressures in congestive heart failure. Am J Cardiol. 2004;94:237–9.
34. Bayes M, Rabasseda X, Prous JR. Gateways to clinical trials. Methods Find Exp Clin Pharmacol. 2003;25:317–40.
35. Tabrizchi R. SLV-306. Solvay. Curr Opin Investig Drugs. 2003;4:329–32.
36. Russell FD, Skepper JN, Davenport AP. Human endothelial cell storage granules: a novel intracellular site for isoforms of the endothelin-converting enzyme. Circ Res. 1998;83:314–21.
37. Maguire JJ, Davenport AP. ETA receptor-mediated constrictor responses to endothelin peptides in human blood vessels in vitro. Br J Pharmacol. 1995;115:191–7.
38. Davenport AP, Maguire JJ. Is endothelin-induced vasoconstriction mediated only by ETA receptors in humans? Trends Pharmacol Sci. 1994;15:9–11.
39. Haynes WG, Webb DJ. Contribution of endogenous generation of endothelin-1 to basal vascular tone. Lancet. 1994;344:852–4.
40. Love MP, Ferro CJ, Haynes WG, et al. Endothelin receptor antagonism in patients with chronic heart failure. Cardiovasc Res. 2000;47:166–72.
41. Kiowski W, Luscher TF, Linder L, Buhler FR. Endothelin-1-induced vasoconstriction in humans. Reversal by calcium channel blockade but not by nitrovasodilators or endothelium-derived relaxing factor. Circulation. 1991;83:469–75.
42. Kurihara Y, Kurihara H, Suzuki H, et al. Elevated blood pressure and craniofacial abnormalities in mice deficient in endothelin-1. Nature. 1994;368:703–10.
43. Fukuroda T, Fujikawa T, Ozaki S, et al. Clearance of circulating endothelin-1 by ETB receptors in rats. Biochem Biophys Res Commun. 1994;199:1461–5.
44. Gasic S, Wagner OF, Vierhapper H, Nowotny P, Waldhausl W. Regional hemodynamic effects and clearance of endothelin-1 in humans: renal and peripheral tissues may contribute to the overall disposal of the peptide. J Cardiovasc Pharmacol. 1992;19:176–80.
45. Plumpton C, Ferro CJ, Haynes WG, Webb DJ, Davenport AP. The increase in human plasma immunoreactive endothelin but not big endothelin-1 or its C-terminal fragment induced by systemic administration of the endothelin antagonist TAK-044. Br J Pharmacol. 1996;119:311–4.
46. Davenport AP, Russell FD. Endothelin converting enzymes and endothelin receptor localisation in human tissues. Handb Exp Pharmacol. 2001;152:209–37.
47. Kuc RE, Karet FE, Davenport AP. Characterization of peptide and nonpeptide antagonists in human kidney. J Cardiovasc Pharmacol. 1995;26 Suppl 3:S373–5.
48. Gariepy CE, Ohuchi T, Williams SC, Richardson JA, Yanagisawa M. Salt-sensitive hypertension in endothelin-B receptor-deficient rats. J Clin Invest. 2000;105:925–33.
49. Gandhi CR, Stephenson K, Olson MS. Endothelin, a potent peptide agonist in the liver. J Biol Chem. 1990;265:17432–5.
50. Pitts KR. Endothelin receptor antagonism in portal hypertension. Expert Opin Investig Drugs. 2009;18:135–42.
51. Rockey DC, Weisiger RA. Endothelin induced contractility of stellate cells from normal and cirrhotic rat liver: implications for regulation of portal pressure and resistance. Hepatology. 1996;241:233–40.

52. Hinterhuber L, Graziadei IW, Kahler CM, Jaschke W, Vogel W. Endothelin-receptor antagonist treatment of portopulmonary hypertension. Clin Gastroenterol Hepatol. 2004;2:1039–42.
53. Reichen J, Gerbes AL, Steiner MJ, Sagesser H, Clozel M. The effect of endothelin and its antagonist Bosentan on hemodynamics and microvascular exchange in cirrhotic rat liver. J Hepatol. 1998;28:1020–30.
54. Kojima H, Yamao J, Tsujimoto T, Uemura M, Takaya A, Fukui H. Mixed endothelin receptor antagonist, SB209670, decreases portal pressure in biliary cirrhotic rats in vivo by reducing portal venous system resistance. J Hepatol. 2000;32:43–50.
55. Feng HQ, Weymouth ND, Rockey DC. Endothelin antagonism in portal hypertensive mice: implications for endothelin receptor-specific signaling in liver disease. Am J Physiol Gastrointest Liver Physiol. 2009;297:G27–33.
56. Johnstrom P, Harris NG, Fryer TD, et al. [^{18}F]-Endothelin-1, a positron emission tomography (PET) radioligand for the endothelin receptor system: radiosynthesis and in vivo imaging using microPET. Clin Sci. 2002;103 Suppl 48:4S–8.
57. Kelland NF, Kuc RE, McLean DL, et al. Endothelial cell-specific ETB receptor knockout: autoradiographic and histological characterisation and crucial role in the clearance of endothelin-1. Can J Physiol Pharmacol. 2010;88:644–51.
58. Bagnall AJ, Kelland NF, Gulliver-Sloan F, et al. Deletion of endothelial cell endothelin B receptors does not affect blood pressure or sensitivity to salt. Hypertension. 2006;48:286–93.
59. Maguire JJ, Johnson CM, Mockridge JW, Davenport AP. Endothelin converting enzyme (ECE) activity in human vascular smooth muscle. Br J Pharmacol. 1997;122:1647–54.
60. Maguire JJ, Davenport AP. Increased response to big endothelin-1 in atherosclerotic human coronary artery: functional evidence for up-regulation of endothelin-converting enzyme activity in disease. Br J Pharmacol. 1998;125:238–40.
61. Dive V, Chang CF, Yiotakis A, Sturrock ED. Inhibition of zinc metallopeptidases in cardiovascular disease–from unity to trinity, or duality? Curr Pharm Des. 2009;15:3606–21.
62. Battistini B, Daull P, Jeng AY. CGS 35601, a triple inhibitor of angiotensin converting enzyme, neutral endopeptidase and endothelin converting enzyme. Cardiovasc Drug Rev. 2005;23: 317–30.
63. Kuc RE, Ashby MJ, Seed A, et al. The ECE/NEP inhibitor SLV306 (Daglutril), inhibits systemic endogenous conversion of infused big endothelin-1 in human volunteers. Proc Br Pharmacol Soc. 2005. doi:http://www.pA2online.org/abstracts/Vol3Issue4abst029P.pdf.
64. Johnstrom P, Fryer TD, Richards HK, et al. Positron emission tomography of [18F]-big endothelin-1 reveals renal excretion but tissue-specific conversion to [18F]-endothelin-1 in lung and liver. Br J Pharmacol. 2010;159:812–9.
65. Naruse K, Naruse M, Watanabe Y, et al. Molecular form of immunoreactive endothelin in plasma and urine of normal subjects and patients with various disease states. J Cardiovasc Pharmacol. 1991;17 Suppl 7:S506–8.
66. Yanagisawa H, Hammer RE, Richardson JA, et al. Disruption of ECE-1 and ECE-2 reveals a role for endothelin-converting enzyme-2 in murine cardiac development. J Clin Invest. 2000;105:1373–82.
67. D'Orleans-Juste P, Houde M, Rae GA, Bkaily G, Carrier E, Simard E. Endothelin-1 (1–31): from chymase-dependent synthesis to cardiovascular pathologies. Vasc Pharmacol. 2008;49:51–62.
68. Fecteau MH, Honore JC, Plante M, Labonte J, Rae GA, D'Orleans-Juste P. Endothelin-1 (1–31) is an intermediate in the production of endothelin-1 after big endothelin-1 administration in vivo. Hypertension. 2005;46:87–92.
69. Maguire JJ, Kuc RE, Davenport AP. Vasoconstrictor activity of novel endothelin peptide, ET-1(1–31), in human mammary and coronary arteries in vitro. Br J Pharmacol. 2001;134: 1360–6.
70. O'Reilly G, Charnock-Jones DS, Morrison JJ, Cameron IT, Davenport AP, Smith SK. Alternatively spliced mRNAs for human endothelin-2 and their tissue distribution. Biochem Biophys Res Commun. 1993;193:834–40.

71. Plumpton C, Ashby MJ, Kuc RE, O'Reilly G, Davenport AP. Expression of endothelin peptides and mRNA in the human heart. Clin Sci. 1996;90(1):37–46.
72. Howard PG, Plumpton C, Davenport AP. Anatomical localization and pharmacological activity of mature endothelins and their precursors in human vascular tissue. J Hypertens. 1992;10:1379–86.
73. Matsumoto H, Suzuki N, Kitada C, Fujino M. Endothelin family peptides in human plasma and urine: their molecular forms and concentrations. Peptides. 1994;15:505–10.
74. Bacon CR, Davenport AP. Endothelin receptors in human coronary artery and aorta. Br J Pharmacol. 1996;117:986–92.
75. Chang I, Bramall, A, Baynash AG et al. Genetic study of ET-2 in mice. Eleventh international conference on endothelin, Montréal; 2009. p. 15.
76. Marciniak SJ, Plumpton C, Barker PJ, et al. Localization of immunoreactive endothelin and proendothelin in the human lung. Pulm Pharmacol. 1992;5:175–82.
77. Davenport AP, Hoskins SL, Kuc RE, Plumpton C. Differential distribution of endothelin peptides and receptors in human adrenal gland. Histochem J. 1996;28:779–89.
78. Davenport AP, Kuc RE, Mockridge JW. Endothelin-converting enzyme in the human vasculature: evidence for differential conversion of big endothelin-3 by endothelial and smooth-muscle cells. J Cardiovasc Pharmacol. 1998;31 Suppl 1:S1–3.
79. Maguire JJ, Davenport AP. No alteration in vasoconstrictor endothelin-B-receptor density or function in human coronary artery disease. J Cardiovasc Pharmacol. 2000;36(5 Suppl 1):S380–1.
80. Halcox JP, Nour KR, Zalos G, Quyyumi AA. Endogenous endothelin in human coronary vascular function: differential contribution of endothelin receptor types A and B. Hypertension. 2007;49:1134–41.
81. McCulloch KM, Docherty CC, Morecroft I, MacLean MR. EndothelinB receptor-mediated contraction in human pulmonary resistance arteries. Br J Pharmacol. 1996;119:1125–30.
82. Davie N, Haleen SJ, Upton PD, et al. ET(A) and ET(B) receptors modulate the proliferation of human pulmonary artery smooth muscle cells. Am J Respir Crit Care Med. 2002;165:398–405.
83. Davenport AP, Kuc RE. Radioligand binding and molecular imaging techniques for the quantitative analysis of established and emerging orphan receptors systems. Methods Mol Biol. 2005;306:93–120.
84. Wu-Wong JR, Chiou WJ, Hoffman DJ, et al. Endothelins and endothelin receptor antagonists: binding to plasma proteins. Life Sci. 1996;58:1839–47.
85. Williams Jr DL, Jones KL, Alves K, Chan CP, Hollis GF, Tung JS. Characterization of cloned human endothelin receptors. Life Sci. 1993;53:407–14.
86. Molenaar P, O'Reilly G, Sharkey A, et al. Characterization and localization of endothelin receptor subtypes in the human atrioventricular conducting system and myocardium. Circ Res. 1993;72:526–38.
87. Aramori I, Nirei H, Shoubo M, et al. Subtype selectivity of a novel endothelin antagonist, FR139317, for the two endothelin receptors in transfected Chinese hamster ovary cells. Mol Pharmacol. 1993;43:127–31.
88. Peter MG, Davenport AP. Characterization of the endothelin receptor selective agonist, BQ3020 and antagonists BQ123, FR139317, BQ788, 50235, Ro462005 and bosentan in the heart. Br J Pharmacol. 1996;117:455–62.
89. Peter MG, Davenport AP. Selectivity of [125I]-PD151242 for human, rat and porcine endothelin ETA receptors in the heart. Br J Pharmacol. 1995;114:297–302.
90. Doherty AM, Cody WL, He JX, et al. In vitro and in vivo studies with a series of hexapeptide endothelin antagonists. J Cardiovasc Pharmacol. 1993;22 Suppl 8:S98–102.
91. Breu V, Loffler BM, Clozel M. In vitro characterization of Ro 46-2005, a novel synthetic non-peptide endothelin antagonist of ETA and ETB receptors. FEBS Lett. 1993;334:210–4.
92. Clozel M, Breu V, Gray GA, et al. Pharmacological characterization of bosentan, a new potent orally active nonpeptide endothelin receptor antagonist. J Pharmacol Exp Ther. 1994;270: 228–35.

93. Wu C, Chan MF, Stavros F, et al. Discovery of TBC11251, a potent, long acting, orally active endothelin receptor-A selective antagonist. J Med Chem. 1997;40:1690–7.
94. Webb ML, Bird JE, Liu EC, et al. BMS-182874 is a selective, nonpeptide endothelin ETA receptor antagonist. J Pharmacol Exp Ther. 1995;272:1124–34.
95. Elliott JD, Lago MA, Cousins RD, et al. 1,3-Diarylindan-2-carboxylic acids, potent and selective non-peptide endothelin receptor antagonists. J Med Chem. 1994;37:1553–7.
96. Johnstrom P, Fryer TD, Richards HK, et al. In vivo imaging of cardiovascular endothelin receptors using the novel radiolabelled antagonist [18F]-SB209670 and positron emission tomography (microPET). J Cardiovasc Pharmacol. 2004;44 Suppl 1:S34–8.
97. Patt WC, Edmunds JJ, Repine JT, et al. Structure-activity relationships in a series of orally active gamma-hydroxy butenolide endothelin antagonists. J Med Chem. 1997;40:1063–74.
98. Maguire JJ, Kuc RE, Davenport AP. Affinity and selectivity of PD156707, a novel nonpeptide endothelin antagonist, for human ET(A) and ET(B) receptors. J Pharmacol Exp Ther. 1997;280:1102–8.
99. Walsh TF, Fitch KJ, Chakravarty PK, et al. The discovery of L-749329, a highly potent, orally active antagonist of endothelin receptors. Abstr Pap Am Chem Soc. 1994;298:145.
100. Fujimoto M, Mihara S, Nakajima S, Ueda M, Nakamura M, Sakurai K. A novel non-peptide endothelin antagonist isolated from bayberry, Myrica cerifera. FEBS Lett. 1992;305:41–4.
101. Mihara S, Nakajima S, Matumura S, Kohnoike T, Fujimoto M. Pharmacological characterization of a potent nonpeptide endothelin receptor antagonist, 97–139. J Pharmacol Exp Ther. 1994;268(3):1122–8.
102. Davenport AP, O'Reilly G, Kuc RE. Endothelin ETA and ETB mRNA and receptors expressed by smooth muscle in the human vasculature: majority of the ETA sub-type. Br J Pharmacol. 1995;114:1110–6.
103. Davenport AP, Kuc RE, Fitzgerald F, Maguire JJ, Berryman K, Doherty AM. [125I]-PD151242: a selective radioligand for human ETA receptors. Br J Pharmacol. 1994;111: 4–6.
104. Maguire JJ, Bacon CR, Fujimoto M, Davenport AP. Myricerone caffeoyl ester (50–235) is a non-peptide antagonist selective for human ETA receptors. J Hypertens. 1994;12(6): 675–80.

Non-Pharmacological Treatment of Peripheral Vascular Disease

11

Janice Tsui and George Hamilton

11.1 Introduction

Peripheral vascular disease (PVD) due to atherosclerosis of the lower limb arteries is an increasing problem in Western societies. Epidemiological studies suggest that the prevalence of asymptomatic PVD is approximately 7–15% in the middle-aged and elderly population [1]. About 15% of these patients develop lower limb symptoms within 5–7 years [2], with 1 in 2,500 of the population developing critical limb ischemia each year [3].

Risk factors for PVD are those of atherosclerosis: smoking, diabetes, hypertension, hyperlipidemia, hypercoagulable states, and sedentary lifestyle. While aggressive control of risk factors, smoking cessation, antiplatelet and statin therapy, and physical exercise are important in patients with PVD whose cardiovascular risks are significantly increased [4], this chapter provides an overview on non-pharmacological therapy in the treatment of lower limb symptoms in these patients.

11.2 Treatment Aims in PVD

Patients with symptomatic PVD present with either intermittent claudication (IC) or critical limb ischemia (CLI). IC describes pain in affected muscle groups on exercise. In CLI, patients suffer from pain at rest and may develop ulcers and gangrene. The viability of the limb is threatened with a significant risk of limb loss. The treatment aims and strategies are different for these two modes of presentation.

J. Tsui (✉) • G. Hamilton
Royal Free Vascular Unit, Royal Free Hampstead NHS Trust,
London, UK
e-mail: janice.tsui@ucl.ac.uk; g.hamilton@ucl.ac.uk

D. Abraham et al. (eds.), *Translational Vascular Medicine*,
DOI 10.1007/978-0-85729-920-8_11, © Springer-Verlag London Limited 2012

11.2.1 Intermittent Claudication

Patients with IC experience pain on exercise when the blood supply to the lower limb muscles is unable to meet metabolic requirements. The muscle groups affected depend on the arterial lesions present and the pain occurs at consistent walking distances with rapid relief on resting. While the lower limb outcomes of these patients are generally good, with less than 10% deteriorating sufficiently to merit revascularization over time [5], they are at significantly higher risks of cardiovascular complications due to atherosclerosis in other vascular territories [6]. The treatment aims in these patients are therefore to reduce their risks of cardiovascular events and to improve quality of life by increasing walking distance. Aggressive risk factor management is crucial to reduce cardiovascular events, but despite available evidence, risk factors generally remain suboptimally monitored and managed in patients with PVD [7–9]. Exercise therapy in the form of supervised exercise training programs is effective in improving walking ability and functional outcomes but long-term compliance is a problem [10, 11]. As interventions are not without risks, in this group of patients where limb viability is not threatened, intervention is only considered for patients with disabling symptoms which are significantly affecting their day-to-day life.

11.2.2 Critical Limb Ischemia

In CLI, patients suffer from pain at rest, which is worse on elevation. Patients describe relentless pain particularly at night and may resort to sleeping in a chair. The viability of the limb is threatened and ulceration and/or gangrene may occur. Revascularization by endovascular, surgical, or a combination of techniques is required to prevent limb loss.

11.3 Revascularization Procedures in PVD

Non-pharmacological treatments of PVD are mainly divided into endovascular and surgical revascularization procedures. Since most patients who require intervention have CLI, the treatment options related to this group of patients will be discussed. The goal of revascularization is to re-establish in-line flow to the foot. In CLI, this generally involves treating multiple arterial segments; inflow disease is addressed prior to treating outflow lesions. While treating proximal lesions may improve symptoms, establishment of uninterrupted flow in at least one infrapopliteal vessel to the foot is usually required where tissue loss is present [12].

11.3.1 Endovascular Treatment of CLI

Many institutions have a strategy of using endovascular intervention as the initial choice of treatment since most patients with CLI have significant comorbidities and high short-term mortality [12, 13]. Endovascular procedures are generally less prolonged than surgical bypasses and can be performed under local anesthetic in high-risk patients. In addition, prior endovascular procedures do not preclude subsequent surgery and in patients with diseased distal vessels and lack of suitable conduit for bypass, endovascular treatment may be the only option for limb salvage.

Recent evidence from the BASIL trial, however, suggests that surgery is the more durable option and should be considered as initial treatment in some patients. In this trial, 452 patients with CLI due to infrainguinal disease were randomized to either surgery first ($n=228$) or angioplasty first ($n=224$). At 5.5-year follow-up, 248 (55%) patients were alive without amputation, 38 (8%) were alive with amputation, 36 (8%) were dead after amputation, and 130 (29%) were dead without amputation. After 6 months, there was no significant difference in amputation-free survival or health-related quality-of-life measurements between the two treatment arms. However, hospital costs were higher in the surgery first group than the angioplasty first group [14]. Beyond 2 years however, the surgery first group did better, suggesting that for patients who have a life expectancy of 2 years or more, surgery may be the more durable option. Moreover, surgical outcomes were worse after an initial failed angioplasty [15].

11.3.1.1 Suprainguinal Intervention

Endovascular procedures to improve inflow include percutaneous transluminal angioplasty (PTA) with or without stenting of the iliac arteries. In CLI, stent placements improve the success rates of treating iliac stenoses and occlusions compared to PTA alone, with primary patency rates of 90%, 74%, and 69% at 1, 3, and 5 years respectively. Women and patients with chronic renal insufficiency had poorer outcomes [16].

11.3.1.2 Infrainguinal Intervention

For infrainguinal lesions, PTA with or without subintimal angioplasty or adjunctive stenting is associated with reasonable rates of limb salvage. In subintimal angioplasty, the wire is intentionally directed subintimally and then redirected within the true lumen, enabling long diffuse stenoses or occlusions to be crossed and treated [17, 18].

In the femoral and popliteal arteries, PTA is associated with lower patency rates than for iliac arteries. In one study of femoro-politeal angioplasty with provisional stent placement in patients with CLI, patency rates at the end of 2 years were 65% but with limb salvage rates of 97% [19]. A meta-analysis of 19 studies reported 3-year patency rates of 30–43% following angioplasty and 60–65% following additional stent placement [20].

Angioplasty of infrapopliteal vessels in CLI is reported to have limb salvage rates of between 92% and 95% [21] (Fig. 11.1). In a series of 235 patients with CLI, tibio-peroneal angioplasty had an overall success rate of 92% with limb salvage rate of 95% at 5 years [22]. Bare metal stents have been used successfully in infrainguinal vessels but with similar success rates to angioplasty alone [23]. Drug-eluting stents are also under investigation to reduce restenosis rates [24].

As increasingly challenging lesions are treated, different approaches have been described including retrograde puncture of pedal vessels and combined antegrade and retrograde approaches [25]. Long-term outcomes of these procedures from large studies are awaited.

11.3.1.3 Adjunctive Endovascular Devices

While the overall technical success rates of angioplasty and stents in the treatment of lower limb arterial lesions are relatively high, immediate failure is most commonly related to lesions that are difficult to cross or dilate, or difficulty in re-entering the distal lumen, due to long lesions, total occlusions, calcified vessels, and diffuse distal disease. Several devices have been developed to try and overcome these problems. These include cutting balloon catheters with embedded atherotomes on the exterior of the balloon aimed at treating heavily calcified arteries which are difficult to dilate; laser-assisted angioplasty devices which aim to debulk atheromatous plaques which are difficult to cross; and directional atherectomy catheters which allow plaque to be excised and removed. Currently, there is inadequate evidence of their efficacy and cost-effectiveness in the treatment of CLI [26].

Late failure occurs due to intimal hyperplasia leading to restenosis of the native vessel as well as in-stent restenoses. A cryoplasty balloon catheter was developed with the aim of delivering cold thermal energy to the vessel wall to induce apoptosis and reduce restenosis, but again, there is little evidence to support its use in CLI. Other adjunctive devices include self-expandable stents which may be useful in the superficial femoral arteries or tortuous or tapering arteries; drug-eluting stents based

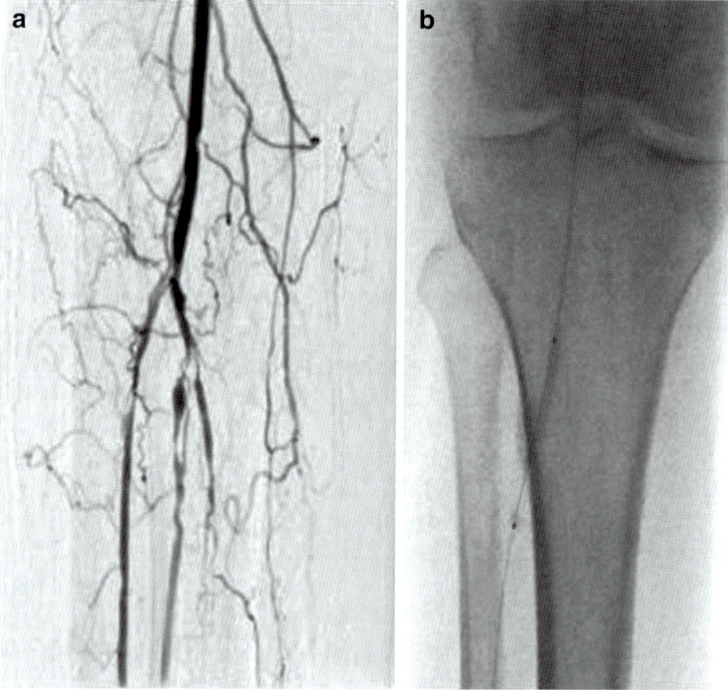

Fig. 11.1 Infrapopliteal angioplasty. (**a**) Tight stenoses at origins of the right infrapopliteal arteries with subsequent occlusion of the posterior tibial artery (*arrow*). (**b**) Angioplasty of the anterior tibial artery. (**c**, **d**) Following successful angioplasty of the anterior tibial and peroneal arteries, in-line flow is restored to the foot

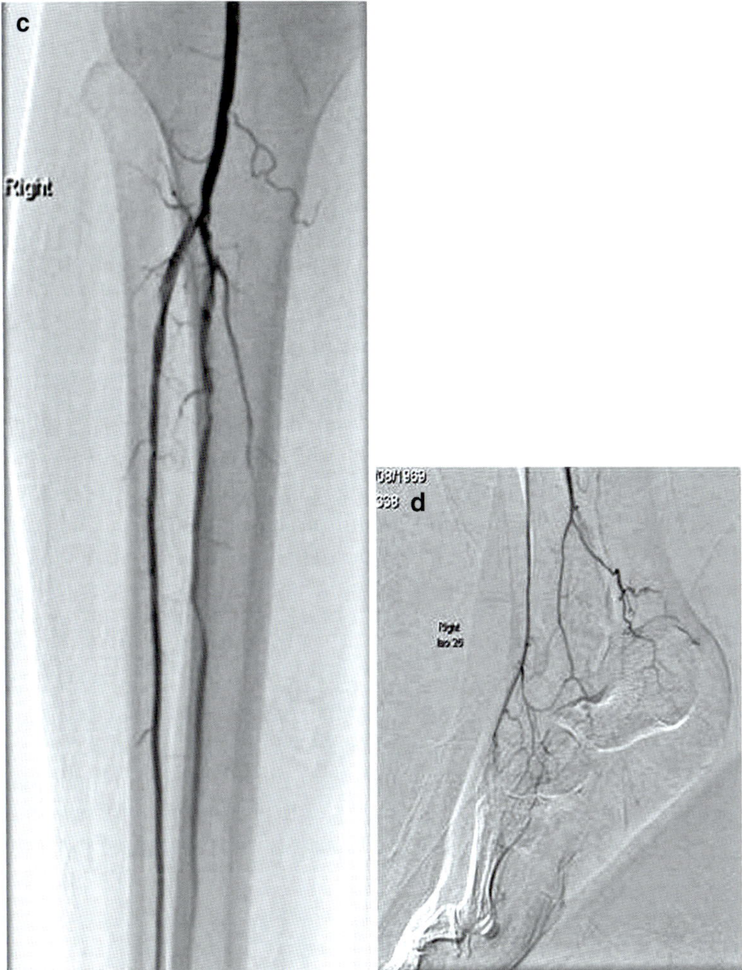

Fig. 11.1 (continued)

on their use in the coronary circulation; covered stents and bioabsorbable stents [26]. While some of these warrant further investigation, they are currently not recommended for routine use in CLI.

11.3.1.4 Complications of Endovascular Procedures

Endovascular procedures are not without complications, particularly in patients with CLI who have calcified vessels and multilevel disease [27].

Puncture site complications are the most common with reported incidence of 2–6% [28]. These include bleeding, hematoma, false aneurysms, arteriovenous fistulae formation, and infection. Bleeding complications usually resolve with conservative management but may require surgical intervention and may be life-threatening such as in cases of massive retroperitoneal hematomas. False aneurysms at the groin can usually be treated with thrombin injection [29].

Complications related to the target vessel include vessel rupture, local dissection, thrombosis, and distal embolization. These are uncommon and can mostly be treated endovascularly with covered stents, further balloon inflations, and local thrombolysis, but may require surgical intervention. Serious complications leading to limb loss have been reported to occur following 2.2% of angioplasties for CLI [27].

In addition, with increasing use of devices, device-specific complications such as device migration and deployment failure may occur.

11.3.2 Surgical Treatment of CLI

In many centers, surgical revascularization is reserved for patients with lesions that are deemed unsuitable for endovascular treatment or for those where endovascular treatment has failed. Younger patients with prolonged life expectancy are more commonly considered for surgical revascularization due to its more durable results, particularly in light of recent evidence from the BASIL trial as discussed above.

11.3.2.1 Suprainguinal Procedures

Aortoiliac disease is treated with either aortic reconstruction or with an extra-anatomical bypass (e.g., axillofemoral, axillobifemoral, femorofemoral bypass). While aorto-bifemoral bypasses have good patency rates of 80% and 72% at 5 and 10 years respectively, operative mortality is on average 3.3%, rising to 8% in patients with significant comorbidities [30]. For patients who are less fit, extra-anatomical bypasses using externally supported Dacron or PTFE grafts are alternatives with 5-year limb salvage rates of 60–90% [31].

11.3.2.2 Infrainguinal Procedures

Infrainguinal procedures include common femoral endarterectomy and profundaplasty which may used to improve inflow prior to an infrainguinal bypass procedure. However, there is evidence that isolated common femoral endarterectomy is sufficient to salvage limbs in some patients with CLI [32].

Infrainguinal bypass procedures are usually taken from the common femoral artery to the above- or below-knee popliteal artery or to the tibial or peroneal arteries. Autologous vein rather than synthetic grafts should be used where possible, particularly in infrageniculate bypasses. A meta-analysis had shown primary patencies of 66% for vein at any level compared to 47% for above-knee PTFE and 33% for below-knee PTFE at 5 years [33]. While the long saphenous vein is the most commonly used autologous conduit, the short saphenous vein, arm veins, and deep leg veins can be used. Different strategies have been used to try and improve the outcome of vein grafts [34], but the lack of a suitable autologous conduit is usually the problem. Vein cuffs performed at the distal anastomosis or less commonly formation of an arterio-venous fistula to an adjacent vein at the distal anastomosis have been used to improve patency of infrageniculate bypasses using synthetic grafts [35, 36]. With advances in the fields of biomaterials and tissue engineering, novel grafts exploiting these innovative technologies are likely to be the solution to improved outcomes of these procedures [37, 38].

11.3.3 Hybrid Procedures in CLI

Hybrid procedures combining endovascular and open procedures are increasingly used to revascularize patients with CLI. These may offer less extensive procedures, reduced perioperative complications, and better outcomes [39, 40]. For example, iliac angioplasty and stenting may be combined with an infrainguinal bypass procedure; angioplasty of an iliac stenosis may then allow a femorofemoral crossover graft to treat bilateral iliac disease; superficial femoral artery (SFA) angioplasty of a focal lesion may enable a shorter SFA-distal bypass graft to be performed particularly where availability of vein is limited. The procedures can be done as staged procedures or increasingly commonly as concomitant procedures [41] (Fig. 11.2).

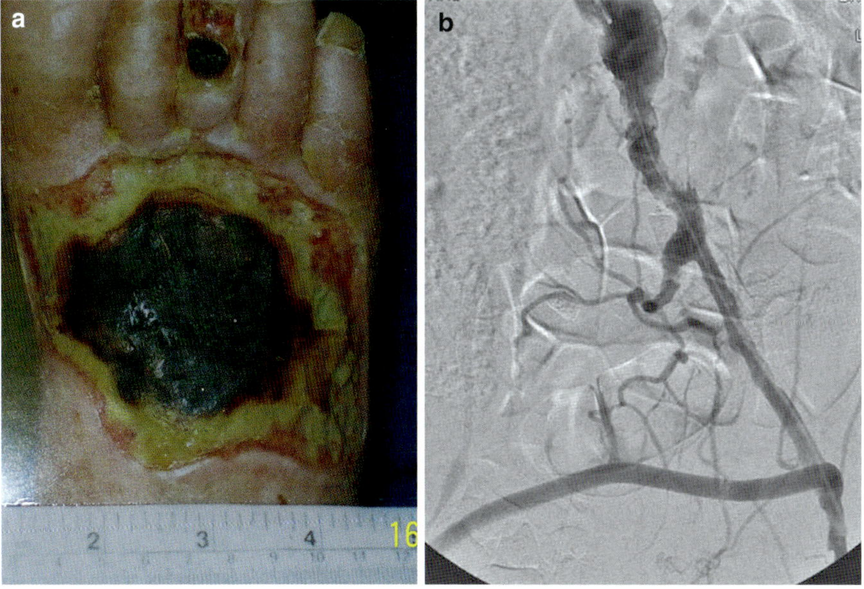

Fig. 11.2 Iliac angioplasty and femorofemoral crossover graft in a 70-year-old man with CLI. (**a**) Gangrenous ulcer on dorsum of right foot at presentation. A left-to-right femorofemoral crossover graft was initially performed which failed 2 weeks later due to inadequate inflow (diseased left iliac artery) and poor outflow (occluded right superficial femoral artery). (**b**) A left iliac angioplasty, crossover graft thrombectomy, and femoro-distal bypass were successfully performed. (**c**) Ulcer healing following revasuclarization. (**d**) Residual ulcer of approximately 1 cm in diameter

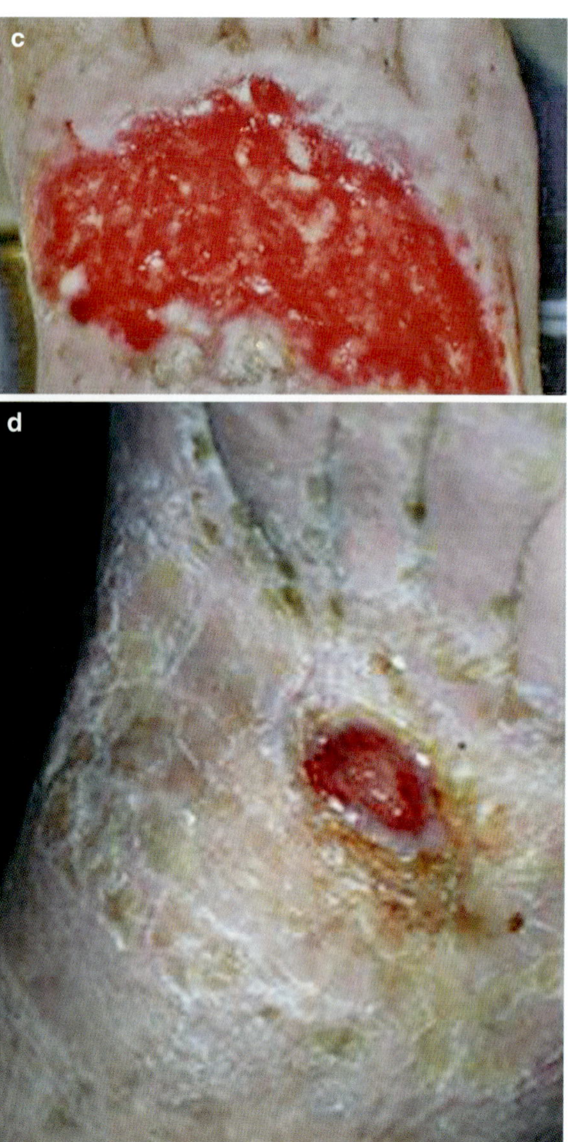

Fig. 11.2 (continued)

11.4 Other Treatment Strategies

In patients with CLI who have non-reconstructable disease, other treatment strategies are offered in an attempt to avoid or delay amputation. They aim to reduce ischemic rest pain, promote ulcer healing, and prevent further deterioration.

11.4.1 Sympathectomy

Lumbar sympathectomy may be offered to patients with non-reconstructable disease. Open surgical sympathectomy has largely been replaced by percutaneous chemical sympathectomy [42] or the laparoscopic approach [43]. These techniques have been shown to be of some benefit in terms of pain relief and limb salvage in patients with CLI [44].

11.4.2 Spinal Cord Stimulation

Spinal cord stimulation employs low-voltage electrical impulses from a subcutaneous pulse generator which are delivered to the epidural space by electrodes placed at the L3/4 level. The exact mechanism of action is unknown but is thought to improve microcirculatory blood flow. Early controlled studies suggested a benefit in pain control although there was no improvement in ulcer healing, limb salvage, or mortality rates [45, 46]. However, most of these studies had small patient numbers and a recent systematic review found no significant benefit of the technique above best medical therapy [47]. Moreover, spinal cord stimulation is more expensive [48].

11.4.3 Therapeutic Angiogenesis

Therapeutic angiogenesis using gene- and cell-based therapies to stimulate new vessel formation has been investigated over the past decade. While initial preclinical studies were promising [49], clinical trials have been less convincing. A meta-analysis identified six phase II randomized, controlled trials in therapeutic angiogenesis for PVD: four in patients with IC and two in CLI [50]. While the meta-analysis of the pooled data from these studies concluded that therapeutic angiogenesis was beneficial in patients with CLI with only a slight increase in side effects of edema, hypotension, and proteinuria, this was mainly due to the results of two of the trials. The TRAFFIC trial showed that intra-arterial recombinant basic fibroblast growth factor (bFGF) improved peak walking time in patients with IC [51] while the TACT trial demonstrated that autologous bone marrow mononuclear cell implantation significantly improved rest pain in CLI patients [52]. Since then, the TAMARIS trial which was a randomized controlled study designed to evaluate the efficacy of NV1FGF, a non-viral plasmid-based gene delivery system for FGF-1, in CLI patients has failed to meet its primary endpoint of prevention of major amputation or death at 12 months [53]. Overall, the results of these studies are disappointing and probably reflect the complexity of angiogenesis which is unlikely to be easily manipulated by the administration of single factors.

11.5 Amputation

Major amputation may be required in patients where no revascularization option is available or where treatment has failed. In some patients, primary amputation may be considered due to an unsalvageable limb, fixed flexion contractures of the leg or in patients who are already bed-bound due to comorbidities. The goals of amputation are pain relief, removal of nonviable or infected tissue, and the formation of a well-healed stump that maximizes chances of rehabilitation. The level of amputation therefore depends on healing and rehabilitation potential as well as prosthetic considerations.

11.5.1 Levels of Amputations

In patients with PAD, minor amputations, i.e., amputations of the foot, are generally performed only following successful revascularization to remove necrotic and/or infected tissue. Without revascularization, minor amputations are unlikely to heal and the patient is then faced with multiple procedures with increased risks and delayed rehabilitation.

The commonest levels of amputation in PAD patients are perigeniculate: below-knee (transtibial), through-knee (knee disarticulation and Gritti–Stokes amputation), and above-knee (transfemoral) amputations. In patients who are likely to achieve prosthetic walking, the knee joint should be preserved if possible and a below-knee amputation performed if healing at this level is likely. If healing is likely to be compromised, an above-knee amputation still allows prosthetic fitting. In patients who are likely to remain chair- or bed-bound, through-knee amputations avoid development of fixed flexion contractures and subsequent difficulties with transfers while providing longer lever lengths and larger surface areas for improved balance. Patients who have already developed a fixed flexion deformity of the knee, an above-knee amputation is more realistic.

Hip disarticulation and hindquarter amputations are extensive procedures which fortunately are rarely performed [54].

In order to provide these patients with optimal care, the involvement of a dedicated multidisciplinary team including specialist physiotherapists, occupational therapists, pain specialists, prosthetists, and social workers, from preoperative planning stages through to rehabilitation, is essential [55, 56].

11.6 Prognosis of Patients with PVD

As mentioned above, lower limb outcomes for patients with IC are generally good. About 50% of patients will remain stable or experience some improvement in their symptoms over a 5-year period; 25% will deteriorate and only 1–2% will require major amputation. However, 2–4% of these patients will have a non-fatal cardiovascular event within the first year of diagnosis with a 1–3% annual incidence thereafter. Overall, patients with PAD have a 25% greater mortality risk than those without [57].

In patients with CLI, overall long-term prognosis is poor with amputation rates of 10–40% and mortality rates of 20% at 1 year and 40–70% at 5 years [58]. While there is data from Denmark [59]and Finland[60] showing reduced amputation rates with more aggressive revascularization policies, the number of major amputations performed each year in the UK for vascular diseases has remained at approximately 3,000 from 2000 to 2007 [54]. In 1999, Dormandy et al. reported that within 2 years, 15% of below-knee amputees required a contralateral major amputation and 30% were dead [61]. More recently, Dillingham et al. reported similar figures: 10% of below-knee amputees had a contralateral amputation and over a third died within 1 year [62].

11.7 Conclusions

In conclusion, PVD is a prevalent disease and with an aging population and an increase in cardiovascular diseases, it will continue to be a challenging healthcare issue. In order to improve the outcomes of these patients, they must be managed by vascular specialists within multidisciplinary teams who are able to combine endovascular and surgical techniques to revascularize threatened limbs and to reduce the impact of their overall cardiovascular risks. Novel therapeutic therapies are urgently required particularly to improve the quality of life of patients with disease that is not amenable to revascularization and also as adjuncts to improve the results of currently available treatment options.

References

1. Fowkes FG, Housley E, Cawood EH, Macintyre CC, Ruckley CV, Prescott RJ. Edinburgh artery study: prevalence of asymptomatic and symptomatic peripheral arterial disease in the general population. Int J Epidemiol. 1991;20:384–92.
2. Hooi JD, Kester AD, Stoffers HE, Overdijk MM, van Ree JW, Knottnerus JA. Incidence of and risk factors for asymptomatic peripheral arterial occlusive disease: a longitudinal study. Am J Epidemiol. 2001;153:666–72.
3. Critical limb ischaemia: management and outcome. Report of a national survey. The Vascular Surgical Society of Great Britain and Ireland. Eur J Vasc Endovasc Surg. 1995;10:108–13.
4. Criqui MH, Denenberg JO, Langer RD, Fronek A. The epidemiology of peripheral arterial disease: importance of identifying the population at risk. Vasc Med. 1997;2:221–6.
5. Leng GC, Lee AJ, Fowkes FG, Whiteman M, Dunbar J, Housley E, et al. Incidence, natural history and cardiovascular events in symptomatic and asymptomatic peripheral arterial disease in the general population. Int J Epidemiol. 1996;25:1172–81.
6. Criqui MH, Langer RD, Fronek A, Feigelson HS, Klauber MR, McCann TJ, et al. Mortality over a period of 10 years in patients with peripheral arterial disease. N Engl J Med. 1992;326:381–6.
7. Bianchi C, Montalvo V, Ou HW, Bishop V, Abou-Zamzam Jr AM. Pharmacologic risk factor treatment of peripheral arterial disease is lacking and requires vascular surgeon participation. Ann Vasc Surg. 2007;21:163–6.

8. Khan S, Flather M, Mister R, Delahunty N, Fowkes G, Bradbury A, et al. Characteristics and treatments of patients with peripheral arterial disease referred to UK vascular clinics: results of a prospective registry. Eur J Vasc Endovasc Surg. 2007;33:442–50.
9. Zeymer U, Parhofer KG, Pittrow D, Binz C, Schwertfeger M, Limbourg T, et al. Risk factor profile, management and prognosis of patients with peripheral arterial disease with or without coronary artery disease: results of the prospective German REACH registry cohort. Clin Res Cardiol. 2009;98:249–56.
10. Regensteiner JG, Steiner JF, Hiatt WR. Exercise training improves functional status in patients with peripheral arterial disease. J Vasc Surg. 1996;23:104–15.
11. Gardner AW, Katzel LI, Sorkin JD, Bradham DD, Hochberg MC, Flinn WR, et al. Exercise rehabilitation improves functional outcomes and peripheral circulation in patients with intermittent claudication: a randomized controlled trial. J Am Geriatr Soc. 2001;49:755–62.
12. Hirsch AT, Haskal ZJ, Hertzer NR, Bakal CW, Creager MA, Halperin JL, et al. ACC/AHA 2005 Practice Guidelines for the management of patients with peripheral arterial disease (lower extremity, renal, mesenteric, and abdominal aortic): a collaborative report from the American Association for Vascular Surgery/Society for Vascular Surgery, Society for Cardiovascular Angiography and Interventions, Society for Vascular Medicine and Biology, Society of Interventional Radiology, and the ACC/AHA Task Force on Practice Guidelines (Writing Committee to Develop Guidelines for the Management of Patients with Peripheral Arterial Disease): endorsed by the American Association of Cardiovascular and Pulmonary Rehabilitation; National Heart, Lung, and Blood Institute; Society for Vascular Nursing; TransAtlantic Inter-Society Consensus; and Vascular Disease Foundation. Circulation. 2006;113:e463–654.
13. Varty K, Nydahl S, Butterworth P, Errington M, Bolia A, Bell PR, et al. Changes in the management of critical limb ischaemia. Br J Surg. 1996;83:953–6.
14. Forbes JF, Adam DJ, Bell J, Fowkes FG, Gillespie I, Raab GM, et al. Bypass versus Angioplasty in Severe Ischaemia of the Leg (BASIL) trial: health-related quality of life outcomes, resource utilization, and cost-effectiveness analysis. J Vasc Surg. 2010;51(5 Suppl):43S–51.
15. Bradbury AW, Adam DJ, Bell J, Forbes JF, Fowkes FG, Gillespie I, et al. Multicentre randomised controlled trial of the clinical and cost-effectiveness of a bypass-surgery-first versus a balloon-angioplasty-first revascularisation strategy for severe limb ischaemia due to infrainguinal disease. The Bypass versus Angioplasty in Severe Ischaemia of the Leg (BASIL) trial. Health Technol Assess. 2010;14:1–210, i–iv.
16. Timaran CH, Stevens SL, Freeman MB, Goldman MH. Predictors for adverse outcome after iliac angioplasty and stenting for limb-threatening ischemia. J Vasc Surg. 2002;36(3): 507–13.
17. Tisi PV, Mirnezami A, Baker S, Tawn J, Parvin SD, Darke SG. Role of subintimal angioplasty in the treatment of chronic lower limb ischaemia. Eur J Vasc Endovasc Surg. 2002;24:417–22.
18. Myers SI, Myers DJ, Ahmend A, Ramakrishnan V. Preliminary results of subintimal angioplasty for limb salvage in lower extremities with severe chronic ischemia and limb-threatening ischemia. J Vasc Surg. 2006;44:1239–46.
19. Conrad MF, Cambria RP, Stone DH, Brewster DC, Kwolek CJ, Watkins MT, et al. Intermediate results of percutaneous endovascular therapy of femoropopliteal occlusive disease: a contemporary series. J Vasc Surg. 2006;44:762–9.
20. Muradin GS, Bosch JL, Stijnen T, Hunink MG. Balloon dilation and stent implantation for treatment of femoropopliteal arterial disease: meta-analysis. Radiology. 2001;221:137–45.
21. Dorros G, Lewin RF, Jamnadas P, Mathiak LM. Below-the-knee angioplasty: tibioperoneal vessels, the acute outcome. Cathet Cardiovasc Diagn. 1990;19:170–8.
22. Dorros G, Jaff MR, Dorros AM, Mathiak LM, He T. Tibioperoneal (outflow lesion) angioplasty can be used as primary treatment in 235 patients with critical limb ischemia: five-year follow-up. Circulation. 2001;104:2057–62.
23. Feiring AJ, Wesolowski AA, Lade S. Primary stent-supported angioplasty for treatment of below-knee critical limb ischemia and severe claudication: early and one-year outcomes. J Am Coll Cardiol. 2004;44:2307–14.

24. Commeau P, Barragan P, Roquebert PO. Sirolimus for below the knee lesions: mid-term results of SiroBTK study. Catheter Cardiovasc Interv. 2006;68:793–8.
25. Gandini R, Pipitone V, Stefanini M, Maresca L, Spinelli A, Colangelo V, et al. The "Safari" technique to perform difficult subintimal infragenicular vessels. Cardiovasc Intervent Radiol. 2007;30:469–73.
26. Arain SA, White CJ. Endovascular therapy for critical limb ischemia. Vasc Med. 2008;13: 267–79.
27. Axisa B, Fishwick G, Bolia A, Thompson MM, London NJ, Bell PR, et al. Complications following peripheral angioplasty. Ann R Coll Surg Engl. 2002;84:39–42.
28. Belli AM, Cumberland DC, Knox AM, Procter AE, Welsh CL. The complication rate of percutaneous peripheral balloon angioplasty. Clin Radiol. 1990;41:380–3.
29. Vlachou PA, Karkos CD, Bains S, McCarthy MJ, Fishwick G, Bolia A. Percutaneous ultrasound-guided thrombin injection for the treatment of iatrogenic femoral artery pseudoaneurysms. Eur J Radiol. 2011;77:172–4.
30. de Vries SO, Hunink MG. Results of aortic bifurcation grafts for aortoiliac occlusive disease: a meta-analysis. J Vasc Surg. 1997;26:558–69.
31. McDaniel MD, Macdonald PD, Haver RA, Littenberg B. Published results of surgery for aortoiliac occlusive disease. Ann Vasc Surg. 1997;11:425–41.
32. Desai M, Tsui J, Davis M, Myint F, Wilson A, Baker DM, et al. Isolated endarterectomy of femoral bifurcation in critical limb ischemia: is restoration of inline flow essential? Angiology. 2011;62:119–25.
33. Hunink MG, Wong JB, Donaldson MC, Meyerovitz MF, Harrington DP. Patency results of percutaneous and surgical revascularization for femoropopliteal arterial disease. Med Decis Making. 1994;14:71–81.
34. Tsui JC, Dashwood MR. Recent strategies to reduce vein graft occlusion: a need to limit the effect of vascular damage. Eur J Vasc Endovasc Surg. 2002;23(3):202–8.
35. Stonebridge PA, Prescott RJ, Ruckley CV. Randomized trial comparing infrainguinal polytetrafluoroethylene bypass grafting with and without vein interposition cuff at the distal anastomosis. The Joint Vascular Research Group. J Vasc Surg. 1997;26:543–50.
36. Hamsho A, Nott D, Harris PL. Prospective randomised trial of distal arteriovenous fistula as an adjunct to femoro-infrapopliteal PTFE bypass. Eur J Vasc Endovasc Surg. 1999;17:197–201.
37. Ravi S, Chaikof EL. Biomaterials for vascular tissue engineering. Regen Med. 2010;5:107–20.
38. de Mel A, Bolvin C, Edirisinghe M, Hamilton G, Seifalian AM. Development of cardiovascular bypass grafts: endothelialization and applications of nanotechnology. Expert Rev Cardiovasc Ther. 2008;6:1259–77.
39. Melliere D, Cron J, Allaire E, Desgranges P, Becquemin JP. Indications and benefits of simultaneous endoluminal balloon angioplasty and open surgery during elective lower limb revascularization. Cardiovasc Surg. 1999;7:242–6.
40. Miyahara T, Miyata T, Shigematsu H, Shigematsu K, Okamoto H, Nakazawa T, et al. Long-term results of combined iliac endovascular intervention and infrainguinal surgical revascularization for treatment of multilevel arterial occlusive disease. Int Angiol. 2005;24:340–8.
41. Dougherty MJ, Young LP, Calligaro KD. One hundred twenty-five concomitant endovascular and open procedures for lower extremity arterial disease. J Vasc Surg. 2003;37:316–22.
42. Pieri S, Agresti P, Ialongo P, Fedeli S, Di Cesare F, Ricci G. Lumbar sympathectomy under CT guidance: therapeutic option in critical limb ischaemia. Radiol Med. 2005;109:430–7.
43. Nemes R, Surlin V, Chiutu L, Georgescu E, Georgescu M, Georgescu I. Retroperitoneoscopic lumbar sympathectomy: prospective study upon a series of 50 consecutive patients. Surg Endosc 2011, Surg Endosc 2011;25:3066–70.
44. Holiday FA, Barendregt WB, Slappendel R, Crul BJ, Buskens FG, van der Vliet JA. Lumbar sympathectomy in critical limb ischaemia: surgical, chemical or not at all? Cardiovasc Surg. 1999;7:200–2.
45. Jivegard LE, Augustinsson LE, Holm J, Risberg B, Ortenwall P. Effects of spinal cord stimulation (SCS) in patients with inoperable severe lower limb ischaemia: a prospective randomised controlled study. Eur J Vasc Endovasc Surg. 1995;9:421–5.

46. Klomp HM, Spincemaille GH, Steyerberg EW, Habbema JD, Van Urk H. Spinal-cord stimulation in critical limb ischaemia: a randomised trial. ESES Study Group. Lancet. 1999;353: 1040–4.
47. Klomp HM, Steyerberg EW, Habbema JD, Van Urk H. What is the evidence on efficacy of spinal cord stimulation in (subgroups of) patients with critical limb ischemia? Ann Vasc Surg. 2009;23:355–63.
48. Klomp HM, Steyerberg EW, Van Urk H, Habbema JD. Spinal cord stimulation is not cost-effective for non-surgical management of critical limb ischaemia. Eur J Vasc Endovasc Surg. 2006;31:500–8.
49. Isner JM, Asahara T. Angiogenesis and vasculogenesis as therapeutic strategies for postnatal neovascularization. J Clin Invest. 1999;103:1231–6.
50. De Haro J, Acin F, Lopez-Quintana A, Florez A, Martinez-Aguilar E, Varela C. Meta-analysis of randomized, controlled clinical trials in angiogenesis: gene and cell therapy in peripheral arterial disease. Heart Vessels. 2009;24:321–8.
51. Lederman RJ, Mendelsohn FO, Anderson RD, Saucedo JF, Tenaglia AN, Hermiller JB, et al. Therapeutic angiogenesis with recombinant fibroblast growth factor-2 for intermittent claudication (the TRAFFIC study): a randomised trial. Lancet. 2002;359:2053–8.
52. Tateishi-Yuyama E, Matsubara H, Murohara T, Ikeda U, Shintani S, Masaki H, et al. Therapeutic angiogenesis for patients with limb ischaemia by autologous transplantation of bone-marrow cells: a pilot study and a randomised controlled trial. Lancet. 2002;360:427–35.
53. Haitt WR, Baumgartner I, Nikol S, Van Belle E, Driver V, Norgren L, et al. NV1FGF gene therapy on amputation-free survival in critical limb ischaemia – phase 3 randomised double-blind placebo controlled trial (TAMARIS). Circulation. 2010;122:2217.
54. Information Services Division NHS Scotland on behalf of National Amputee Statistical Database for the UK (NASDAB). The Amputee Statistical Database for UK Annual Report 2006/2007. http://www.limbless-statistics.org/documents/Report2006-07.pdf.
55. Kaplow M, Muroff F, Fish W, Stillwell D, Mitchell N. The dysvascular amputee: multidisciplinary management. Can J Surg. 1983;26:368–9.
56. Hakimi KN. Pre-operative rehabilitation evaluation of the dysvascular patient prior to amputation. Phys Med Rehabil Clin N Am. 2009;20:677–88.
57. Adam DJ, Bradbury AW. TASC II document on the management of peripheral arterial disease. Eur J Vasc Endovasc Surg. 2007;33:1–2.
58. Dormandy J, Heeck L, Vig S. The fate of patients with critical leg ischemia. Semin Vasc Surg. 1999;12:142–7.
59. Ebskov LB, Schroeder TV, Holstein PE. Epidemiology of leg amputation: the influence of vascular surgery. Br J Surg. 1994;81:1600–3.
60. Luther M. The influence of arterial reconstructive surgery on the outcome of critical leg ischaemia. Eur J Vasc Surg. 1994;8:682–9.
61. Dormandy J, Heeck L, Vig S. Major amputations: clinical patterns and predictors. Semin Vasc Surg. 1999;12:154–61.
62. Dillingham TR, Pezzin LE, Shore AD. Reamputation, mortality, and health care costs among persons with dysvascular lower-limb amputations. Arch Phys Med Rehabil. 2005;86:480–6.

Surgical Approaches to Abdominal Aortic Aneurysm Repair

12

Matt Thompson, Peter Holt, Rob Hinchliffe, and Ian Loftus

12.1 Introduction

In the last decade, there have been dramatic changes to the management of abdominal aortic aneurysms (AAA) in the UK, and further progress is likely in the next few years. The central strategy in managing abdominal aortic aneurysms is to detect these lesions before they rupture and perform an elective repair with low morbidity and mortality. Rupture of an abdominal aortic aneurysm is a catastrophic event which carries a community mortality in excess of 90%.

Historically, the UK has one of the worst mortality rates for aneurysm surgery in the world. The reasons for this are multi-factorial but involve late diagnosis, high rates of comorbidity, low uptake of endovascular technology, and fragmented service organization. This chapter will focus on three specific aspects of abdominal aortic aneurysm surgery which will have a major impact on UK practice in the next few years: the introduction of a national screening program, the use of endovascular techniques to reduce operative mortality, and the urgent need for centralization of aortic surgery.

12.2 Screening for AAA

Community-based ultrasound screening is a noninvasive, cheap, and accurate method of detecting AAA. Large-scale population screening trials have shown that it is effective in men aged 65–75 years [1], and reduces the rate of aneurysm rupture and aneurysm-related mortality. On this basis, the UK Secretary of State for Health announced a UK national screening program for abdominal aortic aneurysms in

M. Thompson (✉) • P. Holt • R. Hinchliffe • I. Loftus
Vascular Surgery, St George's Vascular Institute, St George's Hospital,
London, UK
e-mail: matt.thompson@stgeorges.nhs.uk

D. Abraham et al. (eds.), *Translational Vascular Medicine*,
DOI 10.1007/978-0-85729-920-8_12, © Springer Verlag London Limited 2012

January 2008. The primary aim of the program is to reduce AAA-related mortality by providing a systematic population-based screening program for the male population during their 65th year, and on request for men over 65.

There are still many other practical aspects relating to screening programs that require further work. These include techniques to optimize the uptake of screening, whether to use internal or external aortic diameters, cost-effective surveillance intervals, and the management of anxiety and cardiovascular risk factors within the screened population with small aneurysms. The national quality assurance framework and audit processes in place within the UK screening program may help to clarify some of these issues in due course.

The UK program is being rolled out in a staged process. The first six centers commenced screening in 2009 as "early implementation sites," namely St Georges London, Leicester, Manchester, South Devon, Gloucester and West Sussex. By 2012/13, the aim is to have 60 centers operational around the country covering a population of 270,000 men aged 65. Sites are based on a minimum 800,000 total screening population, working within established vascular networks and able to demonstrate acceptable perioperative aneurysm mortality through submission to the National Vascular Database.

Subjects will receive an invitation to a single ultrasound scan during their 65th year, performed within community healthcare facilities. If the aorta measures less than 3 cm, no further recall scans will be arranged. For aortas measuring 3.0–4.4 cm, a follow-up scan will be arranged for 1 year, and for aortas measuring 4.5–5.4 cm, a further scan is arranged for 3 months. Above this size, an automatic referral is generated to the screening center or local network hospital.

Currently, internal aortic diameters are used though there has been debate about the use of external diameters. Ultrasound has high sensitivity and specificity if performed with adequate quality assurance, particularly for internal diameters. Ultrasound can reliably image the aorta in 99% of subjects. If the aorta is not visualized, the subject should be rescanned by an experienced sonographer. The incidence of false-positive scans is uncertain but is small and of little clinical consequence.

The decision to introduce a national program was based on four randomized trials, namely the Chichester trial in the UK [2], the Viborg trial in Denmark [3], the Western Australia trial [4], and the UK Multi-centre Aneurysm Screening Study (MASS) [1]. In each study, individuals were randomized either to an offer of aneurysm screening, or to no offer of screening. In all four trials, screening was shown to reduce aneurysm-related mortality for men. In the MASS trial, the screening group demonstrated a 43% reduction in overall mortality from aneurysm disease. Overall the odds ratio in favor of screening for men was 0.60 [95%CI 0.47–0.78]. The individual characteristics of the trials are summarized in Table 12.1.

There is no good evidence to support aneurysm screening in women. In the only screening trial conducted in women, there was no reduction in the incidence of aneurysm rupture at 5 or 10 years [5]. Smoking and family history are important independent risk factors for aneurysm development, and though not specifically targeted within the program, these increased risks should be highlighted. Within the UK program, there exists the opportunity for lifestyle advice and cardiovascular risk factor assessment, particularly for those with small aneurysms. For example,

Table 12.1 Summary of the population-based randomized screening trials

Trial	Chichester UK	Viborg Denmark	MASS[a] UK	Western Australia
Number randomized	15,775	12,628	67,800	41,000
People	Men and women	Men	Men	Men
Age (years)	65–80	65–73	65–74	65–79
Dates recruited	1988–90	1994–8	1997–9	1996–8
Date published	1995	2002	2002	2004
% accepting screening	68	76	80	70
Detection rate	4% (7.6% in men)	4%	4.9%	7.2%
Intervention policy	At 6 cm	At 5 cm	At 5.5 cm	None
Mean follow-up (months)	30.5	61	49	43
AAA-mortality, odds ratio Screened vs. not (95%CI)*	0.59 (men only) (0.27–1.29)	0.31 (0.13–0.79)	0.58 (0.42–0.78)	0.72 (0.39–1.32)
All-cause mortality, odds ratio Screened vs. not (95%CI)**	1.07 (men only) (0.93–1.22)		0.97 (0.93–1.02)	0.98 (0.91–1.04)

*Pooled odds ratio over all four trials is strongly in favor of screening, OR 0.57 (0.45–0.74), with a halving of the incidence of aneurysm rupture in screened populations

**Pooled odds ratio trend in favor of screening, OR 0.98 (0.95–1.02)

[a]The MASS trial recently published 10-year follow-up, demonstrating the cost-effectiveness of screening and a significant all-cause mortality benefit, but a rising incidence of AAA rupture in the screened group

statins may reduce aneurysm growth rates by about 50% and smoking cessation appears to reduce growth rate by 20–30%.

While the benefits of screening are clear from the population-based studies, the possibility of causing harm must also be considered. Detection of a small aneurysm with the potential for unpredictable expansion and rupture is likely to create anxiety. Both the MASS and Viborg trials demonstrated a decreased quality of life for a short period after positive screening, though the effects resolved within a few months [6]. More importantly, there is the mortality risk associated with intervention. If screening is to be conducted safely, the referral centers must have an audited low mortality for both open and endovascular repair. For elective open repair, the operative mortality should be less than 5%, as in the Chichester, Viborg, and MASS trials, and for EVAR less than 2%. The early advantage of EVAR is unlikely to result in a greater survival advantage for population screening if the "catch-up" in all-cause mortality demonstrated in the EVAR trials is sustained in contemporary practice [7]. Again, the effectiveness of a national program has been based on the available trial data, though the UK program has set an upper limit perioperative mortality of 7% as the standard for screening centers. The current UK mortality for aneurysm surgery is higher than other European countries, and there is a drive from the Vascular Society of Great Britain and Ireland to reduce this by 50%.

It should be noted that in all studies, an age range up to 75 or 80 was screened. Calculations and projections on the benefits and cost-effectiveness of a national program have been based on these data. The initial detection rate in the UK program is likely to be lower due to the lower age at screening. Self-invitation in the 65–75-year age group has been estimated to only recruit around 2% of the target population, while the trials demonstrate a 68–80% acceptance rate from formal invitation. Early figures from the early implementation centers within the UK national program suggest that screen uptake rates will be below 80%. There is also some concern about late aortic ruptures in men screened early from the MASS trial [8]. Further detailed analysis will be required to see how the decision to limit invited screening to 65-year olds, and the effect of an aging UK male population, influences the success of the national program.

12.3 Endovascular Aneurysm Repair

Since its inception in the early 1990s, endovascular repair of AAA has assumed an increasingly important role in the management of elective and ruptured AAA. It might be argued now that endovascular repair might be considered the first-line therapy for the treatment of AAA, with many units reporting that over 90% of infra-renal aneurysms are treated with this technology.

Endovascular aneurysm repair involves the exclusion of an aneurysm from the circulation using a stent graft that is delivered to the aneurysm via the femoral arteries – often using a totally percutaneous approach (Fig. 12.1 a,b). Most endovascular grafts are designed as a modular reconstruction that is assembled within the aneurysm sac. Standard endovascular aneurysm repair can only be performed if there is an adequate fixation zone between the renal arteries and the start of the aneurysm to allow adequate anchorage of the stent. Most series report that 50–70% of AAA are anatomically suitable for endovascular repair, but this percentage has increased in recent years with newer stent graft systems and the advent of fenestrated and branched endografts (Figs. 12.2 and 12.3).

The potential advantages of endovascular aneurysm repair relate to the minimally invasive nature of the procedure with both a laparotomy and aortic cross clamping being avoided. The reduction in operative severity leads to a reduction in physiological stress with cardiac, respiratory, metabolic, and renal parameters being improved in comparison to conventional surgical procedures. It was hoped that the reduction in physiological stress associated with endovascular repair would translate to a reduction in the mortality and morbidity of elective aneurysm repair.

Endovascular aneurysm repair has arguably evolved into the first-line therapy for patients with infra-renal abdominal aortic aneurysms. The evidence for this position has been derived from a number of sources. At present, there is a substantial body of evidence that suggests an endovascular first strategy is reasonable. Data from the randomized clinical trials has demonstrated that endovascular repair has a significant advantage over an open strategy with regard to operative mortality rates. Data from the randomized trials comparing endovascular with open surgery (EVAR-1 [7], DREAM [9] and OVER [10]) (Table 12.2) suggest an odds ratio for endovascular repair in the order of approximately 0.3 [11]. These data are backed up by large population-based registry figures, which suggest a similar advantage for endovascular repair over all age ranges [12]. It might be expected that as the technology improves, endovascular repair might be performed with a mortality of less than 1%, in contrast to open surgery, which will still most likely have a mortality of 3–5%.

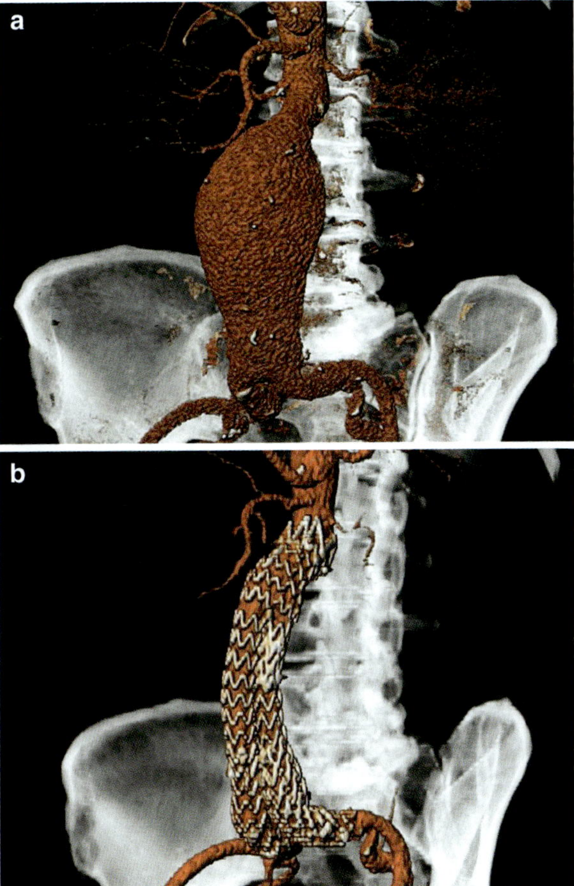

Fig. 12.1 (**a**) 3D reconstruction of an infra-renal abdominal aortic aneurysm with a good landing zone between renal arteries and start of the aneurysm. (**b**) 3D reconstruction of an infra-renal abdominal aortic aneurysm repaired using an endovascular stent graft

Fig. 12.2 3D reconstruction of a juxta-renal aneurysm that has been repaired with a fenestrated endograft. In these cases, the endograft is manufactured with fenestrations that are designed to individual patient anatomy. In the illustrated case, the graft has three fenestrations, one each for the renal arteries and the superior mesenteric artery

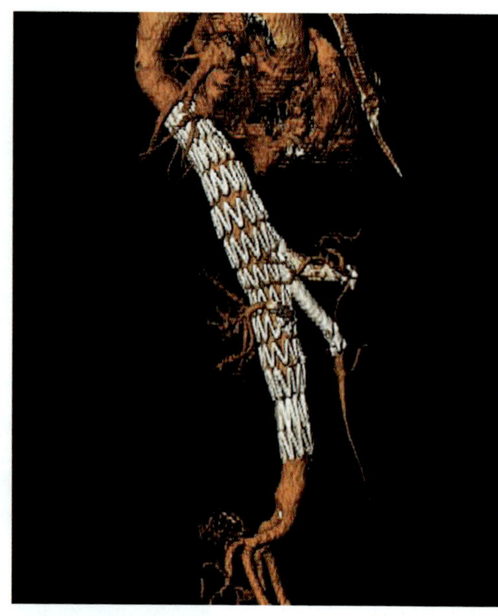

Fig. 12.3 3D reconstruction of thoracoabdominal aneurysm that has been repaired with a branched endograft. In these cases, the endograft is manufactured with branches designed to individual patient anatomy. In the illustrated case, the graft has two fenestrations for the renal arteries and two branches for the visceral vessels

Table 12.2 Mortality rates for open and endovascular surgery (EVR) in three randomized trials

	30-d mortality EVR	30-d mortality open
EVAR-1	1.7	4.7
DREAM	1.2	4.6
OVER1.7	0.2	2.3

Recently performed patient preference studies have also demonstrated that patients express a preference for endovascular repair over open procedures. Winterborne et al. [13] demonstrated that in a population of screened patients, 84% would prefer EVR. Postoperative mortality and morbidity were more important than need for surveillance or long-term problems with EVR. Similarly, Reise et al. [14] demonstrated a clear patient preference for endovascular repair in a cohort of patients given information regarding both procedures.

In the current healthcare climate, preferences for new technology need to be underpinned by a cost-effectiveness analysis. In the UK, the National Institute for Health and Clinical Excellence has recently concluded "endovascular stent grafts are recommended as a treatment option for patients with unruptured infra-renal abdominal aortic aneurysms, for whom surgical intervention is considered appropriate. The decision on whether endovascular aneurysm repair is preferred over open surgical repair should be made jointly by the patient and their clinician after assessment of a number of factors including, aneurysm size and morphology, patient age, general life expectancy and fitness for open surgery, the short and long term benefits and risks of the procedures including aneurysm related mortality and operative mortality." This decision was based on an economic analysis that estimated the incremental cost-effectiveness ratio of endovascular procedures to be in the range of £12,000.

Despite the positive data presented above, endovascular procedures are associated with some disadvantages. Most reports suggest that aneurysm-related reinterventions after endovascular procedures are significantly greater than open repair. It must also be remembered that endovascular procedures are not applicable to all aneurysms, and that the percentage of patients treated by current commercially available endografts may be as low as 40%, if the indications for use are followed.

Endografts have continued to evolve since their inception and there is evidence from several studies that the newer generation of endografts perform better than early generations [10, 15]. In designing new endografts, several features have become increasingly desirable:

- The ability to treat a higher proportion of patients with infra-renal aneurysms. In particular designs need to incorporate features to allow fixation and seal in difficult proximal neck anatomy and narrow, tortuous iliac access
- An ability to reduce intraoperative complication rate should be incorporated with less reliance on adjunctive procedures which are known to affect outcome [16, 17]
- A reduction in the postoperative intervention rate with a reduction in endoleak and limb thrombosis rates

The evolution of endovascular aneurysm repair has been rapid and now approaches the treatment of first choice for many patients with AAA. Improvements in the design and follow-up protocols remain likely over the next few years.

12.4 Centralization of Aortic Surgery

It seems a paradox that, in the modern healthcare climate, vascular professionals and commissioners continue to debate whether complex surgical interventions with high morbidity and mortality should be performed in centers of proven excellence with an adequate caseload, or whether they should remain in a greater number of more local, low-volume providers with little proof of safety. The evidence for centralization appears robust and incontrovertible, but aortic services in the UK have not been rationalized into large volume centers.

12.4.1 The Volume–Outcome Relationship for Elective Aneurysm Repair

There is a strong evidence base that suggests that mortality from elective aneurysm surgery is significantly less in centers with a high caseload than in units that perform a lower number of procedures. A meta-analysis of the existing literature [18] reviewed studies containing 421,299 elective aneurysm repairs and reported a weighted odds ratio of 0.66 in favor of higher volume centers dichotomized at 43 cases per year. This result echoes meta-analyses of most complex surgical interventions and should be regarded as definitive and highly informative.

However, although robust, meta-analyses can be criticized due to publication bias, heterogeneity, and the predominance of data from certain countries. Additional information may be gathered by analyzing national administrative data. A typical "volume–outcome" curve is illustrated in Fig. 12.4 [19] for elective aneurysm repair in the UK between 2001 and 2005. These data demonstrated that the mean mortality for an elective repair was 7.4%, and that 80% of all aneurysm repairs were carried out in units performing less than 33 cases annually (Table 12.3). Importantly, the mortality rate in the units with lowest caseload was 8.5% as compared to the 5.9% reported by units with a higher workload. Even more worrying are the many small volume centers where the elective mortality may often exceed 20% (region A in Fig. 12.4). These data provide the strongest possible inditement of the organization of vascular services.

Individual hospital performance from administrative datasets can be assessed by safety plots [20]. In a safety analysis of UK data, 30 of 410 hospitals performing elective aneurysm surgery had a mortality rate significantly above the national

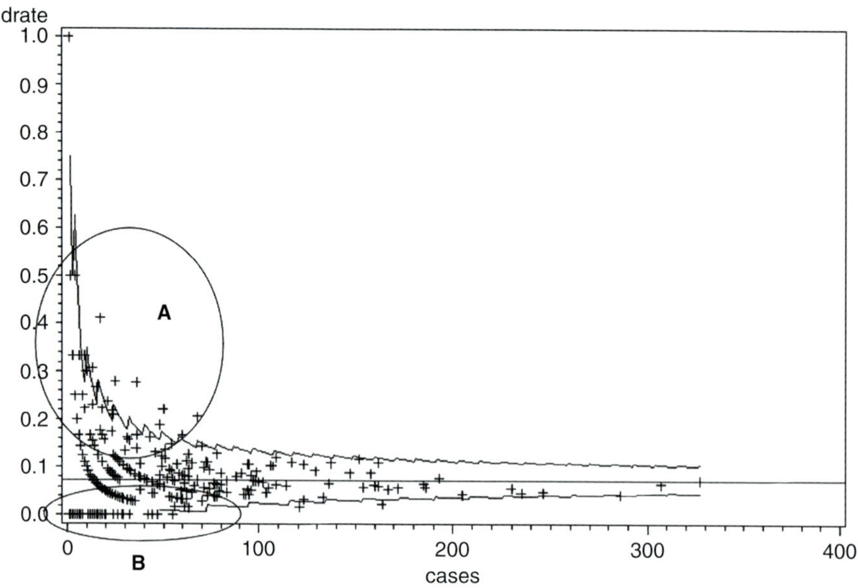

Fig. 12.4 Figure demonstrating mortality plotted against number of aneurysm repairs over a 5-year period (2000–2005)

Table 12.3 Organization of elective aneurysm services as derived from HES data for the years 2000–2005

Quintile	Quintile volume	No. of cases	No. of deaths	Mortality (%)	No. of hospitals
1	0–7.2	3,149	269	8.5	272
2	7.3–12.6	3,070	234	7.6	60
3	12.7–19.4	3,126	225	7.2	38
4	19.5–32	2,943	227	7.7	25
5	>32	3,227	190	5.9	15

average. All of these units with high mortality rates were at the low end of the volume spectrum. Additionally, to statistically demonstrate a record of safe surgery (below the national average), an annual volume of at least 39 elective cases was required with a mean national mortality of 7.4%. If the national mean mortality were to be lower (as might be expected with EVAR or different service configuration), then a greater number of cases would be needed in order to prove safety.

Data from alternative sources [21, 22] confirms that elective and ruptured aneurysm repair is performed with lower mortality rates in units with a large caseload, that services are currently inappropriately organized in a mass of small volume centers, and that units with low volumes cannot demonstrate evidence of safety.

Vascular surgery has been curiously reluctant to recognize the importance of the volume–outcome relationship, with an attendant excess mortality under current service configurations, and centralize aneurysm services. A number of theoretical objections to centralization have been raised which will be discussed below.

12.4.2 Is the Magnitude of Absolute Difference in Mortality Sufficient to Justify Centralization?

It might be argued that the 3–4% absolute mortality difference between the lowest volume and highest volume units does not justify centralization of aneurysm services. Irrespective of the absolute mortality differences in elective surgery, the mortality differences in the emergency setting are more dramatic. In a study of ruptured AAA in the UK between 2003 and 2008, the absolute mortality differences between hospitals in the lowest and highest volume quintiles reached 24% [23].

In addition, relying on operative mortality will minimize differences in outcome, as case mix and patients considered "unfit" for surgery must also be considered. In these areas, there is evidence to suggest disparate practices, with no surgical intervention being offered to over 50% of emergency patients in lowest quintile units as compared to approximately 20% in the highest volume centers [23].

12.4.3 What About Low-Volume Centers with No Mortality?

In any volume–outcome plot, there are a number of relatively low-volume units that have an elective aneurysm mortality of 0% (region B in Fig. 12.4). It is tempting to speculate that these units should not be part of any centralization due to their apparent good results. This zero mortality paradox was investigated by Dimick and Welch [24] who studied hospitals that had reported a zero mortality between 1997 and 1999. When the outcomes for these hospitals in 2000 were compared with the rest of the Medicare data, the "zero mortality" hospitals had a lower caseload (4 vs. 13) and higher mortality (6.3% vs. 5.8%). The finding of zero mortality in this study was therefore not reflective of superior results, just a function of low case volume. None of these hospitals would be able to demonstrate statistical evidence of safety.

12.4.4 Are Volume–Outcome Data Applicable to the Endovascular Era?

The majority of data investigating the effect of caseload on elective aneurysm surgery have been derived by analysis of patients undergoing open repair. Clearly, the advent of endovascular surgery will change this relationship. Two recent studies have investigated the effect of endovascular repair on the volume–outcome relationship for elective aneurysm surgery. The studies demonstrated that:

- The volume–outcome relationship was maintained for endovascular surgery, open surgery, and the combined cohort [25]. There was a significant difference between endovascular mortality between the lowest and highest quintile providers (6.88% vs. 2.88%), and a 77% reduction in mortality was observed for every 100 endovascular repairs performed.
- Higher volume hospitals were more likely to adopt endovascular therapy (44% in high-volume hospitals vs. 18% in low-volume hospitals) [21].
- Hospital volume was an independent predictor of mortality.
- Results were defined by the total aneurysm caseload rather than either endovascular or open cohorts alone, i.e., hospitals with a large, predominantly endovascular, caseload also reported better than average results from open aneurysm repair.

The data from both studies suggested that, if anything, the relationship between hospital caseload and outcome becomes even more important if endovascular technology is incorporated into the analysis.

12.4.5 Travel Times and Patient Preferences

The most important aspect defining the provision of aneurysm (or any other) services must be the acceptability to patients. There is a clear trade-off between the advantages associated with a high-volume center and the difficulties caused by prolonged travel times for both patients and relatives. In a modeling exercise, Holt et al. [26] defined the increased travel times that would be associated with a centralized model of care for aneurysm surgery in the UK. If aneurysm surgery was performed in centers with a record of demonstrable safety and a relatively low-volume threshold of 33 procedures per year, the number of hospitals performing aneurysm repair fell from 242 to 48 and travel times increased by 28 min relative to the nearest hospital.

The acceptability of increased travel times was assessed in a study of 262 patients [27]. Patients were asked to complete a questionnaire that was calibrated against the time an individual was willing to travel to access specific attributes of an aneurysm service. Approximately 92% of individuals stated a willingness to travel for at least 1 h beyond their nearest hospital in order to access services with a lower perioperative mortality, lower nonfatal complication rates, a high annual caseload of aneurysm repairs, and routine availability of endovascular repair. This study demonstrated that patients' preference to access safe, modern surgery in a high-volume center outweighed their concerns over travel.

12.4.6 Centralization Implies Poor Surgeon Performance in Low-Volume Units

Undoubtedly, discussion of centralization has been made more difficult by the feeling that stopping aneurysm surgery at an institution implies that surgeons in these centers are performing poorly. While there is a relationship between individual surgical caseload and outcome [28], it is the institutional experience which is the most important facet of delivering good quality care. The importance of the institutional component was recently emphasized by Ghaferi et al. [29] who studied 84,730 inpatients undergoing vascular or general surgery. The study reported that complication rates after surgery were not different between high- and low-volume institutions but that mortality following major complications was much higher in the low-volume units (21.4% vs. 12.5%). This study gives credence to the impression that outcomes may be defined by the institutional facilities, protocols, and familiarity with challenging management of complex interventions.

The data presented above would imply that aneurysm services should be performed in high-quality, high-volume providers with a proven record of safety. There appear to be no convincing arguments for maintaining aneurysm repair in low-volume hospitals.

Perhaps the most pertinent unresolved question is how to define high- and low-volume centers. The available literature utilizes differing thresholds according to study design with many studies merely dividing caseload data into quartiles or quintiles to demonstrate the nature of the relationship. Exact volume thresholds will differ in various healthcare systems where there is disparate organization of services. However, it is important to note that the volume–outcome relationship is continuous with improvements in outcome seen with increasing volume. Clearly a pragmatic approach to defining an appropriate threshold is mandated. It might be suggested that aneurysm repair should not be undertaken in centers performing less than 50 cases per year, and ideally the annual caseload.

12.5 Concluding Remarks

The UK has been notoriously poor in managing patients with abdominal aneurysms with mortality rates inferior to most international comparators. Initiatives from commissioning bodies to centralize services will continue the trends toward better management that have been stimulated in recent years by the adoption of modern technology and institution of screening programs.

References

1. Ashton HA et al. The Multicentre Aneurysm Screening Study (MASS) into the effect of abdominal aortic aneurysm screening on mortality in men: a randomised controlled trial. Lancet. 2002;360(9345):1531–9.
2. Scott RA et al. Influence of screening on the incidence of ruptured abdominal aortic aneurysm: 5-year results of a randomized controlled study. Br J Surg. 1995;82(8):1066–70.

3. Lindholt JS et al. Screening for abdominal aortic aneurysms: single centre randomised controlled trial. BMJ. 2005;330(7494):750.
4. Norman PE et al. Population based randomised controlled trial on impact of screening on mortality from abdominal aortic aneurysm. BMJ. 2004;329(7477):1259.
5. Scott RA, Bridgewater SG, Ashton HA. Randomized clinical trial of screening for abdominal aortic aneurysm in women. Br J Surg. 2002;89(3):283–5.
6. Marteau TM et al. Poorer self assessed health in a prospective study of men with screen detected abdominal aortic aneurysm: a predictor or a consequence of screening outcome? J Epidemiol Community Health. 2004;58(12):1042–6.
7. EVAR trial participants. Endovascular aneurysm repair versus open repair in patients with abdominal aortic aneurysm (EVAR trial 1): randomised controlled trial. Lancet. 2005;365(9478):2179–86.
8. Thompson SG, Multicentre Aneurysm Screening Study Group, et al. Screening men for abdominal aortic aneurysm: 10 year mortality and cost effectiveness results from the randomised Multicentre Aneurysm Screening Study. BMJ. 2009;338:1538–41, b2307.
9. Blankensteijn JD et al. Two-year outcomes after conventional or endovascular repair of abdominal aortic aneurysms. N Engl J Med. 2005;352(23):2398–405.
10. Lederle FA et al. Outcomes following endovascular vs open repair of abdominal aortic aneurysm: a randomized trial. JAMA. 2009;302(14):1535–42.
11. Drury D et al. Systematic review of recent evidence for the safety and efficacy of elective endovascular repair in the management of infrarenal abdominal aortic aneurysm. Br J Surg. 2005;92(8):937–46.
12. Schermerhorn ML et al. Endovascular vs. open repair of abdominal aortic aneurysms in the Medicare population. N Engl J Med. 2008;358(5):464–74.
13. Winterborn RJ et al. Preferences for endovascular (EVAR) or open surgical repair among patients with abdominal aortic aneurysms under surveillance. J Vasc Surg. 2009;49(3):576–81. e3.
14. Reise JA et al. Patient preference for surgical method of abdominal aortic aneurysm repair: postal survey. Eur J Vasc Endovasc Surg. 2010;39(1):55–61.
15. Franks SC et al. Systematic review and meta-analysis of 12 years of endovascular abdominal aortic aneurysm repair. Eur J Vasc Endovasc Surg. 2007;33(2):154–71.
16. Choke E et al. Outcomes of endovascular abdominal aortic aneurysm repair in patients with hostile neck anatomy. Cardiovasc Intervent Radiol. 2006;29(6):975–80.
17. Biasi L et al. Intra-operative Dyna CT. J Vasc Surg. 2009;49(2):288–95.
18. Holt PJ et al. Meta-analysis and systematic review of the relationship between volume and outcome in abdominal aortic aneurysm surgery. Br J Surg. 2007;94(4):395–403.
19. Holt PJ et al. Epidemiological study of the relationship between volume and outcome after abdominal aortic aneurysm surgery in the UK from 2000 to 2005. Br J Surg. 2007;94(4): 441–8.
20. Holt PJ et al. Demonstrating safety through in-hospital mortality analysis following elective abdominal aortic aneurysm repair in England. Br J Surg. 2008;95(1):64–71.
21. Dimick JB, Upchurch Jr GR. Endovascular technology, hospital volume, and mortality with abdominal aortic aneurysm surgery. J Vasc Surg. 2008;47(6):1150–4.
22. Giles KA et al. Population-based outcomes following endovascular and open repair of ruptured abdominal aortic aneurysms. J Endovasc Ther. 2009;16(5):554–64.
23. Holt PJ et al. Propensity scored analysis of outcomes after ruptured abdominal aortic aneurysm. Br J Surg. 2010;97(4):496–503.
24. Dimick JB, Welch HG. The zero mortality paradox in surgery. J Am Coll Surg. 2008;206(1): 13–6.
25. Holt PJ et al. Effect of endovascular aneurysm repair on the volume-outcome relationship in aneurysm repair. Circ Cardiovasc Qual Outcomes. 2009;2(6):624–32.
26. Holt PJ et al. Model for the reconfiguration of specialized vascular services. Br J Surg. 2008;95(12):1469–74.

27. Holt PJ et al. Screened individuals' preferences in the delivery of abdominal aortic aneurysm repair. Br J Surg. 2010;97(4):504–10.
28. Young EL et al. Meta-analysis and systematic review of the relationship between surgeon annual caseload and mortality for elective open abdominal aortic aneurysm repairs. J Vasc Surg. 2007;46(6):1287–94.
29. Ghaferi AA, Birkmeyer JD, Dimick J. Variation in hospital mortality associated with inpatient surgery. N Engl J Med. 2009;361:1368–75.

Section IV

Clinical and Translational Aspects of Pulmonary Vascular Disease

Understanding the Pathobiology of Pulmonary Vascular Disease

13

Kristin B. Highland

13.1 Introduction

Pulmonary arterial hypertension (PAH) is characterized by a progressive increase in pulmonary arterial pressure in association with variable degrees of pulmonary vascular remodeling, vasoconstriction, and in situ thrombosis. This leads to increased pulmonary vascular resistance and eventual right heart failure and death. A greater understanding of the complex pathobiology of PAH is essential for the future development of new therapeutic options. The following is a brief review of this pathobiology.

13.2 What Is Normal?

Normal pulmonary arteries have a thin media of circular muscle whose thickness is less than 5% of the diameter of the vessel [1] (Fig. 13.1). Consequently, under physiological conditions, the pulmonary circulation is characterized by low pressure and low vascular resistance. An exhaustive systemic review of the literature [2] that included data from 1887 healthy individuals enrolled in 47 studies from 13 countries revealed that the mean pulmonary artery pressure (mPAP) at rest was 14.0 ± 3.3 mmHg, and this was independent of sex and ethnicity and only slightly influenced by age (<30 years: 12 ± 3.1 mmHg, >50 years: 14.7 ± 4.0 mmHg).

Therefore, if the upper limit of normal is defined by the mean plus two times the standard deviation, then the upper limit for the mPAP at rest in healthy subjects is 20.6 mmHg; this is considerably lower than the established definition for pulmonary hypertension of >25 mmHg. This same systematic review [2] showed that the mPAP

K.B. Highland
Pulmonary Hypertension Program, Medical University of South Carolina, Charleston, SC, USA
e-mail: highlakb@musc.edu

D. Abraham et al. (eds.), *Translational Vascular Medicine*,
DOI 10.1007/978-0-85729-920-8_13, © Springer-Verlag London Limited 2012

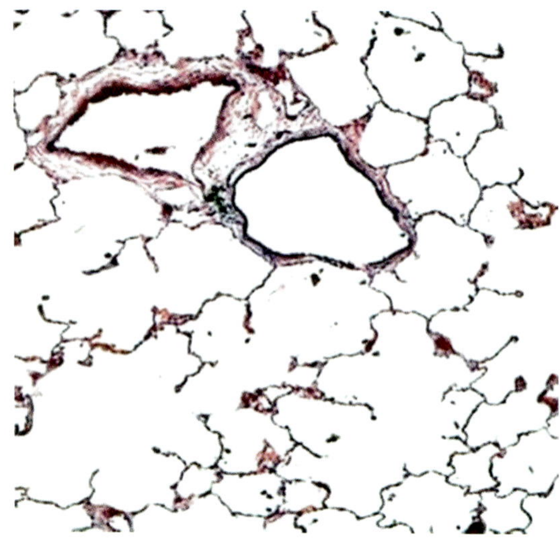

Fig. 13.1 Normal pulmonary arteriole (Van Giesen elastic stain) flanked by a normal bronchiole (the latter at the 11 o'clock position) courtesy of Ellen Reimer, M.D., J.D., Assistant Professor of Pathology and Laboratory Medicine, Medical University of South Carolina.

with exercise was dependent on age, exercise type, and exercise intensity, making it difficult to establish a threshold value that would accurately define exercise-induced pulmonary hypertension. As a result, the former exercise criterion (>30 mmHg) was abandoned during the Fourth World Symposium on Pulmonary Hypertension in Dana Point [3]. Although modestly elevated mPAPs in the setting of chronic lung diseases are often associated with a poor prognosis [4–7], the significance of a "borderline" mPAP (20–25 mmHg) in subjects that are otherwise healthy remains unclear. This highlights the importance of the clinical assessment and the need for early biomarkers as compared to a focus on hemodynamics alone, especially since these data suggest that the prevalence of individuals with an mPAP >25 mmHg will be substantially higher than the known prevalence of PAH [2, 8].

13.3 The Pathologic Lesion

The histologic findings in PAH are characterized by variable intimal hyperplasia, medial hypertrophy, adventitial proliferation, and fibrosis culminating in concentric obliterative lesions (Fig. 13.2) that occur in close proximity to plexiform lesions (Fig. 13.3). The plexiform lesion results from neo-intimal proliferation and progresses from a cellular to fibrotic lesion with advanced disease [9]. It is made up of a predominance of endothelial cells in different stages of vascular organization. The endothelial cells in a plexiform lesion express growth factors typically seen in angiogenesis (vascular endothelial growth factor and hypoxia inducible factor). Therefore, the disease state might represent an abnormal form of angiogenesis [10]. Pulmonary vascular remodeling has also been associated with in situ thrombosis and infiltration by inflammatory and progenitor cells [9, 11]. In idiopathic PAH (IPAH), these histologic abnormalities are heterogeneous in their distribution and prevalence within the

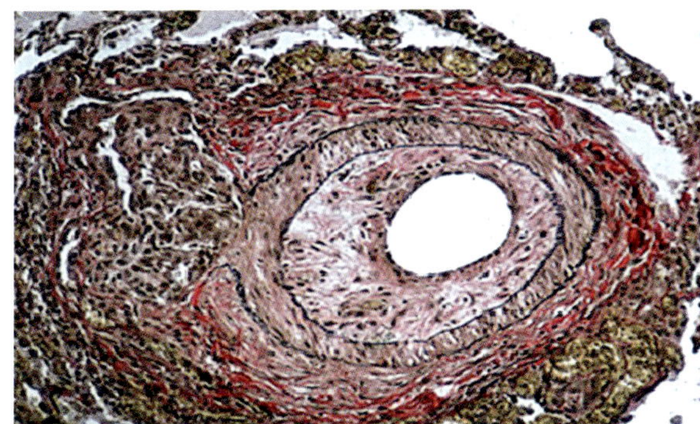

Fig. 13.2 Concentric obliterative lesion characteristic of PAH. There is intimal proliferation with encroachment on the lumen. Note a plexiform lesion to the left of the artery (at the 9 o'clock position) (Van Giesen elastic stain, 40× objective) courtesy of Russel Harley, M.D., Professor of Pathology and Laboratory Medicine, MUS and Chairman, Dept. of Pulmonary and Mediastinal Pathology, AFIP

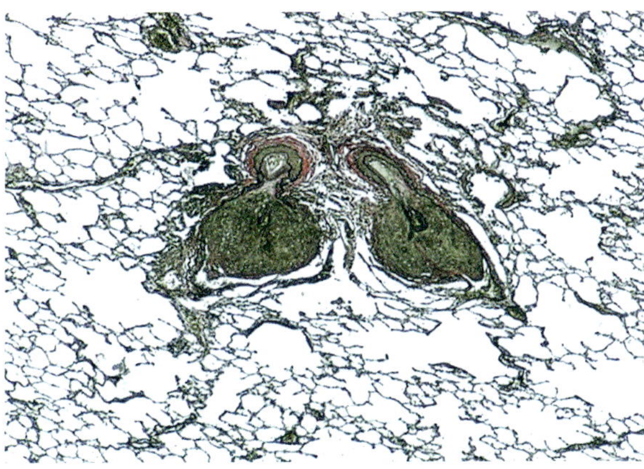

Fig. 13.3 Plexiform lesions characteristic of PAH (Van Giesen elastic stain, 10× objective) courtesy of Russel Harley, M.D., Professor of Pathology and Laboratory Medicine, MUS and Chairman, Dept. of Pulmonary and Mediastinal Pathology, AFIP

lungs and typically spare the airway, veins, bronchial circulation, capillaries, and systemic vasculature [12].

13.4 PAH is a Disease of Resistance Resulting in Right Heart Failure

As the vascular pathology progresses, the pulmonary vascular resistance (PVR) increases and pulmonary artery pressure rises in concert in order to maintain cardiac output. As long as the right ventricle is able to compensate for the resistance, the pressure continues to increase as the PVR increases. When the contractile reserve of the right ventricle (RV) is exhausted, right ventricular systolic failure ensues. A varying degree of right ventricular diastolic dysfunction is also present in pulmonary hypertension and is related to RV muscle mass and after-load and correlates with parameters of disease severity. The combination of reduced RV output and diastolic dysfunction enhances diastolic interdependence, severely impairing left ventricular filling and ultimately resulting in hemodynamic deterioration [13]. Consequently, prognostic indicators are generally related to right ventricular function and include: clinical and echocardiographic findings of right ventricular failure/dysfunction, exercise tolerance, functional class, serum concentrations of B-type natriuretic peptide, and hemodynamics (right atrial pressure, cardiac index) [14].

With longstanding PAH, the right ventricle attempts to revert to the fetal/neonatal phenotype and becomes hypertrophied (RVH) allowing for ejection against an increased pulmonary vascular resistance. For example, in RVH, phosphodiesterase type-5 (PDE-5), which was expressed in the fetal RV, is selectively reexpressed [15]. In addition, there appears to be a metabolic switch to glycolysis with increased expression of the glucose transporter type 4 (GLUT4) and increased activation of adenosine monophosphate (AMP)–activated protein kinase and pyruvate dehydrogenase kinase [16].

13.5 Classification

Pulmonary hypertension was previously classified into two categories: primary pulmonary hypertension or secondary pulmonary hypertension, depending on the absence or the presence of identifiable causes or risk factors. The diagnosis of primary pulmonary hypertension was one of exclusion after ruling out all other causes for PH. Subsequent classification schemes have attempted to create categories of PH that share pathologic and clinical features as well as similar therapeutic options. These classification schemes have allowed investigators to conduct clinical trials in well-defined groups of PAH patients with a shared underlying pathogenesis resulting in nine approved therapies. The more inclusive category of PAH has also afforded increased opportunities for treatment of some rare forms of PAH that were previously too rare for individual treatment studies. The most recent classification scheme was a product of the 4th World Symposium on PH held in 2008 in Dana Point, California [17] (Table 13.1).

Unfortunately, a limitation of these classification schemes is the fact that the many patients with PH have "multifactorial pulmonary hypertension." The clinician is thus faced with treating PH patients with a variety of clinical scenarios that often include features from more than one of the WHO groups. For example, there may be some elevation of pulmonary venous pressures, some obstructive or restrictive lung diseases, or some valvular heart disease that under usual clinical presentations would not account for PH severity. These "out of proportion" PH patients are not included in clinical trials; therefore, there is a paucity of data pertaining to the safety and efficacy of conventional PAH therapies in this population. There are also different survival curves for different types of PAH. For example, patients with congenital heart disease typically have an improved survival compared to patients with IPAH, whereas patients with connective tissue disease have a worse survival [18]. This highlights another limitation of the classification scheme. Although the different types of PAH share similar pathobiology, there are key differences that include different responses of the right ventricle (congenital heart disease), differences in the pathologic lesion (absence or reduced presence of the plexiform lesion seen in the connective tissue diseases), concomitant left

Table 13.1 Updated clinical classification of pulmonary hypertension

1. Pulmonary arterial hypertension (PAH)
1.1 Idiopathic PAH
1.2 Heritable
1.2.1 BMPR2
1.2.2 ALK1, endoglin
1.2.3 Unknown
1.3 Drug or toxin-induced
1.4 Associated with
1.4.1 Connective tissue diseases
1.4.2 HIV infection
1.4.3 Portal hypertension
1.4.4 Congenital heart diseases
1.4.5 Schistosomiasis
1.4.5 Chronic hemolytic anemia
1.5 Persistent pulmonary hypertension of the newborn
1′. Pulmonary veno-occlusive disease and/or pulmonary capillary hemangiomatosis
2. Pulmonary hypertension owing to left heart disease
3. Pulmonary hypertension owing to lung disease and/or hypoxia
4. Chronic thromboembolic pulmonary hypertension (CTEPH)
5. Pulmonary hypertension with multifactorial mechanisms

Adapted from Simonneau et al. [17]

ventricular diastolic dysfunction and/or pulmonary fibrosis (commonly seen in scleroderma), differing response to vasodilatation, hyperdynamic/high flow states (congenital heart disease, portopulmonary hypertension, and chronic hemolysis), and other comorbidities seen in all of the "associated" forms of PAH. Unfortunately, what is known about the pathobiology of PAH largely stems from research on patients with IPAH or animal models that are meant to represent IPAH.

13.6 Pathobiology

The pathobiology of PAH is thought to result from a multiple-hit hypothesis [19] involving the interaction of a predisposing state interacting with an inciting stimulus. This results in the alteration of various pathways and mediators (Table 13.2) that lead to vascular constriction, cellular proliferation, and a pro-thrombotic state, ultimately leading to the pathologic lesion of PAH and its clinical sequelae.

13.6.1 Genetics

Several genotypes have been associated with heritable PAH. These include mutations in bone morphogenetic protein receptor II (BMPR2), active-like kinase type-1 (ALK-1), and endogolin [20].

BMPR2 mutations are seen in 70–80% of patients with heritable PAH, but are relatively uncommon in patients with associated PAH [20, 21]. Fortunately, pene-

Table 13.2 Mediators and pathways in PAH

Increased activity	Decreased activity
Endothelin-1	Prostacyclin
Serotonin	Prostacyclin synthase
Thromoxane A_2	Nitric oxide
Angiopoietin-a	Nitric oxide synthase
Plasminogen activator inhibitor-1	Vasoactive intestinal peptide
Growth factors	Voltage-gated potassium channels
Oxidant stress	Fibrinolysis
Inflammation	

trance is low, and only approximately 25% of carriers will go on to develop PAH [22]. The mechanism of BMPR2 mutations is felt to be largely a result in defective SMAD signaling, which results in vascular proliferation and suppression of apoptosis.

Like BMPR2, Activin-like kinase type-1 and endogolin are also members of the transforming growth factor-beta (TGF-β) super-family and are located on endothelial cells. Mutations in ALK-1 and/or endogolin are associated with the autosomal dominant disorder hereditary hemorrhagic telangiectasia and PAH [23].

Research into epigenetic mechanisms (gene methylation) and single nucleotide polymorphisms with a current focus on the serotonin transporter (SERT) [24], voltage-gated potassium channels (Kv1.5) [25] and TRPC6 (Transient receptor potential cation channel, subfamily C, member 6) [26] is currently underway with hopes to explain other forms of heritable PAH and/or enhanced disease susceptibility.

13.6.2 Cellular Mediators and Pathways

PAH results from an imbalance that favors vasoconstriction, thrombosis, and mitogenesis; restoration of this balance by inhibition of endothelin and thromboxane or augmentation of nitric oxide and prostacyclin forms the basis of today's current therapies. Ongoing research of other mediators and pathways (Table 13.2) promises new targets for novel therapies.

13.6.2.1 Prostacyclin

Prostacyclin (PGI_2) is a product of endothelial cells as a result of the action of prostacyclin synthase on arachidonic acid. Prostacyclin relaxes smooth muscle by increasing intracellular cyclic AMP (cAMP). It is also an inhibitor of platelet aggregation and smooth muscle cell proliferation. Patients with PAH have increased excretion of urinary metabolites of thromboxane and decreased excretion of urinary metabolites of prostacyclin when compared with normal controls [27]. Likewise, there is reduced prostacyclin synthase activity in patients with PAH [28].

13.6.2.2 Endothelin

Endothelin-1 (ET-1) is synthesized and secreted by endothelial cells and is metabolized in the normal lung. It is a potent acute vasodilator and chronically stimulates cellular proliferation and fibrosis. Patients with PAH have increased plasma levels of endothelin-1 and decreased clearance when compared to normal controls; furthermore, levels of ET-1 correlate with severity of PAH and prognosis [29, 30]. Endothelin-1 immunoreactivity is increased in pulmonary arteries of all sizes in subjects with PAH [31] and acts on two different endothelin-1 receptors: ETA and ETB. Both receptors are located on vascular smooth muscle cells. ETB is also expressed on the endothelial cell. Both ETA and ETB receptors mediate vascular smooth muscle proliferation. ETA receptors also mediate vasoconstriction, whereas ETB receptors may have a role in either vasoconstriction via actions on smooth muscle receptors or vasodilation and clearance via actions on endothelial cells [31].

13.6.2.3 Nitric Oxide

Nitric oxide (NO) is a potent vasodilator that is produced by endothelial cells from arginine by nitric oxide synthase and acts on the vascular smooth muscle cells via cyclic guanosine monophosphate (cGMP). Phosphodiesterase-5 degrades cGMP, thus counteracting this vasodilatory pathway. Patients with PAH have decreased plasma levels of nitric oxide metabolites [32]; likewise, endothelial nitric oxide synthase (eNOS) expression is reduced in the pulmonary arteries [33].

13.6.2.4 Serotonin

Serotonin (5-HT) is a smooth muscle mitogen that is transported into cells primarily via serotonin transporter (SERT, 5-HTT). Elevated plasma levels of serotonin and increased SERT function [34, 35] have been observed in patients with PAH. Administration of the selective serotonin reuptake inhibitor (SSRI) fluoxetine, which inhibits SERT uptake of serotonin, results in a decrease in serotonin uptake when compared to controls indicating that the increased uptake of serotonin is through the SERT pathway [34]. The expression of the serotonin receptors is also increased in PAH and mediates vasoconstriction and vascular proliferation [35].

13.6.2.5 Ion Channels

Downregulation of the expression and activity of voltage-gated potassium channels, especially Kv1.5, is common in PAH, particularly in the resistance arteries that are the major site of pathology. Not only do these channels regulate the resting membrane potential important for controlling vascular tone, but through the regulation of intracellular potassium, these channels also affect proliferation and apoptosis and thus vascular remodeling [12]. These channels are inhibited by a number of stimuli including chronic hypoxia and dexfenfluramine, both of which have been implicated in the development of PAH [12].

13.6.2.6 Coagulation

As a result of endothelial dysfunction, abnormalities of the coagulation cascade, and disordered platelet function, a number of procoagulant alterations in patients with PAH have been identified. These include increased levels of von Willebrand factor, plasma fibrinopeptide A, plasminogen activator inhibitor-1, serotonin, and thromboxane and decreased levels of tissue plasminogen activator, thrombomodulin, NO, and PGI_2 [14].

13.6.2.7 Vasoactive Intestinal Peptide

Vasoactive intestinal peptide (VIP) is a member of the glucagon-growth hormone-releasing super-family and increases cardiac output, scavenges oxygen free radical species, inhibits platelet activation, is a potent vasodilator and inhibits the proliferation of pulmonary artery smooth muscle cells [12]. Reduced serum and lung levels of VIP associated with increased VIP receptor expression and receptor-binding affinity in pulmonary artery smooth muscle cells in patients with PAH compared with controls suggests that VIP may be an important mediator [36].

13.6.2.8 Inflammation

There is increasing evidence for the role of inflammation in the pathogenesis of PAH. This includes the presence of perivascular inflammation as well as inflammatory cells within plexiform lesions, autoantibodies to endothelial cells and fibroblasts, and raised cytokine (interleukin-1β and interleukin-6) and chemokine levels [12, 37].

13.7 In Conclusion

In conclusion, PAH is a panvasculopathy that begins in the lumen of the pulmonary artery and extends through the adventitia. The pathobiology includes excesses of vasoconstriction, thrombosis, and mitogenesis, resulting in concentric obliteration of pulmonary arteries, formation of the plexiform lesion, increased pulmonary vascular resistance, right heart failure, and death. PAH may occur as a result of genetic mutations or polymorphisms, coexisting disease, and/or environmental exposures. As a result of a better understanding of the pathobiology of PAH, and the interplay between multiple pathways, novel therapeutic targets and therapeutic strategies are under development that hopefully will lead to a cure.

Acknowledgments Figure 13.1 courtesy of Ellen Reimer, M.D., J.D., Assistant Professor of Pathology and Laboratory Medicine, Medical University of South Carolina.

Figures 13.2 and 13.3 courtesy of Russel Harley, M.D., Professor of Pathology and Laboratory Medicine, MUS and Chairman, Dept. of Pulmonary and Mediastinal Pathology, AFIP.

Disclosures Dr. Highland is on the speaker's bureau and/or has grants/contracts with Actelion Pharmaceuticals, Gilead Sciences, and Pfizer Inc.

References

1. Heath D. Longitudinal muscle in pulmonary arteries. J Pathol Bacteriol. 1963;85:407–12.
2. Kovacs G, Berghold A, Scheidl S, Olschewski H. Pulmonary arterial pressure during rest and exercise in healthy subjects: a systematic review. Eur Respir J. 2009;34:888–94.

3. Badesch DB, Championm HC, Sanchez MA, Hoeper MM, Loyd JE, Manes A, et al. Diagnosis and assessment of pulmonary arterial hypertension. J Am Coll Cardiol. 2009;54(Suppl): S55–66.
4. Behr J, Ryu JH. Pulmonary hypertension in interstitial lung disease. Eur Respir J. 2009;34: 888–94.
5. Chaouat A, Naeije R, Weitzenblum E. Pulmonary hypertension in COPD. Eur Respir J. 2008;31:1357–67.
6. Oswald-Mammosser M, Weitzenblum E, Quoix E, et al. Prognostic factors in COPD patients receiving long-term oxygen therapy. Importance of pulmonary artery pressure. Chest. 1995;107:1193–8.
7. Hamada K, Nagai S, Tanaka S, et al. Significance of pulmonary arterial pressure and diffusion capacity of the lung as prognosticator in patients with idiopathic pulmonary fibrosis. Chest. 2007;131:650–6.
8. Hoeper MM. The new definition of pulmonary hypertension. Eur Respir J. 2009;34:790–1.
9. Tuder RM. Pathology of pulmonary arterial hypertension. Semin Respir Crit Care Med. 2009;30:376–85.
10. Tuder RM, Chacon M, Alger LA, et al. Expression of angiogenesis-related molecules in plexiform lesions in severe pulmonary hypertension: evidence for a process of disordered angiogenesis. J Pathol. 2001;195:367–74.
11. Tuder RM, Groves BM, Badesch DB, Voelkel NF. Exuberant endothelial cell growth and elements of inflammation are present in plexiform lesions of pulmonary hypertension. Am J Pathol. 1994;144:275–85.
12. Archer SL, Weir K, Wilkins MR. Basic science of pulmonary arterial hypertension for clinicians. New concepts and experimental therapies. Circulation. 2010;121:2045–66.
13. Bronicki RA, Baden HP. Pathophysiology of right ventricular failure in pulmonary hypertension. Pediatr Crit Care Med. 2010;11(Suppl):S15–22.
14. McLaughlin VV, McGoon MD. Pulmonary arterial hypertension. Circulation. 2006;114: 1417–31.
15. Nagendran J, Archer SL, Soliman D, et al. Phosphodiesterase type 5 is highly expressed in the hypertrophied human right ventricle, and acute inhibition of phosphodiesterase type 5 improves contractility. Circulation. 2007;116:238–48.
16. Sharma S, Taegtmeyer H, Adrogue J, et al. Dynamic changes of gene expression in hypoxia-induced right ventricular hypertrophy. Am J Physiol Heart Circ Physiol. 2004;286: H1185–92.
17. Simonneau G, Robbins IM, Beghetti M, et al. Updated clinical classification of pulmonary hypertension. J Am Coll Cardiol. 2009;54(Suppl):S43–54.
18. McLaughlin VV, Presberg KW, Doyle RL, et al. Prognosis of pulmonary arterial hypertension: ACCP evidence based clinical practice guidelines. Chest. 2004;126(1 Suppl):78S–92.
19. Yuan JXJ, Rubin LJ. Pathogenesis of pulmonary arterial hypertension: the need for multiple hits. Circulation. 2005;111:534–8.
20. Cogan JD, Pauciulo MW, Batchman AP, et al. High frequency of BMPR2 exonic deletions/duplications in familial pulmonary arterial hypertension. Am J Respir Crit Care Med. 2006;174:590–8.
21. Deng Z, Morse JH, Slager SL, et al. Familial primary pulmonary hypertension (gene PPH1) is caused by mutations in the bone morphogenetic protein receptor-II gene. Am J Hum Genet. 2000;67:737–44.
22. Newman JH, Trembath RC, Morse JA, et al. Genetic basis of pulmonary arterial hypertension: current understanding and future directions. J Am Coll Cardiol. 2004;43:33S–9.
23. Trembath R, Thomson J, Machado R, et al. Clinical and molecular genetic features of pulmonary hypertension in patients with hereditary hemorrhagic telangiectasia. N Engl J Med. 2001;345:325–34.
24. Eddahibi S, Chaouat A, Morrell N, et al. Polymoprphism of the serotonin transporter gene and pulmonary hypertension in chronic obstructive pulmonary disease. Circulation. 2003;108:1839–44.
25. Remillard CV, Tigno DD, Platoshyn O, et al. Function of Kv1.5 channels and genetic variations of KCNA5 in patients with idiopathic pulmonary arterial hypertension. Am J Physiol Cell Physiol. 2007;292:C1837–53.

26. Yu Y, Keller SH, Remillard CV, et al. A functional single-nucleotide polymorphism in the TRPC6 gene promoter associated with idiopathic pulmonary arterial hypertension. Circulation. 2009;119:2313–22.
27. Christman BW, McPherson CD, Newman JH, et al. An imbalance between the excretion of thromboxane and prostacyclin metabolites in pulmonary hypertension. N Engl J Med. 1992;327:70–5.
28. Tuder RM, Cool CD, Geraci MW, et al. Pulmonary prostacyclin synthase is decreased in lungs from patients with severe pulmonary hypertension. Am J Respir Crit Care Med. 1999;159:1925–32.
29. Stewart DJ, Levy RD, Cernacek P, Langleben D. Increased plasma endothelin-1 in pulmonary hypertension: marker or mediator of disease? Ann Intern Med. 1991;114:464–9.
30. Rubens C, Ewert R, Halank M, et al. Big endothelin-1 and endothelin-1 plasma levels are correlated with the severity of primary pulmonary hypertension. Chest. 2001;120:1562–9.
31. Giaid A, Yanagisawa M, Langleben D, et al. Expression of endothelin-1 in the lungs of patients with pulmonary hypertension. N Engl J Med. 1993;328:1732–9.
32. Cella G, Bellotto F, Tona F, et al. Plasma markers of endothelial dysfunction in pulmonary hypertension. Chest. 2001;120:1226–30.
33. Giaid A, Saleh D. Reduced expression of endothelial nitric oxide synthase in the lungs of patients with pulmonary hypertension. N Engl J Med. 1995;333:214–21.
34. Eddahibi S, Humbert M, Fadel E, et al. Serotonin transporter overexpression is responsible for pulmonary artery smooth muscle hyperplasia in primary pulmonary hypertension. J Clin Invest. 2001;108:1141–50.
35. Morecroft I, Heeley RP, Prentice HM, et al. 5-Hdroxytryptoamine receptors mediating contraction in human small pulmonary arteries: importance of the 5-HT1b receptor. Br J Pharmacol. 1999;128:730–4.
36. Petkov V, Mosgoeller W, Ziesche R, et al. Vasoactive intestinal peptide as a new drug for treatment of primary pulmonary hypertension. J Clin Invest. 2003;111:1339–46.
37. Lesprit P, Godeau B, Authier FJ, et al. Pulmonary hypertension in POEMS syndrome: a new feature mediated by cytokines. Am J Respir Crit Care Med. 1998;157:907–11.

Inflammation in Pulmonary Arterial Hypertension

14

Frédéric Perros, Sylvia Cohen-Kaminsky, Peter Dorfmüller, Alice Huertas, Marie-Camille Chaumais, David Montani, and Marc Humbert

F. Perros (✉)
Faculté de médecine, Univ. Paris-Sud, Kremlin-Bicêtre, France

Service de Pneumologie et Réanimation Respiratoire, AP-HP, Centre National de Référence de l'Hypertension Pulmonaire Sévère, Hôpital Antoine Béclère, Clamart, France

INSERM U999, Hypertension Artérielle Pulmonaire: Physiopathologie et Innovation Thérapeutique, Le Plessis-Robinson, France

Centre Chirurgical Marie Lannelongue, Le Plessis-Robinson, France

INSERM U999, Centre Chirurgical Marie Lannelongue, Le Plessis-Robinson, France
e-mail: frederic.perros@gmail.com

S. Cohen-Kaminsky • P. Dorfmüller • A. Huertas • D. Montani • M. Humbert
Faculté de médecine, Univ. Paris-Sud, Kremlin-Bicêtre, France

Service de Pneumologie et Réanimation Respiratoire, AP-HP, Centre National de Référence de l'Hypertension Pulmonaire Sévère, Hôpital Antoine Béclère, Clamart, France

INSERM U999, Hypertension Artérielle Pulmonaire: Physiopathologie et Innovation Thérapeutique, Le Plessis-Robinson, France

Centre Chirurgical Marie Lannelongue, Le Plessis-Robinson, France

D. Abraham et al. (eds.), *Translational Vascular Medicine*,
DOI 10.1007/978-0-85729-920-8_14, © Springer-Verlag London Limited 2012

M.-C. Chaumais
Faculté de médecine, Univ. Paris-Sud,
Kremlin-Bicêtre, France

INSERM U999, Hypertension Artérielle Pulmonaire:
Physiopathologie et Innovation Thérapeutique,
Le Plessis-Robinson, France

Centre Chirurgical Marie Lannelongue,
Le Plessis-Robinson, France

Service de pharmacie, Hôpital Antoine Béclère, Assistance Publique des Hôpitaux de Paris,
Clamart, France

The pathophysiology of PAH is not fully elucidated and no curative treatment is yet available. However, the presence of inflammatory cells and the intense release of inflammatory mediators in pulmonary PAH lesions, associated with the high level of pro-inflammatory cytokines and of autoantibodies targeting vascular components in the sera of patients, raise the question of the involvement of inflammation and autoimmunity in the initiation, the perpetuation, and/or the worsening of the disease. This review covers PAH immunopathological aspects with a special emphasis on the role of inflammation on the pulmonary vascular remodeling, the potential immunopathological mechanisms of PAH, the relevance of inflammatory mediators as prognostic and predictive markers in PAH, and on the immunomodulatory properties of current PAH therapies.

14.1 Introduction

Pulmonary arterial hypertension (PAH) belongs to a heterogeneous group of progressive precapillary diseases characterized by an increase in resting mean pulmonary arterial pressure above 25 mmHg. PAH occurs as a consequence of small pulmonary arterial obstruction that leads to an impaired blood flow in the pulmonary vascular bed. The increased pulmonary vascular resistances (PVR) result in a compensatory right ventricular hypertrophy (RV), followed by right cardiac failure in the late and symptomatic phase of this severe disease [1, 2].

PAH can be idiopathic (IPAH), heritable, or associated with other diseases and drugs (connective tissue diseases, congenital heart disease, human immunodeficiency virus (HIV) infection, portal hypertension, drug-induced anorexia, etc.) [3]. Germline mutations in the bone morphogenetic protein receptor type 2 (*BMPR2*) are detected in 10–40% of IPAH and in 58–74% of heritable PAH [4]. The current PAH therapies are essentially focused on decreasing the PVR by stimulating pulmonary vasodilation (prostacyclin analogues, inhibitors of the phosphodiesterase-5, and endothelin receptor antagonists) [2]. These treatments improve disease symptoms and the quality of life in a majority of PAH patients. Nevertheless, new treatments targeting other PAH pathophysiological mechanisms would be useful to slow down the disease progression. One of the novel pathways under evaluation is represented by the tyrosine kinase inhibitors (TKI), such as imatinib and sorafenib, that have been shown to partially reverse PAH in different animal models [5, 6]. One of the beneficial effects of TKI relies on inhibition of the platelet-derived growth factor (PDGF) receptor, one of the signaling pathways linked to growth factors implicated in PAH pathophysiology [7]. Even though the use of TKI has been suggested to have beneficial effects in few clinical cases [8], it has clearly been shown that imatinib and sorafenib might induce cardiac toxicity, leading to serious safety problems in a disease characterized by underlying cardiac failure [9, 10].

Another therapeutic option would be to target PAH immunopathological component. Increasing evidences suggest that inflammatory mechanisms could play a role in human and experimental PAH genesis. Increased serum levels of pro-inflammatory cytokines and chemokines (cytokines involved in chemoattraction of leukocytes) have been measured in IPAH patients, without any underlying inflammatory, infectious, or recognized autoimmune disease by definition [11–13]. In IPAH, the pulmonary vascular lesions are sites of intense chemokine production often associated to inflammatory cells recruitment [14]. Circulating autoantibodies, in particular anti-endothelial cells and anti-fibroblasts, have been reported in 10–40% of IPAH patients [15, 16], suggesting a possible role of autoimmunity in the pathogenesis of PAH pulmonary vascular lesions. The importance of inflammatory mechanisms in PAH pathophysiology has also been highlighted by the kinetics of inflammatory patterns in standard experimental models, such as monocrotaline (MCT)-induced and hypoxia-induced PAH in rats. In these models, it has been clearly shown that inflammation precedes vascular remodeling and PAH. It has also been demonstrated, particularly in MCT-induced PAH, that immunosuppressive therapies prevent PAH development and reverse totally or partially PAH lesions [17–19]. Finally, immune mechanisms are obviously implicated in the etiology of PAH associated with autoimmune diseases or with HIV infection [14], in which PAH develops in a clear inflammatory context. In these cases, immunosuppressive or anti-inflammatory treatments significantly improve hemodynamic and clinical parameters [20, 21], highlighting the role of immune mechanisms in PAH genesis or progression.

This chapter covers PAH immunopathological aspects and their influence on pulmonary vascular remodeling.

14.2 Inflammation and Pulmonary Vascular Remodeling

The classical form of arterial inflammation, that is arteritis with fibrinoid necrosis, as described in the Heath and Edwards' classification of PAH associated with congenital heart diseases, is less frequently observed nowadays [22]. Classical arteritic PAH lesions comprise transmural inflammatory cell infiltrates with focal vessel wall necrosis and fibrinoid insudation, a histological pattern which has been etiologically linked with particularly severe forms of PAH. The histological "inflammatory mark," which is much more frequent if not common, corresponds to perivascular inflammatory infiltrates, mainly constituted of T lymphocytes, of mast cells, and of macrophages [14]. Furthermore, it has been shown that immature dendritic cells are present within the perivascular infiltrates of idiopathic PAH lungs. These dendritic cells could contribute to the immune disorders observed in PAH [23]. Whether these inflammatory infiltrates are involved in the pathobiology of PAH or whether they are only epiphenomena linked to other pathologic mechanisms leading to pulmonary vascular remodeling is still unclear. However, according to the experience gained in our national reference center which gives us access to a large collection of heart-lung samples from severe PAH, inflammatory lesions seem more often associated to "active" and cellular arterial remodeling, rather than cicatricial-like fibrotic modifications, suggesting an early role of inflammation in disease progression.

In the recent past, there has been increasing scientific evidence for inflammatory involvement in the initiation of pulmonary vascular remodeling in pulmonary hypertension. In this field, animal models have demonstrated their efficiency in dissecting the kinetics of events leading to PAH. In the rat PAH model induced by monocrotaline (MCT), a potent vegetal toxin, an early endothelial injury is followed by a marked pulmonary vascular inflammation during the first 2 weeks post-injection. Subsequently, obliterative vascular remodeling and severe PAH are present at 3 weeks post-injection, leading to right heart failure 1 week later [24]. In this model, a number of immunosuppressive and anti-inflammatory approaches have been successful in treating or preventing the development of the disease [17–19, 25]. Hypoxia-induced PAH involves perivascular inflammatory infiltrates, as well, and the importance of this infiltration is proportional to the extent of pulmonary vascular remodeling [25]. PAH also develops spontaneously in transgenic mice overexpressing specifically in the lungs the pro-inflammatory cytokine IL-6 [26]. Knockdown of genes that are crucial for the integrity of the pulmonary vasculature reciprocally leads to perivascular inflammatory infiltration [27, 28]. More generally, pulmonary vasculature is sensitive to inflammation, and remodels frequently in inflammatory conditions, even though it does not necessarily lead to a recognized PAH. For instance, experimental allergic asthma is associated with pulmonary vascular thickening without PAH [29, 30], and infection of macaques with a chimeric viral construct containing the HIV nef gene in a simian immunodeficiency virus (SIV) backbone (SHIV-nef) [31], or injection of *Schistosoma mansoni* eggs into mice [32], induces pulmonary vascular lesions similar to those described in human explanted PAH lungs. However, this vascular remodeling is not associated with hemodynamic alteration in the analyzed time-course (which could have been too short). In this context, it seems that pulmonary vascular smooth muscle cell (SMC) proliferation is a physiological response to inflammatory stimuli. Indeed, it has been shown that these cells are able to proliferate and/or migrate in vitro, in response to some pro-inflammatory cytokines/chemokines [13, 33, 34]. When pulmonary inflammation is dysregulated, one can hypothesize that vascular remodeling switches from an adaptive and asymptomatic form to an obliterative and symptomatic condition. Chronicity and loss of tolerance seem to be key elements responsible for this imbalance. Swain et al. [35] have recently highlighted this point of view, showing that infection of immunocompetent mice with *Pneumocystis pneumonia* leads to a strong pulmonary inflammation associated with a transient PAH linked to a temporary thickening of the pulmonary vasculature. Conversely, when CD4 T cells are temporally depleted in

Pneumocystis-infected mice, and then allowed to recover, the prolonged inflammation results in PAH that persists even after clearance of *Pneumocystis*. A genetic predisposition can also favor the switch from a transient and asymptomatic vascular remodeling to a fixed and symptomatic condition. Indeed, the pulmonary endothelial injury and inflammation caused by exposure to MCT combined with intratracheal instillation of replication-deficient adenovirus expressing 5-lipoxygenase (MCT + Ad5LO) has no hemodynamic effect in wild-type mice (it is known that mice are resistant to MCT-induced PAH) whereas it induced persistent PAH in heterozygous BMPR2-mutant mice [28]. Moreover, Hagen et al. [36] have demonstrated both in vitro and in vivo a complete negative feedback loop between IL-6 and BMP, suggesting that an important consequence of BMPR2 mutations may be poor regulation of cytokines and thus susceptibility to an inflammatory second hit. Hence, individual genetic predisposition associated to a switch from resolved to chronic uncontrolled inflammatory condition could result in persistent pulmonary vascular remodeling and may precipitate the occurrence of PAH.

14.3 PAH Immunopathological Mechanisms

Besides inflammation in the broader sense, fine targeted immune mechanisms are characterized by a specific response to an antigen. These mechanisms are favored by an inflammatory background and refer to adaptive immunity. The effectors of the immune response are the T and B lymphocytes which are selected in the thymus and in the bone marrow, respectively, to react against the non-self antigens, and are activated only in the presence of foreign antigens from pathogens and/or different from self-antigens. When adaptive immunity attacks the self-antigens, there is a breakdown of self-tolerance, and, as a consequence, there is the development of an autoimmune response (i.e., directed against self-antigens) that can give rise to an autoimmune disease. The self-tolerance is controlled in the periphery by a particular population of T lymphocytes called regulatory T lymphocytes (Treg), which develops in the thymus and plays a role in the pathogenesis of several inflammatory and autoimmune diseases. Tregs are involved in the feedback control of the immune response and in the return to homeostasis. They are also known to dampen autoreactive responses and may delay the onset and progression of autoimmune disorders [37, 38]. Reduced Treg cell count and/or defective suppressor function has been observed in humans, namely patients with systemic lupus erythematosus, juvenile idiopathic arthritis, autoimmune type II polyglandular syndrome, and multiple sclerosis [39–44]. Interestingly, the *BMPR2* pathway plays a role in T cell thymic development [45], which could contribute to an intrinsic defect in the function and/or number of Treg in PAH *BMPR2* mutation carriers. Little is known about the role of Treg in pulmonary diseases, particularly in PAH. Two recent studies showed a Treg increase in peripheral blood in PAH patients [46, 47]. Although these studies raise new hypothesis in PAH physiopathology, the data remain descriptive and Treg identification needs to be better defined [48]. In PAH, the Treg function has not yet been explored. It also remains to investigate the presence of these cells in patients' lungs, which

Table 14.1 Immune disorders linked to severe PAH

	Incidence of PAH	References
Autoimmune disorders		
Scleroderma	7–12%	[49]
Mixed connective tissue disease	Rare, <1%	[50]
Systemic lupus erythematosus	Rare and often multifactorial, 0.5–14%	[51]
Sjögren syndrome	<50 reported cases	[52]
Sarcoidosis	5%, Often multifactorial (interstitial pathology, mediastinitis, pulmonary arterial compression)	[53]
Polymyositis/ dermatopolymyositis	Some cases reported	[54]
Autoimmune thyroid disease	Frequent association, 30–50%	[55]
Systemic vasculitis	Rare	[56]
APECED syndrome	<10 cases described	[57]
Chronic infections		
HIV infections	0.1–0.5%	[58]
Bilharziosis	Frequent in severe hepatosplenic forms of the disease	[59]
Castleman's disease associated with HIV infection	Some cases reported	[60]

APECED syndrome: type-1 autoimmune polyendocrinopathy

represent an important inflammatory site where the self-tolerance breaks and autoimmunity can potentially take place.

The hypothesis that autoimmunity participates in PAH pathogenesis is still largely debated. According to current knowledge, it is particularly difficult to assess if such autoimmunity would be cause or effect of the disease. Nevertheless, since almost 50 years, it is recognized that severe PAH is associated with autoimmune disorders or chronic infections, leading to immunodeficiency (Table 14.1). One trait in common between these immune disorders is a confirmed or latent immunodeficiency that could lead to immune dysregulation and activation of pathogenic T and B cells. It is important to note that patients with an associated immune disorder present pulmonary lesions that cannot be discriminated from those encountered in patients with IPAH, and respond to the same treatments, indicating similar effector mechanisms. It is clear that T and B cells are present in vascular lesions [61], and that dendritic cells invade vascular lesions in both experimental PAH and human IPAH [23]. Immunoglobulin G deposits have even been detected – in and around – the plexiform lesions in patients with IPAH [45]. Hence, all the effectors of a local immune response are present around the remodeled vessels in patients with IPAH.

Little progress has been achieved in understanding how immune aggression could contribute in PAH pathogenesis. However, the search of autoantibodies in patients with IPAH, or with PAH associated with an autoimmune disease, has become a growing field of investigation. It is estimated that 30–40% of patients with IPAH present anti-nuclear antibodies, and 10–15% of these patients have anti-phospholipid antibodies [62]. The latter are able to bind to and activate endothelial cells [63]. It has been proposed that antibodies directed to vascular endothelium could promote

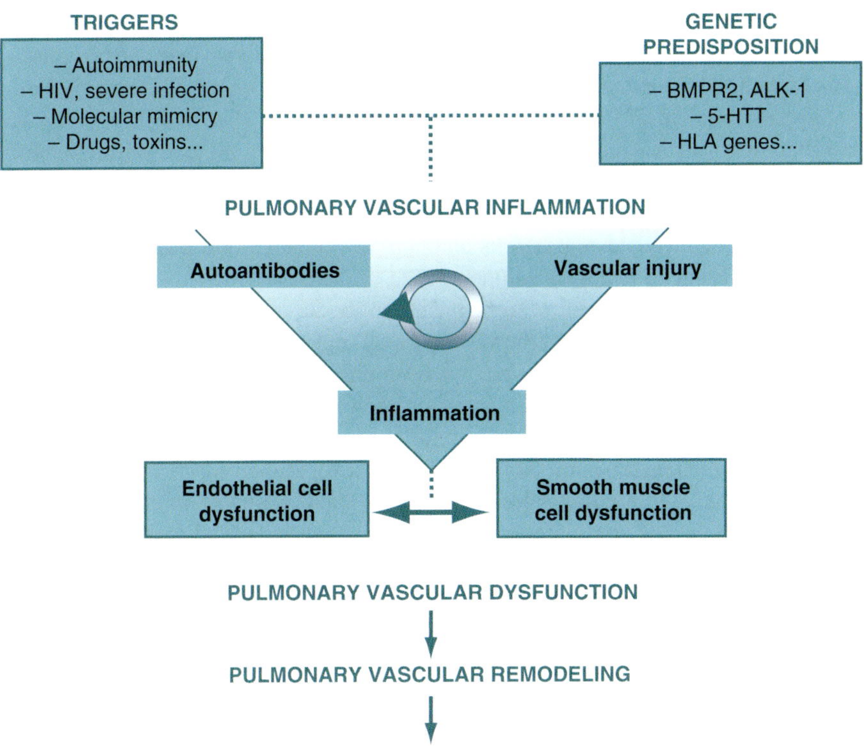

Fig. 14.1 A model of disease progression in PAH. Different triggers intervening in susceptible subjects with genetic predisposition could lead to unresolved pulmonary vascular inflammation. As a consequence, autoantibodies against vascular components are raised and can initiate and/or perpetuate a vicious circle of vascular injury, inflammation, and autoimmunity, leading to endothelial and smooth muscle cell dysfunction. Inflammation and vascular dysfunction promote pulmonary vascular remodeling, increase in the pulmonary vascular resistances and compensatory right heart hypertrophy

endothelial cell apoptosis, and that endothelium aggression could initiate a dysfunction leading to uncontrolled proliferation [45, 64, 65]. Consequently, an altered communication between endothelial cells and smooth muscle cells would lead to the development of the typical vascular lesions found in PAH, to remodeling and vascular dysfunction. Figure 14.1 integrates such an autoimmune mechanism, as central in the pathophysiology of PAH. A similar mechanism could operate in experimental models of PAH using VEGF-R antagonists, which induce early endothelial cell apoptosis, then compensated and relieved by uncontrolled endothelial cell proliferation [66]. Antibodies to endothelial cells are detected in autoimmune diseases associated to PAH, such as lupus and scleroderma [67, 68]. The prevalence of anti-endothelial cell antibodies has recently been estimated to 82% in patients with PAH associated to connective tissue disease [69]. In PAH associated to lupus and to Sjögren syndrome, antibody and complement deposits were localized to the vascular wall [70, 71]. Another pretty favorable condition to local autoimmunity is the presence of mastocytes in and around vascular lesions [72, 73] as a source of IL-4 needed for local B cell expansion and as a link between immune and adaptive immune responses, namely within the context of autoimmunity [74]. All this is on top of the pro-inflammatory environment of PAH, with increased production of IL-1 and IL-6 [11], two proinflammatory cytokines involved in activation, proliferation, and differentiation of B cells. It is worth noting that patients with IPAH or with PAH associated to connective tissue disease do not present autoantibodies directed to BMPR-II or ALK-1, indicating that a mechanism based on autoantibody attack of the BMPR-II pathway does not contribute to PAH development [75].

More recently, systematic search of autoantibodies of the IgG type, directed against different components of the vascular wall (not only endothelial cells but also smooth muscle cells and fibroblasts), has been undertaken, as well as the search of the targets of these autoantibodies using proteomic approaches, in the purpose to identify biomarkers for the diagnosis and follow-up of PAH [15, 16, 76, 77]. Among the 21 targets recognized in the fibroblasts, keys actors involved in cell biology and the maintenance of homeostasis are represented [16]. It is important to note that this approach identifies not only autoantibodies against vascular wall components, but that the differential analysis which is performed reflects in addition pathophysiological changes of the different cell types brought into play. However, among all the autoantibody targets identified, it remains to define which ones are recognized by pathogenic antibodies that would influence the vascular function and/or play a role in remodeling. It is noteworthy that the proteomic approach using bi-dimensional gels does not favor the detection of targets present at the cell membrane, which does not exclude the potential pathogenic role of autoantibodies directed against cytoplasmic or nuclear components. Such autoantibodies would emerge following initial endothelial cell aggression through a toxic compound, inducing endothelial cell apoptosis and neoantigen exposure. Yet, a recent study has confirmed the prevalence of anti-endothelial cell autoantibodies that recognize cell surface components in patients with IPAH (62% prevalence) or associated PAH (prevalence 78%) [65]. The presence of these autoantibodies may indicate the possibility of humoral mechanisms in the pathogenesis of PAH. It remains that such autoantibodies should be considered as a circumstantial observation associated to the disease, and do not constitute a formal proof of autoimmunity. A direct demonstration of pathogenicity is required in experimental models of PAH, and by serum or cell transfer from human to animals.

14.4 Interest in Inflammatory Mediators as Prognostic and Predictive Markers in PAH

Previous chapters highlighted the likely involvement of inflammation and autoimmunity in the progression of PAH, favoring and accompanying pulmonary vascular remodeling, from early stages of disease development to late stages characterized by extensive vascular obstruction and right heart failure requiring lung or heart-lung transplantation.

In clinical practice, these observations find expression in several studies demonstrating a relationship between circulating levels of some inflammatory mediators and patient survival. Quark et al. [78] recently showed that circulating CRP levels were increased in chronic thromboembolic pulmonary hypertension (CTEPH) and PAH patients compared with those in control subjects, and that CRP levels were correlated with PAH severity and patient survival. In additional support to this observation of high CRP levels that could be interpreted as a marker of hemodynamic severity and right heart failure, Soon et al. [79] showed that the circulating levels of interleukin (IL)-1β, -2, -4, -6, -8, -10, -12p70 and TNF-α were increased in IPAH patients and that levels of IL-6, -8, -10, and -12p70 allowed the prediction of patient survival – high levels of these cytokines being associated with poor prognosis, without any correlation with hemodynamic data in these patients. These observations possibly reveal independent markers of right heart function, that are potentially involved in the pathobiology of IPAH. Circulating levels of LIGHT (*Lymphotoxin-like Inducible protein that competes with Glycoprotein D for Herpesvirus entry mediator on T lymphocytes*), a chemokine implicated in vascular inflammation [80], were also associated with PAH patient mortality [81], high levels predicting poor prognosis. In this study, the prothrombotic action of LIGHT on pulmonary vascular endothelium was also highlighted, that could explain the harmful effect of LIGHT on PAH vasculature. Another chemokine, CXCL10/*Interferon gamma-induced protein 10 kDa* (IP-10), which is important for the recruitment of activated T lymphocytes, is also increased in the serum of PAH patients [82]. However, patients presenting the highest circulating levels of CXCL10 do survive better than those with lower levels. As CXCL10 is known to hold anti-angiogenic properties, its rise could be beneficial to counterbalance the aberrant endothelial growth occurring in PAH.

14.5 Immunomodulatory Properties of Current PAH Therapies

Three therapeutic classes are currently used in the treatment of PAH: prostacyclin (epoprostenol) and its analogues (treprostinil, iloprost, beraprost), endothelial receptor antagonists (ERA) (bosentan, ambrisentan), and phosphodiesterase type 5 inhibitors (iPDE5) (sildenafil, tadalafil). These treatments target endothelial dysfunction and induce vasodilation. However, they may also hold immunomodulatory properties, which could contribute to their efficacy.

Prostacyclin and its analogues act on IP_2 receptors, inducing vasodilation and inhibition of platelet aggregation. Moreover, anti-inflammatory properties of prostacyclin have recently been discovered. Iloprost reduces dendritic cell migration and recruitment, and epoprostenol prevents CD4+ Th2 cells' recruitment in animal models of asthma [83, 84]. Prostacyclin analogues inhibit also in vitro pro-inflammatory cytokines production by T lymphocytes [85] and alveolar macrophage activation stimulated by LPS through NF-kappaB [86]. Moreover, prostacyclin analogues diminish in vitro the adhesion between lymphocytes and endothelial cells, and decrease the expression of adhesion molecules and cytokines by a cAMP-dependent mechanism [87]. Circulating levels of VCAM-1 were also decreased with a beraprost treatment in human diabetes mellitus [88]. Finally, treatment with epoprostenol significantly decreases MCP-1 serum levels in PAH patients, a chemokine known to be elevated in this population [89]. Circulating neutrophils of PAH patients release much more inflammatory mediators than those of the control population, and this production is reduced by iloprost treatment [90]. Moreover, endothelial cell activation is decreased in PAH patients treated by prostacyclin associated with ERA, which supports its immunomodulatory role. Indeed, anti-inflammatory properties of prostacyclin and its analogues could constitute an additional benefit in the treatment of PAH. However, due to these anti-inflammatory properties, immunosuppression is not excluded, and increased risk of infection with prostacyclin treatment needs to be considered [91, 92].

As a potent vasoconstrictor, endothelin-1 (ET-1) also holds pro-inflammatory effects through NF-kappaB activation [93], increasing vascular permeability and activation of neutrophils [94]. Dual ET-A and ET-B receptors antagonists bosentan reduces vascular permeability in animal inflammatory models [95]. Bosentan also decreases pro-inflammatory cytokine expression in bronchoalveolar lavages through NF-kappaB inhibition [96]. Moreover, bosentan exposition of CRP-pretreated endothelial cells reduces significantly the expression of adhesion molecules and MCP-1 production [97]. A treatment with a selective antagonist to ET-A receptors, ambrisentan, decreases pro-inflammatory genes expression in ischemia/reperfusion models, leading to a cytoprotective effect on vascular and neuronal microcirculation [98, 99]. ET-1 receptor blockade leads to maturation defect and alteration of antigen-presenting capacity of dendritic cells [100]. In PAH patients, a recent study demonstrated that the reduction of ICAM-1 and of IL-6 plasmatic levels that occurred after bosentan treatment correlated with a hemodynamic improvement [101].

Immunomodulatory effects of iPDE-5 are linked to the cyclic GMP pathway. Treatment by sildenafil decreases inflammation, mucus production, and leukocyte infiltration in animal models of airway inflammation [102, 103]. Moreover, sildenafil restores antitumoral immunity through suppression of arginase 1 and NO synthase inducible expression – two enzymes required to activate immunosuppressor myeloid cells, the *myeloid-derived suppressor cells* (MDSCs) recruited by growing tumors [104]. No study on the immunomodulatory effects of iPDE-5 was reported in PAH patients. However, the potent anti-inflammatory properties of the others iPDE5 (mainly iPDE4), their antiproliferative properties, and restoration capacity of endothelial function support the use of iPDE-5 use as potential treatment for autoimmune disease [105].

The recent implication of PDGF signaling in the pathophysiology of PAH [7] focused attention to tyrosine kinase inhibitors (TKI) as a new therapeutic option in PAH management. Among these TKI, imatinib has anti-inflammatory properties, inhibiting in vitro monocyte/macrophage development from bone marrow progenitors [106], and affecting T lymphocytes and dendritic cells in their capacity to mount a cytotoxic lymphocytic response [107, 108]. Imatinib also has antitumoral properties through activation of a specific type of dendritic cells recently identified as *interferon-producing killer dendritic cells* (IKDC) [109].

In conclusion, current therapeutics of PAH act all along the inflammatory process, blocking adhesion molecules expression on endothelial cells, inhibiting the release of pro-inflammatory cytokines and chemokines, and preventing the activation of effector cells such as lymphocytes and dendritic cells.

14.6 Conclusion

This chapter brings to light a unique sensitivity of the pulmonary vascular bed to inflammatory stimuli. A deregulated and unresolved pulmonary inflammation on the top of a genetic predisposition background could conduct to a persisting vascular remodeling leading to PAH. In this context, some mediators of inflammation are correlated to the survival of patients suffering from this severe condition. Whether autoimmune manifestations are cause or worsening consequence of PAH deserves further in-depth examination. Only circumstantial data on association between the presence of autoantibodies and the disease are currently available. A long road remains to be covered in order to assess the role of autoimmunity in PAH.

Several avenues could be explored:

- A better characterization of inflammatory infiltrates in patients
- The search for deficient immunoregulation, for example, a defect in regulatory T cells
- The formal proof that a tolerance breakdown toward an autoantigen expressed by pulmonary vascular components could conduct to PAH
- The search of pathogenic autoantibodies, and the proof of their mechanism of action, with the possibility to transfer the disease from man to animal
- The discovery of novel experimental models of PAH, involving autoimmune mechanisms
- The formal proof that autoimmunity influences vascular remodeling

In this line, a multifactor appraisal of the pathogenic process in PAH, in which inflammatory mechanisms, namely autoantibodies directed to vascular wall components, play a central role could be proposed (Fig. 14.2). The pulmonary environment being tolerogenic in nature, perturbation agents could act as triggering factors in a given genetic and environmental background. An initial acute inflammation that is normally expected to resolve with return to homeostasis would conduct to the production of autoantibodies against vascular wall components, and would shift to chronic persisting and chronic inflammation, endothelial barrier breakdown, infiltration by immune cells, local and chronic autoimmunity, and vascular remodeling culminating in PAH. Identification of the factors that could trigger this irreversible process remains a major challenge to understanding the mechanisms of PAH, and to the proposal of novel therapeutic targets.

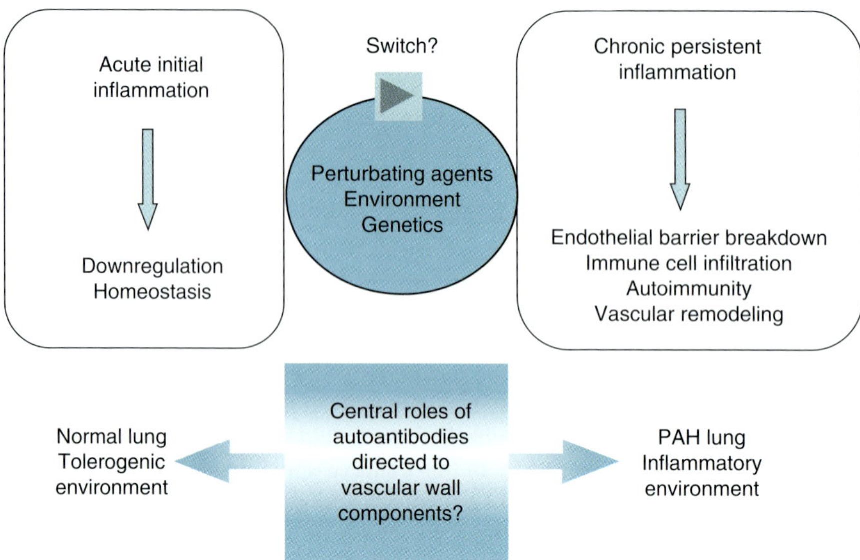

Fig. 14.2 Contribution of inflammatory mechanisms in vascular remodeling. Agents that interfere with the resolution of inflammation and return to homeostasis such as genetic predisposition and environment could favor chronic and persistent pulmonary inflammation, endothelial barrier breakdown, immune cell infiltration, breakdown of self-tolerance and autoimmunity, and vascular remodeling. Autoantibodies to vascular components may have a central role in perpetuating unresolved inflammation and a lung inflammatory environment

Supports Frederic Perros and the team from the INSERM U999 unit are supported by FRM (Fondation pour la Recherche Médicale), team FRM 2010, grant DEQ20100318257.

Peter Dorfmüller and David Montani are supported by the Association HTAPFrance.

References

1. Galie N, Hoeper MM, Humbert M, et al. Guidelines for the diagnosis and treatment of pulmonary hypertension. Eur Respir J. 2009;34(6):1219–63.
2. Humbert M, Sitbon O, Simonneau G. Treatment of pulmonary arterial hypertension. N Engl J Med. 2004;351(14):1425–36.
3. Simonneau G, Robbins IM, Beghetti M, et al. Updated clinical classification of pulmonary hypertension. J Am Coll Cardiol. 2009;54(1 Suppl):S43–54.
4. Girerd B, Montani D, Coulet F, et al. Clinical outcomes of pulmonary arterial hypertension in patients carrying an ACVRL1 (ALK1) mutation. Am J Respir Crit Care Med. 2010;181(8): 851–61.
5. Schermuly RT, Dony E, Ghofrani HA, et al. Reversal of experimental pulmonary hypertension by PDGF inhibition. J Clin Invest. 2005;115(10):2811–21.
6. Klein M, Schermuly RT, Ellinghaus P, et al. Combined tyrosine and serine/threonine kinase inhibition by sorafenib prevents progression of experimental pulmonary hypertension and myocardial remodeling. Circulation. 2008;118(20):2081–90.
7. Perros F, Montani D, Dorfmuller P, et al. Platelet-derived growth factor expression and function in idiopathic pulmonary arterial hypertension. Am J Respir Crit Care Med. 2008;178(1): 81–8.

8. Souza R, Sitbon O, Parent F, Simonneau G, Humbert M. Long term imatinib treatment in pulmonary arterial hypertension. Thorax. 2006;61(8):736.
9. Chen MH, Kerkela R, Force T. Mechanisms of cardiac dysfunction associated with tyrosine kinase inhibitor cancer therapeutics. Circulation. 2008;118(1):84–95.
10. Kerkela R, Grazette L, Yacobi R, et al. Cardiotoxicity of the cancer therapeutic agent imatinib mesylate. Nat Med. 2006;12(8):908–16.
11. Humbert M, Monti G, Brenot F, et al. Increased interleukin-1 and interleukin-6 serum concentrations in severe primary pulmonary hypertension. Am J Respir Crit Care Med. 1995;151(5): 1628–31.
12. Balabanian K, Foussat A, Dorfmuller P, et al. CX(3)C chemokine fractalkine in pulmonary arterial hypertension. Am J Respir Crit Care Med. 2002;165(10):1419–25.
13. Sanchez O, Marcos E, Perros F, et al. Role of endothelium-derived CC chemokine ligand 2 in idiopathic pulmonary arterial hypertension. Am J Respir Crit Care Med. 2007;176(10):1041–7.
14. Dorfmuller P, Perros F, Balabanian K, Humbert M. Inflammation in pulmonary arterial hypertension. Eur Respir J. 2003;22(2):358–63.
15. Tamby MC, Chanseaud Y, Humbert M, et al. Anti-endothelial cell antibodies in idiopathic and systemic sclerosis associated pulmonary arterial hypertension. Thorax. 2005;60(9):765–72.
16. Terrier B, Tamby MC, Camoin L, et al. Identification of target antigens of antifibroblast antibodies in pulmonary arterial hypertension. Am J Respir Crit Care Med. 2008;177(10):1128–34.
17. Price LC, Montani D, Tcherakian C, et al. Dexamethasone reverses monocrotaline-induced pulmonary arterial hypertension in rats. Eur Respir J. 2011;37(4):813–22.
18. Voelkel NF, Tuder RM, Bridges J, Arend WP. Interleukin-1 receptor antagonist treatment reduces pulmonary hypertension generated in rats by monocrotaline. Am J Respir Cell Mol Biol. 1994;11(6):664–75.
19. Ikeda Y, Yonemitsu Y, Kataoka C, et al. Anti-monocyte chemoattractant protein-1 gene therapy attenuates pulmonary hypertension in rats. Am J Physiol Heart Circ Physiol. 2002;283(5): H2021–8.
20. Jouve P, Humbert M, Chauveheid MP, Jais X, Papo T. POEMS syndrome-related pulmonary hypertension is steroid-responsive. Respir Med. 2007;101(2):353–5.
21. Jais X, Launay D, Yaici A, et al. Immunosuppressive therapy in lupus- and mixed connective tissue disease-associated pulmonary arterial hypertension: a retrospective analysis of twenty-three cases. Arthritis Rheum. 2008;58(2):521–31.
22. Heath D, Edwards JE. The pathology of hypertensive pulmonary vascular disease; a description of six grades of structural changes in the pulmonary arteries with special reference to congenital cardiac septal defects. Circulation. 1958;18(4 Part 1):533–47.
23. Perros F, Dorfmuller P, Souza R, et al. Dendritic cell recruitment in lesions of human and experimental pulmonary hypertension. Eur Respir J. 2007;29(3):462–8.
24. Wilson DW, Segall HJ, Pan LC, Dunston SK. Progressive inflammatory and structural changes in the pulmonary vasculature of monocrotaline-treated rats. Microvasc Res. 1989;38(1):57–80.
25. Stenmark KR, Meyrick B, Galie N, Mooi WJ, McMurtry IF. Animal models of pulmonary arterial hypertension: the hope for etiological discovery and pharmacological cure. Am J Physiol Lung Cell Mol Physiol. 2009;297(6):L1013–32.
26. Steiner MK, Syrkina OL, Kolliputi N, Mark EJ, Hales CA, Waxman AB. Interleukin-6 overexpression induces pulmonary hypertension. Circ Res. 2009;104(2):236–44, 28p following 44.
27. Hamidi SA, Prabhakar S, Said SI. Enhancement of pulmonary vascular remodelling and inflammatory genes with VIP gene deletion. Eur Respir J. 2008;31(1):135–9.
28. Song Y, Coleman L, Shi J, et al. Inflammation, endothelial injury, and persistent pulmonary hypertension in heterozygous BMPR2-mutant mice. Am J Physiol Heart Circ Physiol. 2008;295(2):H677–90.
29. Daley E, Emson C, Guignabert C, et al. Pulmonary arterial remodeling induced by a Th2 immune response. J Exp Med. 2008;205(2):361–72.
30. Medoff BD, Okamoto Y, Leyton P, et al. Adiponectin deficiency increases allergic airway inflammation and pulmonary vascular remodeling. Am J Respir Cell Mol Biol. 2009;41(4):397–406.

31. Sehgal PB, Mukhopadhyay S, Patel K, et al. Golgi dysfunction is a common feature in idiopathic human pulmonary hypertension and vascular lesions in SHIV-nef-infected macaques. Am J Physiol Lung Cell Mol Physiol. 2009;297(4):L729–37.
32. Crosby A, Jones FM, Southwood M, et al. Pulmonary vascular remodeling correlates with lung eggs and cytokines in murine schistosomiasis. Am J Respir Crit Care Med. 2010;181(3):279–88.
33. Perros F, Dorfmuller P, Souza R, et al. Fractalkine-induced smooth muscle cell proliferation in pulmonary hypertension. Eur Respir J. 2007;29(5):937–43.
34. Savale L, Tu L, Rideau D, et al. Impact of interleukin-6 on hypoxia-induced pulmonary hypertension and lung inflammation in mice. Respir Res. 2009;10:6.
35. Swain SD, Han S, Harmsen A, Shampeny K, Harmsen AG. Pulmonary hypertension can be a sequela of prior Pneumocystis pneumonia. Am J Pathol. 2007;171(3):790–9.
36. Hagen M, Fagan K, Steudel W, et al. Interaction of interleukin-6 and the BMP pathway in pulmonary smooth muscle. Am J Physiol Lung Cell Mol Physiol. 2007;292(6):L1473–9.
37. Sakaguchi S. Naturally arising CD4+ regulatory t cells for immunologic self-tolerance and negative control of immune responses. Annu Rev Immunol. 2004;22:531–62.
38. Wing K, Sakaguchi S. Regulatory T cells exert checks and balances on self tolerance and autoimmunity. Nat Immunol. 2010;11(1):7–13.
39. Baecher-Allan C, Hafler DA. Suppressor T cells in human diseases. J Exp Med. 2004;200(3):273–6.
40. Crispin JC, Martinez A, Alcocer-Varela J. Quantification of regulatory T cells in patients with systemic lupus erythematosus. J Autoimmun. 2003;21(3):273–6.
41. de Kleer IM, Wedderburn LR, Taams LS, et al. CD4+CD25 bright regulatory T cells actively regulate inflammation in the joints of patients with the remitting form of juvenile idiopathic arthritis. J Immunol. 2004;172(10):6435–43.
42. Kriegel MA, Lohmann T, Gabler C, Blank N, Kalden JR, Lorenz HM. Defective suppressor function of human CD4+ CD25+ regulatory T cells in autoimmune polyglandular syndrome type II. J Exp Med. 2004;199(9):1285–91.
43. Viglietta V, Baecher-Allan C, Weiner HL, Hafler DA. Loss of functional suppression by CD4+CD25+ regulatory T cells in patients with multiple sclerosis. J Exp Med. 2004;199(7):971–9.
44. Matarese G, Carrieri PB, La Cava A, et al. Leptin increase in multiple sclerosis associates with reduced number of CD4(+)CD25+ regulatory T cells. Proc Natl Acad Sci USA. 2005;102(14):5150–5.
45. Nicolls MR, Taraseviciene-Stewart L, Rai PR, Badesch DB, Voelkel NF. Autoimmunity and pulmonary hypertension: a perspective. Eur Respir J. 2005;26(6):1110–8.
46. Ulrich S, Nicolls MR, Taraseviciene L, Speich R, Voelkel N. Increased regulatory and decreased CD8+ cytotoxic T cells in the blood of patients with idiopathic pulmonary arterial hypertension. Respiration. 2008;75(3):272–80.
47. Austin ED, Rock MT, Mosse CA, et al. T lymphocyte subset abnormalities in the blood and lung in pulmonary arterial hypertension. Respir Med. 2010;104(3):454–62.
48. Perros F, Cohen-Kaminsky S, Humbert M. Understanding the role of CD4+CD25 (high) (so-called regulatory) T cells in idiopathic pulmonary arterial hypertension. Respiration. 2008;75(3):253–6.
49. Hachulla E, Gressin V, Guillevin L, et al. Early detection of pulmonary arterial hypertension in systemic sclerosis: a French nationwide prospective multicenter study. Arthritis Rheum. 2005;52(12):3792–800.
50. Haroon N, Nisha RS, Chandran V, Bharadwaj A. Pulmonary hypertension not a major feature of early mixed connective tissue disease: a prospective clinicoserological study. J Postgrad Med. 2005;51(2):104–7. discussion 7–8.
51. Fois E, Le Guern V, Dupuy A, Humbert M, Mouthon L, Guillevin L. Noninvasive assessment of systolic pulmonary artery pressure in systemic lupus erythematosus: retrospective analysis of 93 patients. Clin Exp Rheumatol. 2010;28(6):836–41.

52. Launay D, Hachulla E, Hatron PY, Jais X, Simonneau G, Humbert M. Pulmonary arterial hypertension: a rare complication of primary Sjogren syndrome: report of 9 new cases and review of the literature. Medicine (Baltimore). 2007;86(5):299–315.
53. Nunes H, Humbert M, Capron F, et al. Pulmonary hypertension associated with sarcoidosis: mechanisms, haemodynamics and prognosis. Thorax. 2006;61(1):68–74.
54. Minai OA. Pulmonary hypertension in polymyositis-dermatomyositis: clinical and hemodynamic characteristics and response to vasoactive therapy. Lupus. 2009;18(11):1006–10.
55. Chu JW, Kao PN, Faul JL, Doyle RL. High prevalence of autoimmune thyroid disease in pulmonary arterial hypertension. Chest. 2002;122(5):1668–73.
56. Launay D, Souza R, Guillevin L, et al. Pulmonary arterial hypertension in ANCA-associated vasculitis. Sarcoidosis Vasc Diffuse Lung Dis. 2006;23(3):223–8.
57. Garcia-Hernandez FJ, Ocana-Medina C, Gonzalez-Leon R, Garrido-Rasco R, Sanchez-Roman J. Autoimmune polyglandular syndrome and pulmonary arterial hypertension. Eur Respir J. 2006;27(3):657–8.
58. Sitbon O, Lascoux-Combe C, Delfraissy JF, et al. Prevalence of HIV-related pulmonary arterial hypertension in the current antiretroviral therapy era. Am J Respir Crit Care Med. 2008;177(1):108–13.
59. Lapa M, Dias B, Jardim C, et al. Cardiopulmonary manifestations of hepatosplenic schistosomiasis. Circulation. 2009;119(11):1518–23.
60. Montani D, Achouh L, Marcelin AG, et al. Reversibility of pulmonary arterial hypertension in HIV/HHV8-associated Castleman's disease. Eur Respir J. 2005;26(5):969–72.
61. Tuder RM, Groves B, Badesch DB, Voelkel NF. Exuberant endothelial cell growth and elements of inflammation are present in plexiform lesions of pulmonary hypertension. Am J Pathol. 1994;144(2):275–85.
62. Karmochkine M, Cacoub P, Dorent R, et al. High prevalence of antiphospholipid antibodies in precapillary pulmonary hypertension. J Rheumatol. 1996;23(2):286–90.
63. Riboldi P, Gerosa M, Raschi E, Testoni C, Meroni PL. Endothelium as a target for antiphospholipid antibodies. Immunobiology. 2003;207(1):29–36.
64. Mouthon L, Guillevin L, Humbert M. Pulmonary arterial hypertension: an autoimmune disease? Eur Respir J. 2005;26(6):986–8.
65. Arends SJ, Damoiseaux J, Duijvestijn A, et al. Prevalence of anti-endothelial cell antibodies in idiopathic pulmonary arterial hypertension. Eur Respir J. 2010;35(4):923–5.
66. Taraseviciene-Stewart L, Kasahara Y, Alger L, et al. Inhibition of the VEGF receptor 2 combined with chronic hypoxia causes cell death-dependent pulmonary endothelial cell proliferation and severe pulmonary hypertension. FASEB J. 2001;15(2):427–38.
67. Renaudineau Y, Dugue C, Dueymes M, Youinou P. Antiendothelial cell antibodies in systemic lupus erythematosus. Autoimmun Rev. 2002;1(6):365–72.
68. Negi VS, Tripathy NK, Misra R, Nityanand S. Antiendothelial cell antibodies in scleroderma correlate with severe digital ischemia and pulmonary arterial hypertension. J Rheumatol. 1998;25(3):462–6.
69. Li MT, Ai J, Tian Z, et al. Prevalence of anti-endothelial cell antibodies in patients with pulmonary arterial hypertension associated with connective tissue diseases. Chin Med Sci J. 2010;25(1):27–31.
70. Quismorio Jr FP, Sharma O, Koss M, et al. Immunopathologic and clinical studies in pulmonary hypertension associated with systemic lupus erythematosus. Semin Arthritis Rheum. 1984;13(4):349–59.
71. Nakagawa N, Osanai S, Ide H, et al. Severe pulmonary hypertension associated with primary Sjogren's syndrome. Intern Med. 2003;42(12):1248–52.
72. Heath D, Yacoub M. Lung mast cells in plexogenic pulmonary arteriopathy. J Clin Pathol. 1991;44(12):1003–6.
73. Tucker A, McMurtry IF, Alexander AF, Reeves JT, Grover RF. Lung mast cell density and distribution in chronically hypoxic animals. J Appl Physiol. 1977;42(2):174–8.
74. Benoist C, Mathis D. Mast cells in autoimmune disease. Nature. 2002;420(6917):875–8.

75. Satoh T, Kimura K, Okano Y, Hirakata M, Kawakami Y, Kuwana M. Lack of circulating autoantibodies to bone morphogenetic protein receptor-II or activin receptor-like kinase 1 in mixed connective tissue disease patients with pulmonary arterial hypertension. Rheumatology (Oxford). 2005;44(2):192–6.
76. Tamby MC, Humbert M, Guilpain P, et al. Antibodies to fibroblasts in idiopathic and scleroderma-associated pulmonary hypertension. Eur Respir J. 2006;28(4):799–807.
77. Terrier B, Tamby MC, Camoin L, et al. Antifibroblast antibodies from systemic sclerosis patients bind to {alpha}-enolase and are associated with interstitial lung disease. Ann Rheum Dis. 2010;69(2):428–33.
78. Quarck R, Nawrot T, Meyns B, Delcroix M. C-reactive protein: a new predictor of adverse outcome in pulmonary arterial hypertension. J Am Coll Cardiol. 2009;53(14):1211–8.
79. Soon E, Holmes AM, Treacy CM, et al. Elevated levels of inflammatory cytokines predict survival in idiopathic and familial pulmonary arterial hypertension. Circulation. 2010;122(9):920–7.
80. Otterdal K, Smith C, Oie E, et al. Platelet-derived LIGHT induces inflammatory responses in endothelial cells and monocytes. Blood. 2006;108(3):928–35.
81. Otterdal K, Andreassen AK, Yndestad A, et al. Raised LIGHT levels in pulmonary arterial hypertension: potential role in thrombus formation. Am J Respir Crit Care Med. 2008;177(2): 202–7.
82. Heresi GA, Aytekin M, Newman J, Dweik RA. CXC-chemokine ligand 10 in idiopathic pulmonary arterial hypertension: marker of improved survival. Lung. 2010;188(3):191–7.
83. Idzko M, Hammad H, van Nimwegen M, et al. Inhaled iloprost suppresses the cardinal features of asthma via inhibition of airway dendritic cell function. J Clin Invest. 2007;117(2):464–72.
84. Jaffar Z, Ferrini ME, Buford MC, Fitzgerald GA, Roberts K. Prostaglandin I2-IP signaling blocks allergic pulmonary inflammation by preventing recruitment of CD4+ Th2 cells into the airways in a mouse model of asthma. J Immunol. 2007;179(9):6193–203.
85. Zhou W, Hashimoto K, Goleniewska K, et al. Prostaglandin I2 analogs inhibit proinflammatory cytokine production and T cell stimulatory function of dendritic cells. J Immunol. 2007;178(2):702–10.
86. Raychaudhuri B, Malur A, Bonfield TL, et al. The prostacyclin analogue treprostinil blocks NFkappaB nuclear translocation in human alveolar macrophages. J Biol Chem. 2002;277(36): 33344–8.
87. Zardi EM, Zardi DM, Cacciapaglia F, et al. Endothelial dysfunction and activation as an expression of disease: role of prostacyclin analogs. Int Immunopharmacol. 2005;5(3): 437–59.
88. Goya K, Otsuki M, Xu X, Kasayama S. Effects of the prostaglandin I2 analogue, beraprost sodium, on vascular cell adhesion molecule-1 expression in human vascular endothelial cells and circulating vascular cell adhesion molecule-1 level in patients with type 2 diabetes mellitus. Metabolism. 2003;52(2):192–8.
89. Katsushi H, Kazufumi N, Hideki F, et al. Epoprostenol therapy decreases elevated circulating levels of monocyte chemoattractant protein-1 in patients with primary pulmonary hypertension. Circ J. 2004;68(3):227–31.
90. Rose F, Hattar K, Gakisch S, et al. Increased neutrophil mediator release in patients with pulmonary hypertension–suppression by inhaled iloprost. Thromb Haemost. 2003;90(6):1141–9.
91. Oudiz RJ, Farber HW. Dosing considerations in the use of intravenous prostanoids in pulmonary arterial hypertension: an experience-based review. Am Heart J. 2009;157(4):625–35.
92. Aronoff DM, Peres CM, Serezani CH, et al. Synthetic prostacyclin analogs differentially regulate macrophage function via distinct analog-receptor binding specificities. J Immunol. 2007;178(3):1628–34.
93. Browatzki M, Schmidt J, Kubler W, Kranzhofer R. Endothelin-1 induces interleukin-6 release via activation of the transcription factor NF-kappaB in human vascular smooth muscle cells. Basic Res Cardiol. 2000;95(2):98–105.
94. Helset E, Lindal S, Olsen R, Myklebust R, Jorgensen L. Endothelin-1 causes sequential trapping of platelets and neutrophils in pulmonary microcirculation in rats. Am J Physiol. 1996;271(4 Pt 1):L538–46.

95. Finsnes F, Skjonsberg OH, Tonnessen T, Naess O, Lyberg T, Christensen G. Endothelin production and effects of endothelin antagonism during experimental airway inflammation. Am J Respir Crit Care Med. 1997;155(4):1404–12.
96. Finsnes F, Lyberg T, Christensen G, Skjonsberg OH. Effect of endothelin antagonism on the production of cytokines in eosinophilic airway inflammation. Am J Physiol Lung Cell Mol Physiol. 2001;280(4):L659–65.
97. Verma S, Li SH, Badiwala MV, et al. Endothelin antagonism and interleukin-6 inhibition attenuate the proatherogenic effects of C-reactive protein. Circulation. 2002;105(16):1890–6.
98. Uhlmann D, Gabel G, Ludwig S, et al. Effects of ET(A) receptor antagonism on proinflammatory gene expression and microcirculation following hepatic ischemia/reperfusion. Microcirculation. 2005;12(5):405–19.
99. Hauck EF, Hoffmann JF, Heimann A, Kempski O. EndothelinA receptor antagonist BSF-208075 causes immune modulation and neuroprotection after stroke in gerbils. Brain Res. 2007;1157:138–45.
100. Guruli G, Pflug BR, Pecher S, Makarenkova V, Shurin MR, Nelson JB. Function and survival of dendritic cells depend on endothelin-1 and endothelin receptor autocrine loops. Blood. 2004;104(7):2107–15.
101. Karavolias GK, Georgiadou P, Gkouziouta A, et al. Short and long term anti-inflammatory effects of bosentan therapy in patients with pulmonary arterial hypertension: relation to clinical and hemodynamic responses. Expert Opin Ther Targets. 2010;14(12):1283–9.
102. Wang T, Liu Y, Chen L, et al. Effect of sildenafil on acrolein-induced airway inflammation and mucus production in rats. Eur Respir J. 2009;33(5):1122–32.
103. Toward TJ, Smith N, Broadley KJ. Effect of phosphodiesterase-5 inhibitor, sildenafil (Viagra), in animal models of airways disease. Am J Respir Crit Care Med. 2004;169(2):227–34.
104. Serafini P, Meckel K, Kelso M, et al. Phosphodiesterase-5 inhibition augments endogenous antitumor immunity by reducing myeloid-derived suppressor cell function. J Exp Med. 2006;203(12):2691–702.
105. Shenoy P, Agarwal V. Phosphodiesterase inhibitors in the management of autoimmune disease. Autoimmun Rev. 2010;9(7):511–5.
106. Dewar AL, Domaschenz RM, Doherty KV, Hughes TP, Lyons AB. Imatinib inhibits the in vitro development of the monocyte/macrophage lineage from normal human bone marrow progenitors. Leukemia. 2003;17(9):1713–21.
107. Seggewiss R, Price DA, Purbhoo MA. Immunomodulatory effects of imatinib and second-generation tyrosine kinase inhibitors on T cells and dendritic cells: an update. Cytotherapy. 2008;10(6):633–41.
108. Appel S, Boehmler AM, Grunebach F, et al. Imatinib mesylate affects the development and function of dendritic cells generated from CD34+ peripheral blood progenitor cells. Blood. 2004;103(2):538–44.
109. Taieb J, Chaput N, Menard C, et al. A novel dendritic cell subset involved in tumor immunosurveillance. Nat Med. 2006;12(2):214–9.

Endothelin Receptor Antagonists in Cardiovascular Medicine: Challenges and Opportunities

15

Matthias Barton

15.1 Endothelin: An Endothelium-Derived Vasoconstrictor

In 1980, Robert Furchgott made the seminal observation that endothelial cells modulate vascular tone by releasing a vasodilator factor [1], which was later identified as nitric oxide [2]. Only 1 year later, de Mey and Vanhoutte first reported endothelium-dependent vasoconstriction [3–5]. By the mid-1980s, several investigators independently reported a peptidergic vasoconstrictor activity released from cultured endothelial cells [3, 6–9]. After a combined effort of several Japanese groups led by Tomoh Masaki [10, 11], sequences of the gene and the peptide encoding for endothelin were published in *Nature* in March 1988 [12]. Today, endothelin still represents the most potent and long-lasting vasoconstrictor known in humans [13, 14], being 100-times more potent than noradrenaline [15, 16].

15.2 Molecular Biology and Biological Functions of Endothelin

15.2.1 The Endothelin Peptide Family

Endothelin-1 (ET-1) is a 21-amino acid peptide with a hydrophobic C terminus and two cysteine bridges at the N terminus and the main member of the endothelin peptide family [3, 17]. Two structurally related peptides differing by two and six amino acids were identified and termed endothelin-2 (ET-2) and endothelin-3 (ET-3), respectively; they were identified shortly after the discovery of ET-1 [17].

ET-1 is produced by vascular endothelial [18] and smooth muscle cells, airway epithelial cells, macrophages, fibroblasts, cardiac myocytes, brain neurons, pancreatic islets,

M. Barton
Molecular Internal Medicine, LTK Y44 G22, University of Zurich,
Zürich, Switzerland
e-mail: barton@access.uzh.ch

D. Abraham et al. (eds.), *Translational Vascular Medicine*,
DOI 10.1007/978-0-85729-920-8_15, © Springer-Verlag London Limited 2012

and also by other cells (reviewed in [15, 19]). Endothelial cell–specific overexpression of ET-1 in vascular endothelial cells causes hypertension-associated changes in the vasculature, including hypertrophy and inflammation [20–22]. In contrast, endothelial cell–specific deletion of the preproendothelin gene is associated with hypotension [18, 22]; this blood pressure–lowering effect of gene deletion is similar to that of vascular smooth muscle–specific deletion of the ET_A receptor [18]. In endothelial cell–specific ppET-1 null mice, plasma levels of ET-1 are reduced by about 90%, indicating that endothelial cells are indeed the major source of circulating ET-1 [18]. Moreover, ET-1 tissue levels in organs such as the heart and the lung markedly reduced in these animals consistent with the notion that endothelial cells in these organs largely contribute to endothelin-1 production in tissue [18].

ET-2 is expressed in the ovary and in intestinal epithelial cells, and, among other functions, is involved in the regulation of lung alveolarization, thermoregulation, ovulation, and intestinal epithelial cell homeostasis, and thus possibly for inflammatory bowel disease [23–26].

ET-3 is found in endothelial cells, brain neurons, renal tubular epithelial cells, and intestinal epithelial cells and mediates release of the vasodilators NO and prostacyclin, among others [19].

15.2.2 Endothelin-Converting Enzymes

The endothelin-1 precursors are processed by two proteases to create the mature active forms [19, 27] (Fig. 15.1). The 212-residue preproendothelins are cleaved at dibasic sites by furin-like endopeptidase to form biologically inactive intermediates, namely 37- to 41-amino acid peptides termed pro- or big endothelins (big ETs) (Fig. 15.1). Processing is mediated by a family of membrane-bound zinc metalloproteases from the neprilysin superfamily, termed endothelin-converting enzymes (ECEs) (reviewed in [19]). Depending on the cleaving enzyme (Fig. 15.1), 21-, 31-, or 32-amino acid isoforms with specific receptor activities are formed. In addition to these proteases, other enzymes such as vascular chymase [28–30] and non-ECE metalloproteinase [31] must contribute to the final processing step, since in mice lacking both ECE-1 and ECE-2, the levels of mature endothelin peptides are reduced by only one-third [32]. Recent studies suggest that carboxypeptidase A (cathepsin A) plays an important role in degradation of the ECE product endothelin-1 [33]. The role of ET-1 degradation in physiology and disease, however, remains yet to be studied.

15.2.3 Endothelin Receptors

In humans, two seven-transmembrane domain, G protein–coupled endothelin receptors (ET_A and ET_B) mediate the cellular activities of endothelins (reviewed in [34]) (Fig. 15.1). It is currently not clear whether receptor dimerization into homo- or heterodimers [35] plays a role of endothelin receptor activity and function in vivo such as endothelin effects on diuresis [23, 36], or whether receptor dimerization is

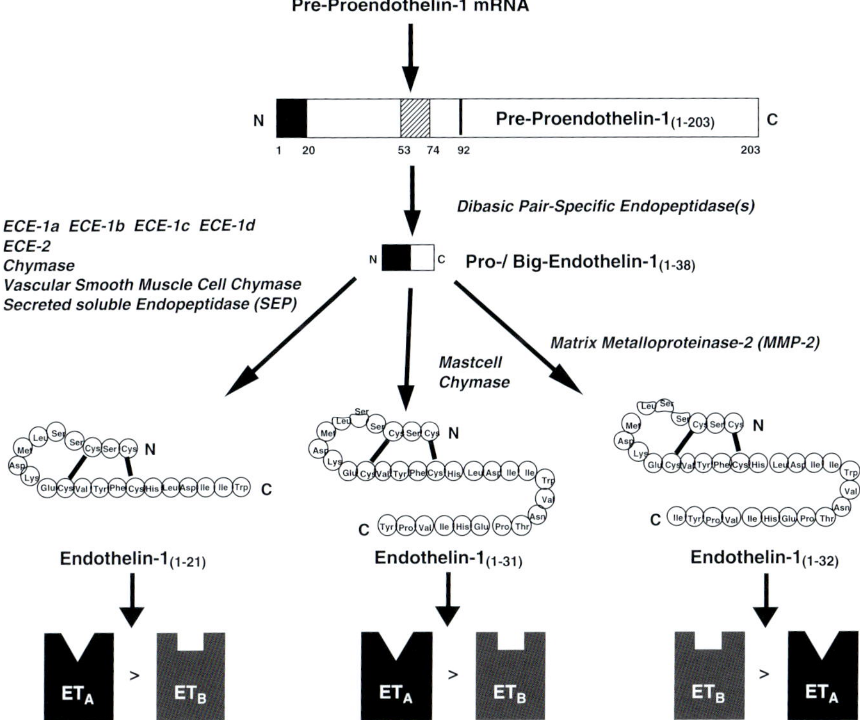

Fig. 15.1 Biosynthetic pathways of endothelin peptides ET-1$_{(1-21)}$, ET$_{(1-31)}$, and ET$_{(1-32)}$ peptides following transcription of preproendothelin-1 mRNA. Enzymes are in *italics*. Prepro-ET mRNA is translated into preproET-1, a 203-amino acid peptide, which is cleaved by furin convertase to the 38-amino acid precursor, big ET-1$_{(1-38)}$. Big ET-1 is processed to ET-1$_{(1-21)}$ by endothelin-converting enzymes (ECEs), mast cell and smooth muscle cell chymases, and a novel non-ECE metalloprotease (secreted soluble endopeptidase, SEP) (*left*). Two novel pathways involve mast cell chymase, yielding the 31-amino acid ET-1$_{(1-31)}$ (*middle*), and matrix metalloproteinase-2 (MMP-2), forming another vasoconstrictor peptide, ET-1$_{(1-32)}$ (*right*)

affected by drug treatment. The ET$_A$ receptor shows sub-nanomolar affinities for ET-1 and ET-2 and 100-fold lower affinity for ET-3 [19]. ET$_B$ has equal sub-nanomolar affinities for all endothelin peptides. The ET$_A$ receptor mediates constriction and growth, whereas the ET$_B$ receptor inhibits cell growth and vascular tone; the ET$_B$ receptor also functions as a "clearance receptor," as indicated by the ET$_B$-dependent inhibition of the accumulation of intravenously administered, radiolabeled ET-1 in tissue [15, 19]. This ET$_B$ receptor–mediated clearance mechanism is particularly important in the lung, which clears about 80% of circulating ET-1 [15]. The ET$_A$ receptor can be considered the principal vasoconstrictor and growth-promoting receptor, whereas the endothelial ET$_B$ receptor generally inhibits cell growth and vasoconstriction in the vascular system, with only few vascular beds expressing contraction-mediating ET$_B$ receptors in the media [15, 27]. The ET$_B$ receptor is important for embryonic development, and its deficiency is associated with a phenotype consistent with Hirschsprung's disease, featuring aganglionosis and megacolon development [37–39]. The endothelial ET_B receptor also functions as a "clearance receptor," because ET_B-selective antagonists inhibit the accumulation of intravenously administered, radio-labeled ET-1 in tissue [15, 19, 40]. This ET_B receptor–mediated clearance mechanism is particularly important in the lung which clears about 80% of circulating ET-1 [41], which however is affected by chronic endothelin receptor antagonist (ERA) treatment distributing clearance to other organs such as the liver [40]. In animals lacking ET_A receptors specifically in vascular smooth muscle, a compensatory upregulation of vasoconstrictor ET_B receptors occurs [18]. In vast majority of tissues, the disease-promoting effects of ET-1 are mediated by activation of the ET_A receptor, and include activity such as inflammation, excessive cell proliferation, contraction, ROS formation, and coagulatory activation (Fig. 15.2) [27, 43, 44].

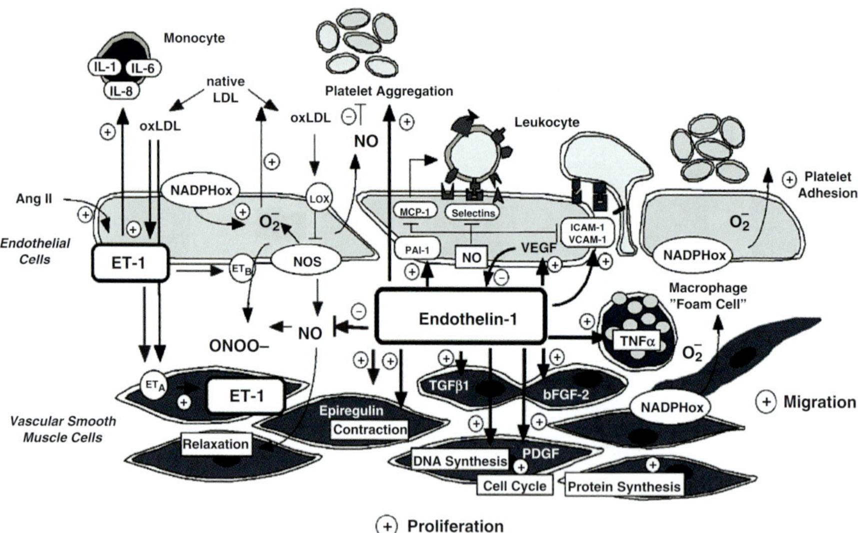

Fig. 15.2 Effects of endothelin (ET-1) in vascular endothelial and smooth muscle cells. ET-1 is generated by endothelial and smooth muscle cells in response to lipoproteins, Ang II, and inflammatory stimuli. Activation of endothelial ET_B receptors increases the release of nitric oxide (NO), whereas ET_A receptors mediate cell proliferation, migration, and contraction. Endothelin induces expression of TNF-α and interleukins in monocytes and vascular adhesion molecule expression, and stimulates leukocyte adherence and platelet aggregation. Endothelin also enhances production and activity of other growth factors, and promotes DNA and protein synthesis and progression of the cell cycle. Abbreviations used in figure: *Ang II* angiotensin II, *ONOO*⁻ peroxynitrite, *ET-1* endothelin-1, ET_A endothelin ET_A receptor, ET_B endothelin ET_B receptor, *NO* nitric oxide, *NOS* nitric oxide synthase, *MCP-1* monocyte chemoattractant protein-1, *ICAM-1* intracellular adhesion molecule-1, *VCAM-1* vascular cell adhesion molecule-1, *LDL* low-density lipoprotein, *oxLDL* oxidized low-density lipoproteins, O_{2-} superoxide anion; *LOX* oxidized LDL receptor, *IL-1* interleukin-1, *IL-6* interleukin-6, *IL-8* interleukin-8, *TNF-α* tumor necrosis factor-α, *TGF-β1* transforming growth factor-β1, *phox* NADPH oxidase, (+) stimulation, (–) inhibition (Reproduced from [42], with permission of the publisher and the American Heart Association)

15.2.4 Biological Functions of Endothelin

Endothelins (ETs), of which ET-1 represents the predominant and biologically most relevant isoform [15, 19], can be considered ubiquitously expressed stress-responsive regulators working in a paracrine and autocrine fashion, with both beneficial and detrimental effects [19]. Endothelins exert a number functions during embryonic development and physiology [45], including neural crest cell development and neurotransmission (reviewed in [19]). In the vascular system, endothelin via activation of ET_A receptors has a basal vasoconstricting role [46] and contributes to the development of vascular disease in hypertension and atherosclerosis [43, 47] (Fig. 15.2). Endothelins are involved in the regulation of myocardial contractility, [14], chronotropy [19], and arrhythmogenesis [48], as well as myocardial remodeling during congestive heart failure [49]. In the lung, the endothelin system regulates the bronchial tone [50] and proliferation of pulmonary airways blood vessels and promotes the development of pulmonary hypertension [51]. Endothelin also controls water and sodium excretion and renal acid–base balance under physiological conditions [27], and promotes the development of glomerulosclerosis [52–55]. In the brain, the endothelin system modulates cardio-respiratory centers and release of hormones [19] and regulates the growth guidance of developing sympathetic neurons (Makita et al. 2008). In addition, endothelins participate in physiologic and pathophysiologic functions of the immune system [44, 56, 57], the liver [58], muscle, adipose tissue, the reproductive system, and are involved in glucose homeostasis [58–60].

15.3 Role of Endothelin in Development and Therapy of Cardiovascular Disease

15.3.1 Arterial Hypertension

The identification of endothelin as a vasoconstrictor [16] and the finding of its release from vascular endothelial cells suggested that this peptide is involved in the pathogenesis of hypertension and vascular disease [61]. Further support for this hypothesis came from case reports of hemangioendothelioma patients that presented with markedly elevated high levels of plasma ET-1 and hypertension and showed normalization of elevated ET-1 levels and blood pressure after tumor removal [62]. In contrast, ET-1 plasma levels are mostly normal in patients with essential hypertension; however, local ET-1 levels increase in the vascular wall in hypertension [63, 64]. In the 1990s, the role of endothelin and experimental hypertension due to high salt or angiotensin II was investigated in several laboratories, results demonstrating potent antihypertensive effects and end-organ protection of endothelin receptor antagonists (reviewed in [47, 64–67]). Endothelial cell–specific overexpression of endothelin in vascular endothelial cells causes hypertension-associated changes in the vasculature, including hypertrophy and inflammation, yet does not cause hypertension [20–22]. On the other hand, endothelial cell–specific deletion of the preproendothelin gene is associated with hypotension [18, 22]; this blood pressure–lowering effect of gene deletion is similar to that of vascular smooth muscle–specific deletion of the ET_A receptor [18]. In mice lacking endothelin in vascular endothelial cells, plasma levels of ET-1 are reduced by about 90%, indicating that endothelial cells are indeed the major source of circulating ET-1 [18]. In human hypertension, ET-1 plasma levels are mostly normal; however, ET-1 levels increase locally in the vascular wall in hypertension [47, 65, 66, 68].

The kidney expresses all components of the endothelin system [69]. Endothelins were shown to be involved in the regulation of renal blood flow, re-absorption of water and sodium, as well as in acid–base balance [23, 69]. The renal vasculature represents one of the most sensitive vascular beds contracting to endothelin concentrations in the picomolar range, and contraction is mainly mediated by ET_A receptors [16, 70]. Renal endothelin has been linked to the development of salt-sensitive hypertension which may involve both the inflammatory NOS (iNOS) [43, 71], which is constitutively expressed in the kidney, [43, 71] and the ET_B receptor [32]. Indeed, blockade of the endothelin system with an ET_A receptor antagonist in genetically salt-sensitive hypertensive Dahl rats increases the abnormally low NO synthase activity in the kidney and markedly attenuates blood pressure induced by salt feeding [43]. Most recent work indicates that NOS also – at least in part – mediates endothelin-1-dependent sodium excretion in the collecting duct and blood pressure [72]. Interestingly, collecting duct–specific deficiency of endothelin abrogated the increase in activity of all three NOS isoforms in the inner medulla in response to sodium loading [72]. Also, genetic deficiency of the ET_B receptor via conditional knockout results in sodium-sensitive hypertension that can be improved by blocking the luminal epithelial sodium channel using amiloride [32]. These studies indicate that salt sensitivity, a common feature of patients with resistant hypertension, involves several underlying pathomechanisms and that it may be particularly accessible to treatment with endothelin antagonists.

Preclinical data on hypertension have been underscored by clinical studies in humans with essential hypertension. Treatment with either the nonselective ET receptor antagonist bosentan [73] or the ET_A receptor-selective antagonist darusentan [74, 75] causes substantial reductions of arterial blood pressure in patients with essential or resistant essential hypertension, darusentan even when added to existing therapy of at least three antihypertensives, including a diuretic [74, 76, 77]. It remains currently unclear whether selective antagonists provide an advantage over nonselective compounds. Selectivity appears to be a crucial issue as blockade of ET_B receptor–mediated effects may attenuate the pressure-lowering effect and interfere with endothelium-dependent dilation [78, 79]. In this regard, the reported selectivity and specificity data of drugs may depend on the assays and cells employed for the selectivity determination and appear to largely vary between drug companies. Indeed, recent studies indicate that selectivity profiles of several endothelin receptor antagonists from different pharmaceutical companies largely differ from the data in published literature depending on whether human or animal cell assays were used [80]. This also dictates caution with the interpretation of results from experimental studies regarding the selectivity of individual components. Resistant hypertension is frequently seen in African American and in obese patients, who are both at increased risk for cardiovascular and renal disease and show elevated plasma ET-1 levels

[81–83]. Long-term clinical studies are required to determine whether treatment with darusentan or other endothelin antagonists has the potential role to lower mortality in these patients, which might involve organ protection beyond that of the pressure-lowering effects of ET receptor blockade [84–86].

15.3.2 Atherosclerosis and Coronary Artery Disease

Expression of endothelin and its receptors is increased in the atherosclerotic plaques of human coronary arteries [87, 88] and both endothelin-1 peptide and ET_A receptors have been causally implicated in the development of atherosclerosis, as inhibition of this pathway inhibits formation of atherosclerotic plaque in animal models [43, 71, 89–91]; moreover, acute blockade of endothelin ET_A receptors ameliorates myocardial ischemia and biochemical changes caused by infarction in mice with coronary atherosclerosis [92] and reduces lipid-induced macrophage activation [89]. Indeed, endothelin has strong growth-promoting activity in the vascular wall and both endothelin and its receptors are widely expressed in macrophages, vascular smooth muscle cells, and fibroblasts (reviewed in [15, 43]) (Fig. 15.2). A common observation made in almost all studies investigating effects of endothelin receptor blockade on vascular function in animal models of hypertension, hypercholesterolemia, or atherosclerosis was that chronic treatment improved endothelium-dependent, NO-mediated vasodilation [43, 71, 93, 94]. This improvement of NO-dependent vasodilation after endothelin ET_A receptor blockade has also been observed in clinical studies and is blocked by ET_B antagonists [79]. Acute blockade of endothelin receptors of isolated internal mammary arteries in vitro obtained from in patients with coronary atherosclerosis improves endothelium-dependent vasodilation [95, 96], and similar findings have been reported from in vivo studies in humans with atherosclerosis [97–101]. A recent study of the effects of 6-month treatment with the ET_A antagonist atrasentan in patients with coronary artery disease also reported improved coronary artery endothelium-dependent vasodilation [102, 103]. Although ACE inhibitors [104] and statins [105] inhibit endothelin expression in vitro, ACE inhibition and statin treatment surprisingly has no effect on the markedly elevated endothelin peptide expression in the mammary artery of patients with coronary atherosclerosis [106], suggesting the need for additional therapies. In contrast, ET_A blockade is effective to completely normalize endothelin peptide levels in atherosclerosis, at least in experimental studies [63, 71].

Environmental cardiovascular risk factors such as cigarette smoking or air pollution have been only recently investigated. Smoking is one of the central risk factors contributing to many cardiovascular deaths [107]. Cigarette smoke enhances inflammatory airway responses [108] and induces ECE-1 peptide expression [21]. Correspondingly, contractile responses to ET-1 in arteries from patients with coronary artery disease are much stronger in smokers than in non-smokers [109]. Cox et al. described protective effects of ERA treatment after smoke inhalation–induced pulmonary injury [110], and preventive effects of ERA treatment on emphysema development, one of the long-term consequences of smoking and COPD, have been reported [111]. As with cigarette smoke, air pollution by car fumes, particularly

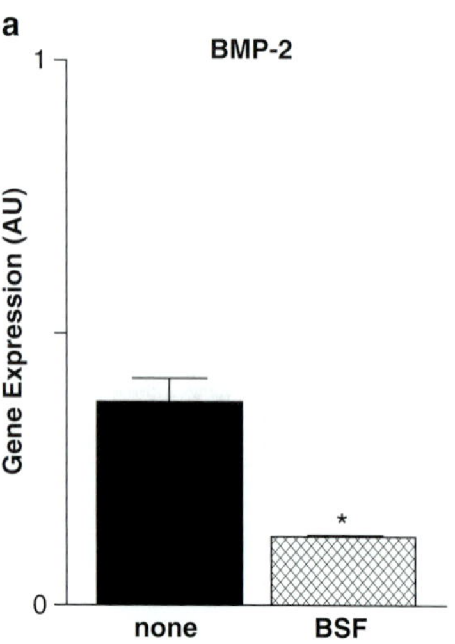

Fig. 15.3 Vascular mRNA expression of bone morphogenetic protein (BMP) receptor II in mice with autoimmune diabetes. Chronic treatment with an ET_A receptor-selective ERA (BSF 431314, a follow-up compound of ambrisentan) for 6 weeks reduced BMP-receptor II expression by almost 80%, indicating its regulation through endothelin (Reproduced from [132], with permission of the publisher)

diesel exhaust, increases cardiovascular morbidity [112]. Exposure to diesel exhaust results in ET_B receptor dysfunction [113] and increases in vasoconstrictor responses to ET-1 [114]; car fumes have also shown to increase vascular endothelin in atherosclerosis [114, 115]. Effects of diesel exhaust on endothelin activation have also been observed in humans [116]. Fine particulate matter as part of air pollution has also been implicated in diseases such as hypertension and airway diseases [117] and causes inflammation [21]. Increases in carbon and particulate matter air pollution increases circulating ET-1 levels [118], and ET_A receptor expression [111, 119]. Importantly, air pollutants may also induce endothelin and endothelin receptors in the absence of any local or systemic inflammation [120].

Obesity is another independent risk factor for atherosclerosis [121, 122]. Six years ago, according to WHO estimates, 1.6 billion adults worldwide were overweight, and 400 million were obese. By 2015, the numbers are expected to increase further to 2.3 billion overweight and 700 million obese, respectively [123]. In both cases, these numbers do not include children and adolescents, in which obesity has also become a worldwide problem [121]. Obesity leads to insulin resistance and subsequently to diabetes, and is associated with activation of the renal but not the pulmonary renin-angiotensin system in an ET_A-dependent manner [121]. Antidiabetic and beneficial structural effects of ERA treatment have been reported in numerous preclinical studies [58, 124–127], and recent clinical data suggest that proteinuria in diabetes may be directly linked to endothelin activation [128–131]. In diet-induced obesity, renal activation of ACE occurs, which is regulated via ET_A receptors, suggesting that under certain conditions ET_A receptors may actually act as ACE inhibitors [43]. ET_A receptors also regulate vascular expression of bone morphogenetic protein (BMP)-2 [132] (Fig. 15.3), an important regulator of vascular calcification and cell growth [133]. In obesity, endothelin and ET_A receptors are increased in the vasculature and kidney [43, 134, 135] ET-mediated vascular tone and metabolic function is abnormal in obesity and diabetes [59, 134–137]. In type 1-diabetes, ET_A receptor blockade also prevents

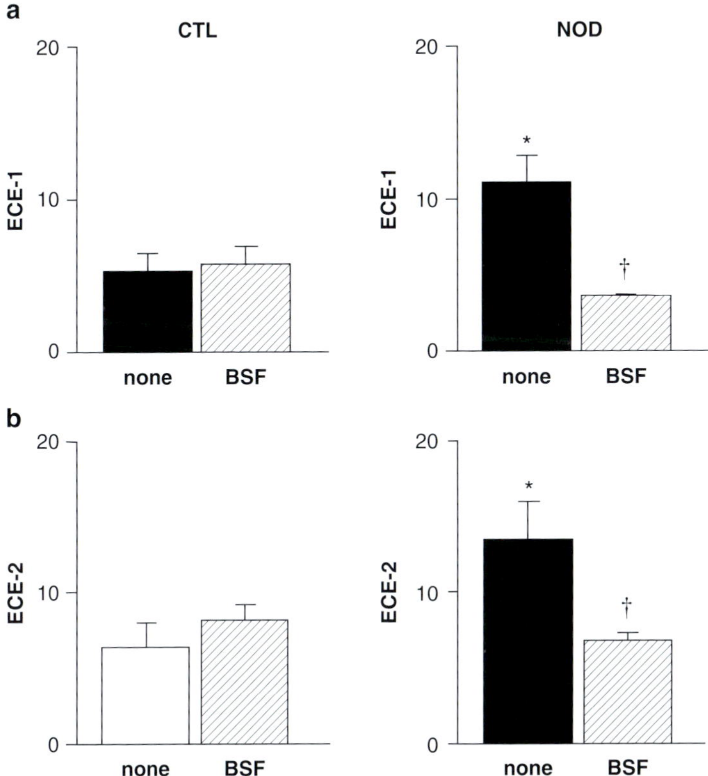

Fig. 15.4 Effects of diabetes and ERA treatment on ECE-1 and ECE-2 gene expression in the arterial vasculature of control (CTL) and non-obese diabetic (NOD) mice, a model of type 1 diabetes. Diabetes increased expression of ECE-1 and ECE-2. This upregulation was completely prevented by concomitant endothelin receptor blockade using the orally active compound BSF/LU461314, indicating that ERA treatment has ECE-inhibitor-like effects under certain pathological conditions (Reproduced from [126])

upregulation of ECE-1 and ECE-2 isoenzymes [126] (Fig. 15.4). Endothelin also directly affects obesity development by regulating adipogenesis and lipolysis [138, 139], and stimulates the release of pro-inflammatory cytokines from adipocytes [25]. Thus, endothelin blockade may be particularly feasible to interfere with cardiovascular and renal disease in obese patients and possibly might also be suitable for treatment of obesity and its associated complications such as insulin resistance [58]. Endothelin inhibits insulin action [140], and accordingly, Pernow and coworkers have recently shown that insulin sensitivity or impaired skeletal muscle glucose uptake in insulin-resistant humans [59] is improved after ERA treatment [141]. Insulin, which facilitates glucose uptake in target tissues [59], not only stimulates endothelin expression and synthesis [142], but insulin secretion is also stimulated by endothelin [143], suggesting a positive feedback loop between these two pathways which may be of therapeutic importance regarding ERA treatment.

Endothelin blockade may thus be particularly feasible to prevent cardiovascular and renal disease in these patients and possibly might also be suitable for treatment of obesity and its associated complications such as insulin resistance [58]. Finally, endothelin contributes to glycemic control and glucose uptake [144–146] and development of type 1 diabetes [126], making metabolic diseases and obesity potential and attractive clinical targets for the application of endothelin receptor antagonists.

15.3.3 Heart Failure

Congestive heart failure is a clinical syndrome with high mortality caused by different etiopathologies, hypertension and coronary artery disease being among the most important ones. In the heart, the endothelin system helps to maintain cardiac function, with ET-1 and the ET_A receptor being the predominant signaling components of the endothelin system [147–149]. In the normal heart, endothelins contribute to inotropy, chronotropy, and arrhythmogenesis, as well as myocardial contractility [19]. Early studies have shown that impairment of cardiac function results in increases in circulating levels of ET-1 or big ET-1 [17] that are reliable prognostic indicators of survival in patients with heart failure [150, 151]. Elevated circulating levels of ET-1 in heart failure are thought to derive from pulmonary congestion, which impairs the clearance function of the lung [152]. The role of endothelin in the post-infarct heart remains controversial. Although a number of experimental prevention studies have demonstrated a benefit of chronic endothelin blockade on survival and left ventricular remodeling in animals of myocardial infarction [149, 153–155], there is currently no evidence for a protective effect of chronic endothelin antagonism in humans with heart failure [23, 156]. It is also important to note that in all experimental studies except for one [157] treatment was begun in animals without preexisting heart failure, i.e., before or immediately after inducing myocardial infarction. In chronic heart failure, i.e., in long-term survivors of experimental infarction, ET_A receptor blockade more or less normalized hypertrophy of the right atrium and ventricle and reduced pulmonary congestion [157] (Fig. 15.5). Moreover, treatment was performed in models that had no ischemic myocardial damage due to coronary artery disease which is present in many heart failure patients. Thus, well-designed experimental studies are still lacking. Studies in humans showed that treatment with the nonselective ET receptor antagonist bosentan over 2 weeks reduced pulmonary and mean arterial pressures, pulmonary and systemic resistance between 10% and 30%, and caused a 13% increase of the cardiac index [158]. Similar effects were seen in heart failure patients who received the ET_A receptor-selective antagonist BQ-123 where treatment reduced pulmonary and arterial pressures, decreased systemic (but not pulmonary) resistance, and increased cardiac index [159]. Although the results of these early studies looked promising, all long-term clinical trials investigating chronic endothelin receptor antagonist treatment in patients with acute or chronic congestive heart failure have been negative without exception [23, 55, 156, 160]. These studies include the ENABLE trial (bosentan), the HEAT-CHF trial (darusentan), the EARTH trial (darusentan), the ENCOR trial (enrasentan), and the RITZ-1 through RITZ-4 trials (tezosentan) [23]. A problem inherent to most of these studies is the fact that

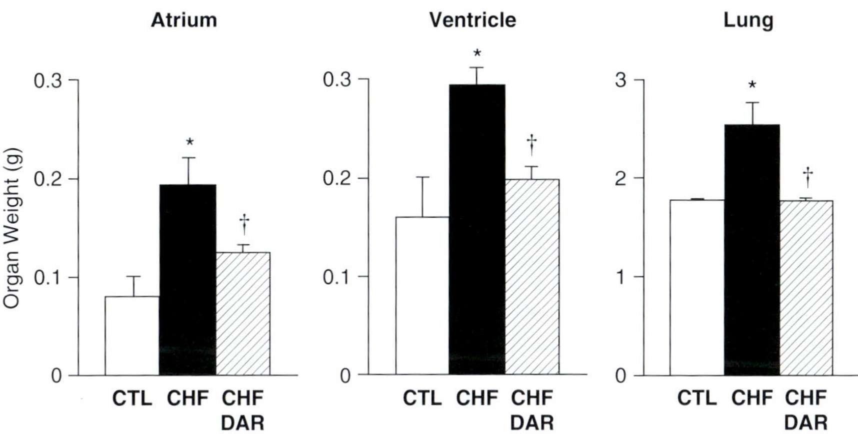

Fig. 15.5 Effects of chronic experimental heart failure and ERA treatment with darusentan on right atrial and ventricular remodeling and pulmonary edema measured by tissue weight. In rats which had survived acute myocardial infarction for 6 months, right atrial and ventricular weights were increased compared to sham controls (CTL), compatible with right heart hypertrophy. Similarly, lung weight, an indicator of pulmonary edema, was increased by approximately 35%. After darusentan treatment (50 mg/kg/d, DAR) of animals with chronic heart failure for 6 weeks, increased myocardial or pulmonary weights were reduced or even normalized. *$p<0.05$ vs. CTL; † $p<0.05$ vs. CHF (Reproduced from [157], with permission of the publisher)

only the results of some of these studies were published and also that if published, not all data were included in the manuscripts or are otherwise available to the scientific community [161]. Also, due to FDA regulations study patients had to be maintained on standard heart failure therapy and received the endothelin antagonist on top of standard treatment, which could be one of the reasons for the disappointing results. It may well be possible that certain predisposing pathological conditions in heart failure patients may be determinants of therapeutic success or failure, such as development of peripheral edema, one of the most often encountered side effects seen in patients on endothelin blocker. In addition, drug dosages [23] and drug toxicity of sulfonamide-based ERAs in patients with heart failure may be a critical problem, particularly in those with right heart failure and subsequent hepatic congestion. Before these data have been fully analyzed and published, no definitive conclusion on whether endothelin antagonists on top of ACE or ARB treatment may be effective remedies in selected patients with heart failure is possible.

15.3.4 Pulmonary Arterial Hypertension (PAH)

Both heart and lungs are important sources and targets of ET-1. Unlike normal subjects, patients with pulmonary hypertension have higher pulmonary arterial vs. venous plasma levels of ET-1, suggesting increased pulmonary ET-1 production and/or decreased lung clearance [19]. In the pulmonary vasculature, ET-1 induces ET_A-dependent vasoconstriction, and perhaps more importantly, acts as a growth factor leading to proliferation of pulmonary artery vascular smooth muscle cells

(Fig. 15.2) [51]. The inhibition of cell growth by ERAs is likely to be one of the important factors contributing to the long-term benefit of endothelin blockade in patients with pulmonary hypertension interfering with pulmonary artery remodeling [162]. Effectiveness of both ET_A receptor antagonists as well as nonselective ET receptor blockade with bosentan has been demonstrated to reduce pulmonary artery pressures, right ventricular hypertrophy, and remodeling of pulmonary arteries in a number of experimental studies [147, 162, 163]. In contrast, ET_B-selective antagonists administered to dogs with pulmonary hypertension increase pulmonary resistance and pressures [19]. This suggests that selective ET_A receptor antagonists might be advantageous in the treatment of pulmonary hypertension.

In the past two decades, the pharmaceutical industry has extensively tested pulmonary hypertension as a clinical target for ET antagonism, and first randomized clinical trials have demonstrated beneficial effects on clinical outcome and quality of life compared with placebo [164, 165]. In 2001, bosentan (*Tracleer*™) was the first endothelin receptor antagonist ever to receive approval for the treatment of patients [166]. This historically important approval was granted for the treatment of primary pulmonary arterial hypertension (PAH), a severe disease with unfavorable prognosis often seen in patients with connective tissue disease, heart failure, or HIV infection [167, 168]. Meanwhile, another ET antagonist has been granted approval by the FDA or by Federal Health Agencies around the world for the treatment of PAH, the ET_A receptor-selective antagonists ambrisentan (*Letairis*™), a follow-up compound of darusentan. Whether selective antagonists are superior over nonselective ones in terms of clinical benefits, side effects, and survival in PAH patients is unknown, and the same holds true for the two different classes of endothelin antagonists (propionic vs. sulfonamide compounds); respective clinical trials are needed, and ongoing trials also include combination therapy of endothelin antagonists as with other pulmonary vasodilators such as sildenafil or prostacyclin [23]. A first study suggests superiority of bosentan over prostacyclin treatment in patients with cirrhosis and pulmonary hypertension [26]. Finally, new treatment options such as aerosol delivery of endothelin antagonists appear to be efficacious and can minimize side effects [169–172]. Endothelin antagonists have now become a standard part of pulmonary hypertension therapy to improve survival in these severely ill patients.

15.3.5 Cardiac Transplantation and Allograft Rejection

Chronic allograft rejection after cardiac transplantation increases endothelin in the graft [173] but also increases the expression of endothelin system components in the host organs such as the liver [174] (Fig. 15.6). An immunomodulatory role of endothelin has also been shown in different models of acute or chronic rejection following solid organ transplantation. Even in the absence of standard immunosuppression from endothelin receptor, ET_A blockade was able to prevent upregulation of circulating interleukin and TNF α levels [175] after cardiac allo-transplantation which again would be compatible with a direct role of endogenous endothelin contributing to the host's immune response. It is thus not surprising that treatment with

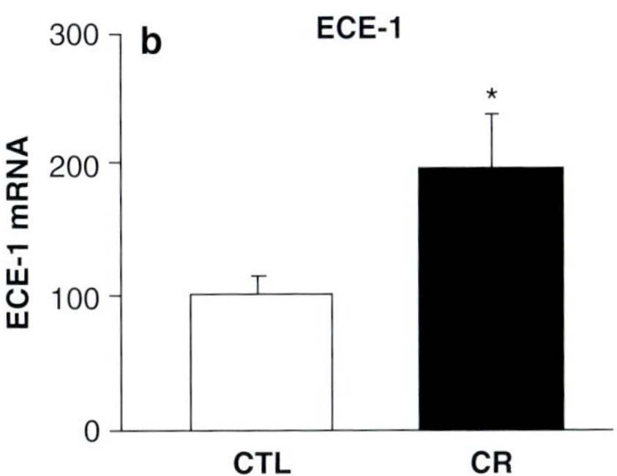

Fig. 15.6 Hepatic mRNA expression of endothelin-converting enzyme (ECE)-1 in a cardiac allograft recipient undergoing chronic rejection (CR). Chronic allograft rejection of the heart was associated with a twofold upregulation of ECE-1 mRNA in the liver of the host (Reproduced from [174], with permission of the publisher)

ERAs very effectively interferes with the development of graft atherosclerosis or the development of fibrosis or glomerulosclerosis-related to solid organ transplantation of the liver, lung, aorta, heart, or kidney, even in the absence of immunosuppression [154, 176–180]. Up to now, no clinical studies have been performed to investigate the therapeutic potential of endothelin receptor antagonists in transplantation medicine, which possibly could also lead to improved donor organ preservation by adding ET antagonists to the preservation solution [181] or to novel combination therapies. Possibly, this would also allow to reduce the amount of immunosuppressant drugs which are responsible for many of the unwanted side effects of solid organ transplantation, such as neurotoxicity [182, 183], development of kidney disease due to nephrotoxicity of drugs like cyclosporin, and secondary, drug-induced hypertension due to immunosuppressive therapy with cyclosporin [184, 185].

15.3.6 Proteinuric Renal Disease

Chronic renal disease and proteinuria are independent risk factors for atherosclerosis and coronary artery disease [52]; in fact, the majority of deaths of patients with renal disease is due to cardiovascular causes [52]. Work from several laboratories in the early 1990s has demonstrated that the endothelin system contributes to the pathological changes leading to glomerulosclerosis in models of hypertension or renal ablation (reviewed in [186–188]). This pro-sclerotic effect of endothelin in the kidney was confirmed by overexpressing human ET-1 in mice which develop glomerulosclerosis even without developing hypertension [54]. A large number of experimental prevention studies have investigated the effects of chronic endothelin blockade on the development of glomerulosclerosis due to hypertension, subtotal nephrectomy, chronic nitric oxide deficiency, diabetic nephropathy, focal segmental glomerulosclerosis, among others [55, 189]. The majority of these studies found pronounced nephroprotective effects that were either in part or even completely independent of systemic blood pressure [186, 190–194]. The mechanisms by which endothelin contributes to glomerular injury following damage of podocytes [192–194], which form the glomerular filtration barrier, include protein that induces endothelin in glomerular podocytes, which in turn causes reorganization of the podocyte actin cytoskeleton [195].

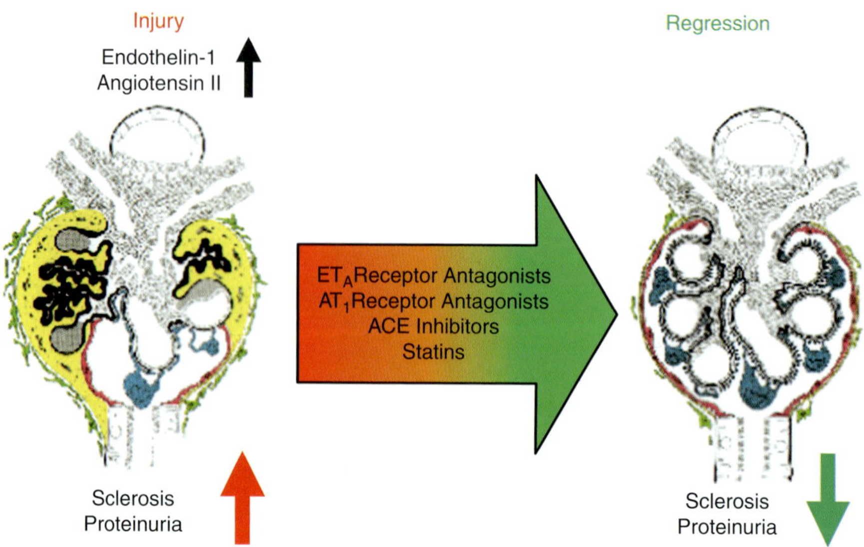

Fig. 15.7 Proposed concept of renal disease regression after inhibition of endothelin action through blockade of endothelin ET_A receptors, RAAS-inhibition through blockade of angiotensin AT_1 receptor blockers (ARBs) or ACE inhibitors, or statin therapy. *Left*: Glomerular renal injury with damage of podocytes (*dark green*) and formation of fibrotic, "sclerotic" tissue (*yellow*) resulting in proteinuria ("Injury"). *Right*: Glomerulosclerosis can be reversed by drug treatment if renal structural injury is less than severe. Disease regression is accompanied by improvements of glomerular architecture and structural improvement of podocytes (*dark green*) and the GBM of the glomerular capillary ("Regression") (Reproduced from [52], with permission of the publisher)

This effect is mediated by ET_A receptors [194, 196]. Interestingly, dietary protein aggravates renal injury by augmenting renal acid production and worsening proteinuria [197]. This effect can be blocked by the ET_A-selective antagonist darusentan but not by the nonselective antagonist bosentan [197], indicating a role for the ET_B receptor which is highly expressed in podocytes. Indeed, this hypothesis is supported by a most recent study indicating sera of patients with proteinuria increase glomerular formation of endothelin and shedding of the podocyte-specific protein nephrin, which can be prevented by an endothelin ET_A receptor antagonist [198]. These studies collectively and strongly indicate that glomerular protein loss, which is caused by and further aggravates podocyte injury, depends on mechanisms that are at least in part endothelin-mediated [52]. This appears to be even more important since proteinuria or albuminuria is a good predictor of future cardiovascular events [189, 199].

Only few studies have investigated the effects of endothelin receptor blockade in conditions in which renal disease was already established [52]. Studies have investigated the anti-proteinuric effect of endothelin receptor antagonists in normotensive or severely hypertensive animal models [192, 194, 200, 201]. In these studies, treatment not only a reversed proteinuria but also lead to a partial healing of the previously injured glomeruli and podocytes (Fig. 15.7). This suggested that renal disease is a particularly relevant area for the clinical application of ERAs with the potential to reverse established glomerular disease [53, 194, 201], as has been demonstrated for experimental vascular disease [202, 203].

15.4 Endothelin Antagonists in Clinical Practice: Current Developments

Within only 4 years after the discovery of endothelin, its receptors were cloned and receptor antagonists had become available [204, 205]. Several hundred compounds are available today of which the majority is orally active [206]. The first clinical trial in patients with congestive heart failure was performed in Zurich, Switzerland, in the early 1990s and results were published in 1995 [158]. Nevertheless, it took a number of years and numerous unsuccessful clinical trials in heart failure patients until endothelin receptor blockade could be established as a new therapeutic concept in clinical medicine [55, 147, 207]. Ten years ago, bosentan (*Tracleer*™) was the first endothelin receptor antagonist to receive approval for clinical application from the U. S. Federal Food and Drug Administration (FDA) [166]. Bosentan, which is a nonselective ET_A/ET_B receptor antagonist, was approved in 2001 for the treatment of patients with primary pulmonary arterial hypertension (PAH) [164, 165]. In 2007, ambrisentan (*Letairis*™), an ET_A receptor-selective antagonist, was also approved by the U.S. FDA for the same indication. A number of other receptor antagonists have been or are being evaluated in clinical studies for indications such as PAH, congestive heart failure–resistant hypertension, cancer, coronary artery disease, or proteinuric renal disease. The highly selective ET_A receptor antagonist sitaxsentan (*Thelin*™) had even approved for treatment of PAH in three continents; yet, after several cases of fatal liver failure, Pfizer withdrew Thelin™ from the market at the end of 2010 [208]. Two years earlier, Speedel announced the discontinuation of the development of avosentan after severe drug-related side effects including heart failure had occurred in diabetics with advanced proteinuric renal disease [208]. These studies will be briefly discussed below. The development of the ET_A receptor antagonist darusentan as an antihypertensive with nephroprotective properties was abandoned by Gilead Sciences, Inc. after the completion of two phase III trials in patients with resistant hypertension at the end of 2009 [74, 76, 77, 208]. According to information available, issues such as short remaining patent life of darusentan and financial risks for further phase III trials were among the reasons for the discontinuation. In addition, ERAs have been evaluated in clinical trials for the treatment of coronary artery disease and atherosclerosis [95–103], cancer [23, 209–211], and autoimmune diseases such as scleroderma [51, 212–218]. More recently, proteinuric renal disease has been revived as target for endothelin antagonism [84, 208]. Treatment with a selective ET_A receptor antagonist atrasentan (*Xinlay*™) for 12 weeks reduces proteinuria and blood pressure [219], and even more impressive reductions in albuminuria reduction were seen using the endothelin ET_A receptor blocker avosentan in patients with diabetic nephropathy in the ASCEND trial [129, 130]. Interestingly, these effects were mostly independent of systemic blood pressure [129, 130]. In many of the patients with chronic renal disease participating in the ASCEND trial, fluid retention and heart failure developed because of which the study was stopped prematurely [129]. It is noteworthy that the effects of proteinuria were seen despite the fact that patients are already receiving ACE inhibitors or AT_1 receptor antagonists, indicating additive and thus independent beneficial effects of both treatments. Thus, ERAs represent a new treatment option to halt and even reverse proteinuric renal disease (Fig. 15.7). Antiproteinuric effects of ERA therapy were also observed with the ET_A receptor antagonist sitaxsentan [216, 220–223]. The ET_A antagonist atrasentan was also recently reported to have anti-proteinuric effects in patients with diabetic nephropathy, without having major side effects [128]. It should be noted that most of the patients studied in randomized endothelin antagonist trials in hypertension and diabetic nephropathy were overweight or obese, conditions known to be associated with impaired renal sodium handling [224]. Thus, it is likely that the side effects of endothelin blockade will depend on the overall health status and comorbidities of the patient receiving the drug. Indeed, in patients with chronic renal disease and normal body weight receiving ET_A receptor antagonists, edema rarely occured [216, 220–223].

15.5 Conclusion and Perspectives

Endothelin – acting predominantly through the ETA receptor – is now recognized as a multifunctional peptide with cytokine-like activity affecting almost all aspects of cardiovascular cell function. Although pharmaceutical companies had rapidly developed drugs blocking endothelin receptors within only a few years after the discovery of endothelin, clinical drug development has been complicated by the fact that both, pharmaceutical industry and clinical investigators, embarked in clinical trials without really knowing endothelin physiology in general and in humans in particular. Also, at the time, the relevance of the endogenous endothelin system in maintaining central organ function in disease (i.e., during heart failure) was not known. It came to no surprise that the largest part of clinical trials were negative, i.e., patient selection (comorbidity burden, stage/severity of disease), excessively high doses of ERAs used, liver toxicity of the sulfonamide-based ERAs in patients with liver comorbidities (hepatic congestion in heart failure, fatty liver disease in obese type II diabetics), and the FDA requirement to study ERAs only if given on top of standard therapy are some of the reasons why so many trials failed. Also, timing of therapy initiation appears to be a critical issue, as is evident from studies in patients with advanced form of cancer, which have been mostly negative [209]. Recent work from Theodorescu's laboratory has shown that endothelin plays a critical role in cancer metastasis [225], which may in part explain the inefficacy of ERA treatment in advanced forms of cancer [209, 210, 226]. It is therefore reasonable to assume that ERA treatment (with its controllable side effect fluid retention) will be effective only if disease is diagnosed early, as has been shown for moderate but not advanced renal failure [128–130].

Some of the clinical indications initially chosen for clinical studies such as heart failure have been not yet shown to benefit from ERA therapy on top of standard

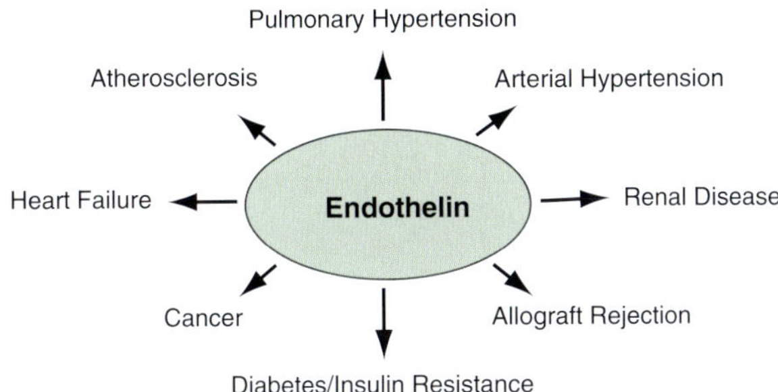

Fig. 15.8 Disease states in which a causative involvement of endothelin receptors has been demonstrated in preclinical and clinical studies (Reproduced from [27], with the permission of the publisher)

treatment due to the lack of adequately designed studies and inadequate patient selection [227]. In contrast, therapeutic efficacy of ERA therapy had been shown in preclinical studies of PAH, which became the first clinical indication for ERAs [166]. Similarly, results from preclinical studies of diseases that are similarly associated with cell growth and/or inflammatory activation such as or resistant arterial hypertension and glomerulosclerosis or immune-mediated disease such as cancer, connective tissue diseases, chronic allograft rejection, or metabolic diseases such as obesity or diabetes (Fig. 15.8) suggest that these conditions could become new indications for endothelin antagonist therapy in the future [27, 208, 227]. Today, more than two decades after the discovery of endothelin and its receptors, only two compounds (bosentan and ambrisentan) are approved and in use for treatment of patients, and only for two indications (PAH and scleroderma-related ulcerations). It is well possible that factors such as the desire to be the first to publish results from clinical studies with these new drugs, lack of knowledge about limitations of patient suitability and disease severity, and perhaps hope for economic reward from the potential sales of "blockbuster" drug candidates pre-marketed to investors have contributed to the unsuccessful clinical drug development of ERAs [227]. Also, some drug companies dropped drug candidates after successful completion of phase III trials for which the remaining patent lives was only a few years. Several hundreds of ERAs have been developed [206], and well and carefully designed clinical studies in correctly selected patients are still warranted to test, verify, or disprove any therapeutic benefit of ERAs for cardiovascular medicine and related fields. That this is possible, without risking severe side effects using carefully selected patients, was recently demonstrated by Kohan and colleagues in patients with moderate proteinuric renal disease [128]. Hope remains that ERA treatment will be available for more than only two indications once the pharmaceutical industry realizes the potential of their drugs and supports investigators in performing correctly done clinical trials.

Acknowledgments This work was supported by grants from the Swiss National Science Foundation (Nr. 108 258 and 122 504). I thank present and former members of my laboratory as well as my collaborators who have contributed to the original research discussed in this chapter.

References

1. Furchgott RF, Zawadzki JV. The obligatory role of endothelial cells in the relaxation of arterial smooth muscle by acetylcholine. Nature. 1980;299:373–6.
2. Ignarro LJ, Buga GM, Wood KS, Byrns RE, Chaudhuri G. Endothelium-derived relaxing factor produced and released from artery and vein is nitric oxide. Proc Natl Acad Sci USA. 1987;84:9265–9.
3. Barton M. The discovery of endothelium-dependent contraction: the legacy of Paul M. Vanhoutte. Pharmacol Res. 2011. doi:S1043-6618(11)00064-8 [pii] 10.1016/j.phrs.2011.02.013.
4. De Mey JG, Vanhoutte PM. Contribution of the endothelium to the response to anoxia in the canine femoral artery. Arch Int Pharmacodyn Ther. 1981;253(2):325–6.
5. De Mey JG, Vanhoutte PM. Heterogeneous behavior of the canine arterial and venous wall. Importance of the endothelium. Circ Res. 1982;51(4):439–47.
6. Gillespie MN, Owasoyo JO, McMurtry IF, O'Brien RF. Sustained coronary vasoconstriction provoked by a peptidergic substance released from endothelial cells in culture. J Pharmacol Exp Ther. 1986;236(2):339–43.
7. Hickey KA, Rubanyi GM, Paul RJ, Highsmith RF. Characterization of a coronary vasoconstrictor produced by cultured endothelial cells. Am J Physiol. 1985;248:C550–6.
8. O'Brien RF, Robbins RJ, McMurtry IF. Endothelial cells in culture produce a vasoconstrictor substance. J Cell Physiol. 1987;132(2):263–70.
9. Rubanyi GM. The discovery of endothelin: the power of bioassay and the role of serendipity in the discovery of endothelium-derived vasoactive substances. Pharmacol Res. 2011. doi:S1043-6618(10)00167-2 [pii] 10.1016/j.phrs.2010.08.004.
10. Masaki T. The discovery of endothelins. Cardiovasc Res. 1998;39(3):530–3.
11. Masaki T. Historical review: endothelin. Trends Pharmacol Sci. 2004;25(4):219–24.
12. Yanagisawa M, Kurihara H, Kimura S, Tomobe Y, Kobayashi M, Mitsui Y, et al. A novel potent vasoconstrictor peptide produced by vascular endothelial cells. Nature. 1988;332(6163):411–5.
13. Hillier C, Berry C, Petrie MC, O'Dwyer PJ, Hamilton C, Brown A, et al. Effects of urotensin II in human arteries and veins of varying caliber. Circulation. 2001;103(10):1378–81.
14. Maguire JJ, Davenport AP. Is urotensin-II the new endothelin? Br J Pharmacol. 2002;137(5): 579–88.
15. Barton M, Carmona R, Krieger JE, Goettsch W, Morawietz H, d'Uscio LV, et al. Endothelin regulates angiotensin-converting enzyme in the mouse kidney. J Cardiovasc Pharmacol. 2000;36(5 Suppl 1):S244–7.
16. Tomobe Y, Miyauchi T, Saito A, Yanagisawa M, Kimura S, Goto K, et al. Effects of endothelin on the renal artery from spontaneously hypertensive and wistar kyoto rats. Eur J Pharmacol. 1988;152(3):373–4.
17. Miyauchi T, Yanagisawa M, Tomizawa T, Sugishita Y, Suzuki N, Fujino M, et al. Increased plasma concentrations of endothelin-1 and big endothelin-1 in acute myocardial infarction. Lancet. 1989;2(8653):53–4.
18. Kisanuki YY, Emoto N, Ohuchi T, Widyantoro B, Yagi K, Nakayama K, et al. Low blood pressure in endothelial cell-specific endothelin 1 knockout mice. Hypertension. 2010;56:121–8. doi:HYPERTENSIONAHA.109.138701 [pii] 10.1161/HYPERTENSIONAHA.109.138701.
19. Kedzierski RM, Yanagisawa M. Endothelin system: the double-edged sword in health and disease. Annu Rev Pharmacol Toxicol. 2001;41:851–76.
20. Amiri F, Ko EA, Javeshghani D, Reudelhuber TL, Schiffrin EL. Deleterious combined effects of salt-loading and endothelial cell restricted endothelin-1 overexpression on blood pressure and vascular function in mice. J Hypertens. 2010. doi:10.1097/HJH.0b013e328338bb8b.
21. Anggrahini DW, Emoto N, Nakayama K, Widyantoro B, Adiarto S, Iwasa N, et al. Vascular endothelial cell-derived endothelin-1 mediates vascular inflammation and neointima formation following blood flow cessation. Cardiovasc Res. 2009;82(1):143–51. doi:cvp026 [pii] 10.1093/cvr/cvp026.

22. Pollock DM. Dissecting the complex physiology of endothelin. New lessons from genetic models. Hypertension. 2010;56:31–3. doi:HYPERTENSIONAHA.109.139758 [pii] 10.1161/HYPERTENSIONAHA.109.139758.
23. Battistini B, Berthiaume N, Kelland NF, Webb DJ, Kohan DE. Profile of past and current clinical trials involving endothelin receptor antagonists: the novel "-sentan" class of drug. Exp Biol Med (Maywood). 2006;231(6):653–95.
24. Bramall AN, Han RN, Deng Y, Yanagisawa M, McInnes RR, Stewart DJ. Endothelin 2 (et-2) plays a critical role in lung alveolarization: novel insight from the et-2-deficient mouse model. Meeting Abstract Book 10th International Conference on Endothelin, Bergamo; September 2007.
25. Chai SP, Chang YN, Fong JC. Endothelin-1 stimulates interleukin-6 secretion from 3t3-l1 adipocytes. Biochim Biophys Acta. 2009;1790(3):213–8. doi:S0304-4165(08)00270-5 [pii] 10.1016/j.bbagen.2008.12.002.
26. Hoeper MM, Seyfarth HJ, Hoeffken G, Wirtz H, Spiekerkoetter E, Pletz MW, et al. Experience with inhaled iloprost and bosentan in portopulmonary hypertension. Eur Respir J. 2007;30(6):1096–102.
27. Barton M, Yanagisawa M. Endothelin: 20 years from discovery to therapy. Can J Physiol Pharmacol. 2008;86(8):485–98. doi:y08-059 [pii] 10.1139/y08-059.
28. Guo C, Ju H, Leung D, Massaeli H, Shi M, Rabinovitch M. A novel vascular smooth muscle chymase is upregulated in hypertensive rats. J Clin Invest. 2001;107(6):703–15. doi:10.1172/JCI9997.
29. Guo C, Rabinovitch M. A novel chymase cDNA cloned from rat pulmonary artery smooth muscle cells with increased vascular expression in spontaneously hypertensive rats. Circulation. 1998;98:I–745 (abstract).
30. Ju H, Gros R, You X, Tsang S, Husain M, Rabinovitch M. Conditional and targeted overexpression of vascular chymase causes hypertension in transgenic mice. Proc Natl Acad Sci USA. 2001;98(13):7469–74. doi:10.1073/pnas.131147598 98/13/7469 [pii].
31. Ikeda K, Emoto N, Raharjo SB, Nurhantari Y, Saiki K, Yokoyama M, et al. Molecular identification and characterization of novel membrane-bound metalloprotease, the soluble secreted form of which hydrolyzes a variety of vasoactive peptides. J Biol Chem. 1999;274(45):32469–77.
32. Gariepy CE, Ohuchi T, Williams SC, Richardson JA, Yanagisawa M. Salt-sensitive hypertension in endothelin-B receptor-deficient rats. J Clin Invest. 2000;105(7):925–33.
33. Seyrantepe V, Hinek A, Junzheng P, Fedjaev M, Ernest S, Kadota Y, et al. Enzymatic activity of lysosomal carboxypeptidase (cathepsin) A is required for proper elastic fiber formation and inactivation of endothelin-1. Circulation. 2008;117:1973–81.
34. Davenport AP. International union of pharmacology. XXIX. Update on endothelin receptor nomenclature. Pharmacol Rev. 2002;54(2):219–26.
35. Evans NJ, Walker JW. Endothelin receptor dimers evaluated by FRET, ligand binding, and calcium mobilization. Biophys J. 2008;95(1):483–92.
36. Watts SW. Endothelin receptors: what's new and what do we need to know? Am J Physiol Regul Integr Comp Physiol. 2010;298. doi:00584.2009 [pii] 10.1152/ajpregu.00584.2009.
37. Attie T, Till M, Pelet A, Amiel J, Edery P, Boutrand L, et al. Mutation of the endothelin-receptor B gene in Waardenburg-Hirschsprung disease. Hum Mol Genet. 1995;4(12):2407–9.
38. Baynash AG, Hosoda K, Giaid A, Richardson JA, Emoto N, Hammer RE, et al. Interaction of endothelin-3 with endothelin-B receptor is essential for development of epidermal melanocytes and enteric neurons. Cell. 1994;79(7):1277–85.
39. Puffenberger EG, Hosoda K, Washington SS, Nakao K, de Wit D, Yanagisawa M, et al. A missense mutation of the endothelin-B receptor gene in multigenic Hirschsprung's disease. Cell. 1994;79(7):1257–66.
40. Burkhardt M, Barton M, Shaw SG. Receptor- and non-receptor-mediated clearance of big-endothelin and endothelin-1: differential effects of acute and chronic ET_A receptor blockade. J Hypertens. 2000;18(3):273–9.

41. de Nucci G, Thomas R, D'Orleans-Juste P, Antunes E, Walder C, Warner TD, et al. Pressor effects of circulating endothelin are limited by its removal in the pulmonary circulation and by the release of prostacyclin and endothelium-derived relaxing factor. Proc Natl Acad Sci USA. 1988;85(24):9797–800.
42. Luscher T, Barton M. Endothelins and endothelin receptor antagonists: therapeutic considerations for a novel class of cardiovascular drugs. Circulation. 2000;102:2434–40.
43. Barton M. Endothelial dysfunction and atherosclerosis: endothelin receptor antagonists as novel therapeutics. Curr Hypertens Rep. 2000;2(1):84–91.
44. Barton M, Nett PC, Amann K, Teixeira MM. Anti-inflammatory effects of endothelin receptor antagonists and their importance for treating human disease. In: Chaudhary I, Ur-Rahman A, editors. Frontiers in cardiovascular drug discovery, vol. 1. Oak Parks: Bentham Science; 2010. p. 236–258.
45. Rubanyi GM, Polokoff MA. Endothelins: molecular biology, biochemistry, pharmacology, physiology, and pathophysiology. Pharmacol Rev. 1994;46(3):325–415.
46. Haynes WG, Webb DJ. Contribution of endogenous generation of endothelin-1 to basal vascular tone. Lancet. 1994;344(8926):852–4.
47. Schiffrin EL. State-of-the-art lecture. Role of endothelin-1 in hypertension. Hypertension. 1999;34(4 Pt 2):876–81.
48. Proven A, Roderick HL, Conway SJ, Berridge MJ, Horton JK, Capper SJ, et al. Inositol 1,4,5-trisphosphate supports the arrhythmogenic action of endothelin-1 on ventricular cardiac myocytes. J Cell Sci. 2006;119(Pt 16):3363–75.
49. Sakai S, Miyauchi T, Kobayashi M, Yamaguchi I, Goto K, Sugishita Y. Inhibition of myocardial endothelin pathway improves long-term survival in heart failure. Nature. 1996;384(6607): 353–5.
50. Uchida Y, Ninomiya H, Saotome M, Nomura A, Ohtsuka M, Yanagisawa M, et al. Endothelin, a novel vasoconstrictor peptide, as potent bronchoconstrictor. Eur J Pharmacol. 1988;154(2): 227–8.
51. Denton CP, Humbert M, Rubin L, Black CM. Bosentan treatment for pulmonary arterial hypertension related to connective tissue disease: a subgroup analysis of the pivotal clinical trials and their open-label extensions. Ann Rheum Dis. 2006;65(10):1336–40.
52. Barton M. Reversal of proteinuric renal disease and the emerging role of endothelin. Nat Clin Pract Nephrol. 2008;4(9):490–501. doi:ncpneph0891 [pii] 10.1038/ncpneph0891.
53. Chatziantoniou C, Dussaule JC. Insights into the mechanisms of renal fibrosis: is it possible to achieve regression? Am J Physiol. 2005;289(2):F227–34.
54. Hocher B, Thöne-Reinecke C, Rohmeiss P, Schmager F, Slowinski R, Burst V, et al. Endothelin-1 transgenic mice develop glomerulosclerosis, interstitial fibrosis, and renal cysts but not hypertension. J Clin Invest. 1997;99:1380–9.
55. Remuzzi G, Perico N, Benigni A. New therapeutics that antagonize endothelin: promises and frustrations. Nat Rev Drug Discov. 2002;1(12):986–1001.
56. Spirig R, Potapova I, Shaw-Boden J, Tsui J, Rieben R, Shaw SG. TLR2 and TLR4 agonists induce production of the vasoactive peptide endothelin-1 by human dendritic cells. Mol immunol 2009;46(15):3178–82.
57. Nett PC, Teixeira MM, Candinas D, Barton M. Recent developments on endothelin antagonists as immunomodulatory drugs–from infection to transplantation medicine. Recent Pat Cardiovasc Drug Discov. 2006;1(3):265–76.
58. Berthiaume N, Carlson CJ, Rondinone CM, Zinker BA. Endothelin antagonism improves hepatic insulin sensitivity associated with insulin signaling in Zucker fatty rats. Metabolism. 2005;54(11):1515–23.
59. Lteif A, Vaishnava P, Baron AD, Mather KJ. Endothelin limits insulin action in obese/insulin-resistant humans. Diabetes. 2007;56(3):728–34.
60. van Harmelen V, Eriksson A, Astrom G, Wahlen K, Naslund E, Karpe F, et al. Vascular peptide endothelin-1 links fat accumulation with alterations of visceral adipocyte lipolysis. Diabetes. 2008;57(2):378–86.

61. Iwasa S, Fan J, Miyauchi T, Watanabe T. Blockade of endothelin receptors reduces diet-induced hypercholesterolemia and atherosclerosis in apolipoprotein E-deficient mice. Pathobiology. 2001;69(1):1–10.
62. Yokokawa K, Tahara H, Kohno M, Murakawa K, Yasunari K, Nakagawa K, et al. Hypertension associated with endothelin-secreting malignant hemangioendothelioma. Ann Intern Med. 1991;114(3):213–5.
63. Barton M, Shaw S, d'Uscio LV, Moreau P, Luscher TF. Angiotensin II increases vascular and renal endothelin-1 and functional endothelin converting enzyme activity in vivo: role of ETA receptors for endothelin regulation. Biochem Biophys Res Commun. 1997;238(3):861–5.
64. Schiffrin EL. Role of endothelin-1 in hypertension and vascular disease. Am J Hypertens. 2001;14(6 Pt 2):83S–9S.
65. Barton M, Kiowski W. The therapeutic potential of endothelin receptor antagonists in cardiovascular disease. Curr Hypertens Rep. 2001;3(4):322–30.
66. Barton M, Luscher TF. Endothelin antagonists for hypertension and renal disease. Curr Opin Nephrol Hypertens. 1999;8(5):549–56.
67. Schiffrin EL. Vascular endothelin in hypertension. Vascul Pharmacol. 2005;43(1):19–29.
68. Schiffrin EL. Endothelin and endothelin antagonists in hypertension. J Hypertens. 1998;16 (12 Pt 2):1891–5.
69. Kohan DE. Endothelins in the normal and diseased kidney. Am J Kidney Dis. 1997;29(1):2–26.
70. Katoh T, Chang H, Uchida S, Okuda T, Kurokawa K. Direct effects of endothelin in the rat kidney. Am J Physiol. 1990;258(2 Pt 2):F397–402.
71. Barton M, Haudenschild CC, d'Uscio LV, Shaw S, Munter K, Luscher TF. Endothelin eta receptor blockade restores no-mediated endothelial function and inhibits atherosclerosis in apolipoprotein E-deficient mice. Proc Natl Acad Sci USA. 1998;95(24):14367–72.
72. Schneider MP, Ge Y, Pollock DM, Pollock JS, Kohan DE. Collecting duct-derived endothelin regulates arterial pressure and Na excretion via nitric oxide. Hypertension. 2008;51:1–6.
73. Krum H, Viskoper RJ, Lacourciere Y, Budde M, Charlon V. The effect of an endothelin-receptor antagonist, bosentan, on blood pressure in patients with essential hypertension. N Engl J Med. 1998;338:784–90.
74. Black HR, Bakris GL, Weber MA, Weiss R, Shahawy ME, Marple R, et al. Efficacy and safety of darusentan in patients with resistant hypertension: results from a randomized, double-blind, placebo-controlled dose-ranging study. J Clin Hypertens. 2007;9(10):760–9.
75. Nakov R, Pfarr E, Eberle S. Darusentan: an effective endothelinA receptor antagonist for treatment of hypertension. Am J Hypertens. 2002;15(7 Pt 1):583–9.
76. Bakris GL, Lindholm LH, Black HR, Krum H, Linas S, Linseman JV, et al. Divergent results using clinic and ambulatory blood pressures: report of a darusentan-resistant hypertension trial. Hypertension. 2010;56(5):824–30. doi:HYPERTENSIONAHA.110.156976 [pii] 10.1161/HYPERTENSIONAHA.110.156976.
77. Weber MA, Black H, Bakris G, Krum H, Linas S, Weiss R, et al. A selective endothelin-receptor antagonist to reduce blood pressure in patients with treatment-resistant hypertension: a randomised, double-blind, placebo-controlled trial. Lancet. 2009;374(9699):1423–31. doi:S0140-6736(09)61500-2 [pii] 10.1016/S0140-6736(09)61500-2.
78. Pollock DM, Schneider MP. Clarifying endothelin type B receptor function. Hypertension. 2006;48(2):211–2.
79. Verhaar MC, Strachan FE, Newby DE, Cruden NL, Koomans HA, Rabelink TJ, et al. Endothelin-A receptor antagonist-mediated vasodilatation is attenuated by inhibition of nitric oxide synthesis and by endothelin-B receptor blockade. Circulation. 1998;97(8):752–6.
80. Green S, Bundey RA, Nunley K, Hartman JC, Melvin LS, Gorczynski RJ, Insel PA, Bristow MR, Pitts KR. Determination of endothelin receptor antagonist affinities and selectivities in human cardiac membranes. Meeting Abstract Book 10th International Conference on Endothelin, Bergamo; September 2007.
81. Ergul A. Hypertension in black patients: an emerging role of the endothelin system in salt-sensitive hypertension. Hypertension. 2000;36(1):62–7.

82. Ergul S, Parish DC, Puett D, Ergul A. Racial differences in plasma endothelin-1 concentrations in individuals with essential hypertension. Hypertension. 1996;28(4):652–5.
83. Parrinello G, Scaglione R, Pinto A, Corrao S, Cecala M, Di Silvestre G, et al. Central obesity and hypertension: the role of plasma endothelin. Am J Hypertens. 1996;9(12 Pt 1): 1186–91.
84. Barton M, Mullins JJ, Bailey MA, Kretzler M. Role of endothelin receptors for renal protection and survival in hypertension: waiting for clinical trials. Hypertension. 2006;48(5): 834–7.
85. Belaidi E, Joyeux-Faure M, Ribuot C, Launois SH, Levy P, Godin-Ribuot D. Major role for hypoxia inducible factor-1 and the endothelin system in promoting myocardial infarction and hypertension in an animal model of obstructive sleep apnea. J Am Coll Cardiol. 2009;53(15):1309–17. doi:S0735-1097(09)00334-9 [pii] 10.1016/j.jacc.2008.12.050.
86. Duru F, Barton M, Luscher TF, Candinas R. Endothelin and cardiac arrhythmias: do endothelin antagonists have a therapeutic potential as antiarrhythmic drugs? Cardiovasc Res. 2001;49(2):272–80.
87. Lerman A, Edwards BS, Hallett JW, Heublein DM, Sandberg SM, Burnett Jr JC. Circulating and tissue endothelin immunoreactivity in advanced atherosclerosis. N Engl J Med. 1991;325(14):997–1001.
88. Winkles JA, Alberts GF, Brogi E, Libby P. Endothelin-1 and endothelin receptor mRNA expression in normal and atherosclerotic human arteries. Biochem Biophys Res Commun. 1993;191(3):1081–8.
89. Babaei S, Picard P, Ravandi A, Monge JC, Lee TC, Cernacek P, et al. Blockade of endothelin receptors markedly reduces atherosclerosis in LDL receptor deficient mice: role of endothelin in macrophage foam cell formation. Cardiovasc Res. 2000;48(1):158–67.
90. Schiffrin EL. Beyond blood pressure: the endothelium and atherosclerosis progression. Am J Hypertens. 2002;15(10 Pt 2):115S–22S.
91. Tepe G, Brehme U, Seeger H, Raschack M, Claussen CD, Duda SH. Endothelin a receptor antagonist LU 135252 inhibits hypercholesterolemia-induced, but not deendothelialization-induced, atherosclerosis in rabbit arteries. Invest Radiol. 2002;37(6):349–55.
92. Caligiuri G, Levy B, Pernow J, Thoren P, Hansson GK. Myocardial infarction mediated by endothelin receptor signaling in hypercholesterolemic mice. Proc Natl Acad Sci USA. 1999;96(12):6920–4.
93. Best PJ, Lerman LO, Romero JC, Richardson D, Holmes Jr DR, Lerman A. Coronary endothelial function is preserved with chronic endothelin receptor antagonism in experimental hypercholesterolemia in vitro. Arterioscler Thromb Vasc Biol. 1999;19(11):2769–75.
94. Best PJ, McKenna CJ, Hasdai D, Holmes Jr DR, Lerman A. Chronic endothelin receptor antagonism preserves coronary endothelial function in experimental hypercholesterolemia. Circulation. 1999;99(13):1747–52.
95. Barton M, Glodny B. Endothelin receptor blockade and nitric oxide bioactivity. Cardiovasc Res. 2001;52(1):161–3.
96. Verma S, Lovren F, Dumont AS, Mather KJ, Maitland A, Kieser TM, et al. Endothelin receptor blockade improves endothelial function in human internal mammary arteries. Cardiovasc Res. 2001;49(1):146–51. doi:S0008-6363(00)00244-3 [pii].
97. Bohm F, Beltran E, Pernow J. Endothelin receptor blockade improves endothelial function in atherosclerotic patients on angiotensin converting enzyme inhibition. J Intern Med. 2005;257(3):263–71. doi:JIM1448 [pii] 10.1111/j.1365-2796.2005.01448.x.
98. Bohm F, Jensen J, Svane B, Settergren M, Pernow J. Intracoronary endothelin receptor blockade improves endothelial function in patients with coronary artery disease. Can J Physiol Pharmacol. 2008;86(11):745–51. doi:y08-081 [pii] 10.1139/y08-081.
99. Bohm F, Pernow J. The importance of endothelin-1 for vascular dysfunction in cardiovascular disease. Cardiovasc Res. 2007;76(1):8–18. doi:S0008-6363(07)00281-7 [pii] 10.1016/j.cardiores.2007.06.004.

100. Kalani M, Pernow J, Bragd J, Jorneskog G. Improved peripheral perfusion during endothelin-A receptor blockade in patients with type 2 diabetes and critical limb ischemia. Diabetes Care. 2008;31(7):e56. doi:31/7/e56 [pii] 10.2337/dc08-0409.
101. Settergren M, Pernow J, Brismar K, Jorneskog G, Kalani M. Endothelin-A receptor blockade increases nutritive skin capillary circulation in patients with type 2 diabetes and microangiopathy. J Vasc Res. 2008;45(4):295–302. doi:000113601 [pii] 10.1159/000113601.
102. Raichlin E, Prasad A, Mathew V, Kent B, Holmes Jr DR, Pumper GM, et al. Efficacy and safety of atrasentan in patients with cardiovascular risk and early atherosclerosis. Hypertension. 2008;52(3):522–8. doi:HYPERTENSIONAHA.108.113068 [pii] 10.1161/HYPERTENSIONAHA. 108.113068.
103. Reriani M, Raichlin E, Prasad A, Mathew V, Pumper GM, Nelson RE, et al. Long-term administration of endothelin receptor antagonist improves coronary endothelial function in patients with early atherosclerosis. Circulation. 2010;122(10):958–66. doi:CIRCULATIONAHA.110.967406 [pii] 10.1161/CIRCULATIONAHA.110.967406.
104. Clavell AL, Mattingly MT, Stevens TL, Nir A, Wright S, Aarhus LL, et al. Angiotensin converting enzyme inhibition modulates endogenous endothelin in chronic canine thoracic inferior vena caval constriction. J Clin Invest. 1996;97(5):1286–92. doi:10.1172/JCI118544.
105. Hernandez-Perera O, Perez-Sala D, Navarro-Antolin J, Sanchez-Pascuala R, Hernandez G, Diaz C, et al. Effects of the 3-hydroxy-3-methylglutaryl-CoA reductase inhibitors, atorvastatin and simvastatin, on the expression of endothelin-1 and endothelial nitric oxide synthase in vascular endothelial cells. J Clin Invest. 1998;101(12):2711–9. doi:10.1172/JCI1500.
106. Sutherland AJ, Nataatmadja MI, Walker PJ, Cuttle L, Garlick RB, West MJ. Vascular remodeling in the internal mammary artery graft and association with in situ endothelin-1 and receptor expression. Circulation. 2006;113(9):1180–8. doi:CIRCULATIONAHA.105.582890 [pii] 10.1161/CIRCULATIONAHA.105.582890.
107. Cesaroni G, Forastiere F, Agabiti N, Valente P, Zuccaro P, Perucci CA. Effect of the Italian smoking ban on population rates of acute coronary events. Circulation. 2008;117(9):1183–8. doi:CIRCULATIONAHA.107.729889 [pii] 10.1161/CIRCULATIONAHA.107.729889.
108. Bhavsar TM, Liu X, Cerreta JM, Liu M, Cantor JO. Endothelin-1 potentiates smoke-induced acute lung inflammation. Exp Lung Res. 2008;34(10):707–16. doi:906804275 [pii] 10.1080/01902140802389701.
109. Muller-Schweinitzer E, Muller SE, Reineke DC, Kern T, Carrel TP, Eckstein FS, et al. Reactive oxygen species mediate functional differences in human radial and internal thoracic arteries from smokers. J Vasc Surg. 2010;51(2):438–44. doi:S0741-5214(09)01928-4 [pii] 10.1016/j.jvs.2009.09.040.
110. Cox RA, Soejima K, Burke AS, Traber LD, Herndon DN, Schmalstieg FC, et al. Enhanced pulmonary expression of endothelin-1 in an ovine model of smoke inhalation injury. J Burn Care Rehabil. 2001;22(6):375–83.
111. Chen Y, Hanaoka M, Droma Y, Chen P, Voelkel NF, Kubo K. Endothelin-1 receptor antagonists prevent the development of pulmonary emphysema in rats. Eur Respir J. 2009. doi:09031936.00003909 [pii] 10.1183/09031936.00003909.
112. Brook RD, Rajagopalan S, Pope 3rd CA, Brook JR, Bhatnagar A, Diez-Roux AV, et al. Particulate matter air pollution and cardiovascular disease: an update to the scientific statement from the American heart association. Circulation. 2010;121(21):2331–78. doi:CIR.0b013e3181dbece1 [pii] 10.1161/CIR.0b013e3181dbece1.
113. Cherng TW, Campen MJ, Knuckles TL, Gonzalez Bosc L, Kanagy NL. Impairment of coronary endothelial cell ET(B) receptor function after short-term inhalation exposure to whole diesel emissions. Am J Physiol Regul Integr Comp Physiol. 2009;297(3):R640–7. doi:90899.2008 [pii]10.1152/ajpregu.90899.2008.
114. Campen MJ, Lund AK, Knuckles TL, Conklin DJ, Bishop B, Young D, et al. Inhaled diesel emissions alter atherosclerotic plaque composition in ApoE(–/–) mice. Toxicol Appl Pharmacol. 2009. doi:S0041-008X(09)00466-9 [pii] 10.1016/j.taap. 2009.10.021.

115. Lund AK, Knuckles TL, Obot Akata C, Shohet R, McDonald JD, Gigliotti A, et al. Gasoline exhaust emissions induce vascular remodeling pathways involved in atherosclerosis. Toxicol Sci. 2007;95(2):485–94. doi:kfl145 [pii] 10.1093/toxsci/kfl145.
116. Langrish JP, Lundback M, Mills NL, Johnston NR, Webb DJ, Sandstrom T, et al. Contribution of endothelin 1 to the vascular effects of diesel exhaust inhalation in humans. Hypertension. 2009;54(4):910–5. doi:HYPERTENSIONAHA.109.135947 [pii] 10.1161/HYPERTENSIONAHA.109.135947.
117. Brook RD, Urch B, Dvonch JT, Bard RL, Speck M, Keeler G, et al. Insights into the mechanisms and mediators of the effects of air pollution exposure on blood pressure and vascular function in healthy humans. Hypertension. 2009;54(3):659–67. doi:HYPERTENSIONAHA.109.130237 [pii] 10.1161/HYPERTENSIONAHA.109.130237.
118. Liu L, Ruddy T, Dalipaj M, Poon R, Szyszkowicz M, You H, et al. Effects of indoor, outdoor, and personal exposure to particulate air pollution on cardiovascular physiology and systemic mediators in seniors. J Occup Environ Med. 2009;51(9):1088–98. doi:10.1097/JOM.0b013e3181b35144.
119. Xie YH, Wang SW, Zhang Y, Edvinsson L, Xu CB. Up-regulation of G-protein-coupled receptors for endothelin and thromboxane by lipid-soluble smoke particles in renal artery of rat. Basic Clin Pharmacol Toxicol. 2010; Epub April 22. doi:PTO585 [pii] 10.1111/j.1742-7843.2010.00585.x.
120. Upadhyay S, Stoeger T, Harder V, Thomas RF, Schladweiler MC, Semmler-Behnke M, et al. Exposure to ultrafine carbon particles at levels below detectable pulmonary inflammation affects cardiovascular performance in spontaneously hypertensive rats. Part Fibre Toxicol. 2008;5:19. doi:1743-8977-5-19 [pii] 10.1186/1743-8977-5-19.
121. Barton M, Carmona R, Ortmann J, Krieger JE, Traupe T. Obesity-associated activation of angiotensin and endothelin in the cardiovascular system. Int J Biochem Cell Biol. 2003;35(6):826–37.
122. Meyer MR, Clegg DJ, Prossnitz ER, Barton M. Obesity, insulin resistance and diabetes: sex differences and role of oestrogen receptors. Acta Physiol (Oxf). 2011;203(1):259–69. doi:10.1111/ j.1748-1716.2010.02237.x.
123. Malik SM, Popkin BM, Bray GA, Despres JP, Hu FB. Sugar-sweetened beverages, obesity, type 2 diabetes mellitus, and cardiovascular disease risk. Circulation. 2010;121:1256–364.
124. Balsiger B, Rickenbacher A, Boden PJ, Biecker E, Tsui J, Dashwood M, et al. Endothelin a-receptor blockade in experimental diabetes improves glucose balance and gastrointestinal function. Clin Sci (Lond). 2002;103 Suppl 48:430S–3.
125. Harris AK, Elgebaly MM, Li W, Sachidanandam K, Ergul A. Effect of chronic endothelin receptor antagonism on cerebrovascular function in type 2 diabetes. Am J Physiol Regul Integr Comp Physiol. 2008;294(4):R1213–9. doi:00885.2007 [pii] 10.1152/ajpregu.00885.2007.
126. Ortmann J, Nett PC, Celeiro J, Traupe T, Tornillo L, Hofmann-Lehmann R, et al. Endothelin inhibition delays onset of hyperglycemia and associated vascular injury in type I diabetes: evidence for endothelin release by pancreatic islet beta-cells. Biochem Biophys Res Commun. 2005;334(2):689–95.
127. Sachidanandam K, Elgebaly MM, Harris AK, Hutchinson JR, Mezzetti EM, Portik-Dobos V, et al. Effect of chronic and selective endothelin receptor antagonism on microvascular function in type 2 diabetes. Am J Physiol Heart Circ Physiol. 2008;294(6):H2743–9. doi:91487.2007 [pii] 10.1152/ajpheart.91487.2007.
128. Kohan DE, Pritchett Y, Molitch M, Wen S, Garimella T, Audhya P, et al. Addition of atrasentan to renin-angiotensin system blockade reduces albuminuria in diabetic nephropathy. J Am Soc Nephrol. 2011;22(4):763–72. doi:ASN.2010080869 [pii] 10.1681/ASN.2010080869.
129. Mann JF, Green D, Jamerson K, Ruilope LM, Kuranoff SJ, Littke T, et al. Avosentan for overt diabetic nephropathy. J Am Soc Nephrol. 2010;21(3):527–35. doi:ASN.2009060593 [pii] 10.1681/ASN.2009060593.
130. Wenzel RR, Littke T, Kuranoff S, Jurgens C, Bruck H, Ritz E, et al. Avosentan reduces albumin excretion in diabetics with macroalbuminuria. J Am Soc Nephrol. 2009;20(3):655–64. doi:ASN.2008050482 [pii] 10.1681/ASN.2008050482.

131. Zanatta CM, Gerchman F, Burttet L, Nabinger G, Jacques-Silva MC, Canani LH, et al. Endothelin-1 levels and albuminuria in patients with type 2 diabetes mellitus. Diabetes Res Clin Pract. 2008;80:299–304.
132. Nett PC, Ortmann J, Celeiro J, Haas E, Hofmann-Lehmann R, Tornillo L, et al. Transcriptional regulation of vascular bone morphogenetic protein by endothelin receptors in early autoimmune diabetes mellitus. Life Sci. 2006;78(19):2213–8.
133. Emmanuele L, Ortmann J, Doerflinger T, Traupe T, Barton M. Lovastatin stimulates human vascular smooth muscle cell expression of bone morphogenetic protein-2, a potent inhibitor of low-density lipoprotein-stimulated cell growth. Biochem Biophys Res Commun. 2003;302(1):67–72. doi:S0006291X03001098 [pii].
134. Mundy AL, Haas E, Bhattacharya I, Widmer CC, Kretz M, Hofmann-Lehmann R, et al. Fat intake modifies vascular responsiveness and receptor expression of vasoconstrictors: implications for diet-induced obesity. Cardiovasc Res. 2007;73(2):368–75.
135. Traupe T, Lang M, Goettsch W, Munter K, Morawietz H, Vetter W, et al. Obesity increases prostanoid-mediated vasoconstriction and vascular thromboxane receptor gene expression. J Hypertens. 2002;20(11):2239–45.
136. Cardillo C, Campia U, Bryant MB, Panza JA. Increased activity of endogenous endothelin in patients with type II diabetes mellitus. Circulation. 2002;106(14):1783–7.
137. Cardillo C, Campia U, Iantorno M, Panza JA. Enhanced vascular activity of endogenous endothelin-1 in obese hypertensive patients. Hypertension. 2004;43(1):36–40.
138. Eriksson AK, van Harmelen V, Stenson BM, Astrom G, Wahlen K, Laurencikiene J, et al. Endothelin-1 stimulates human adipocyte lipolysis through the ETA receptor. Int J Obes. 2009;33(1):67–74. doi:ijo2008212 [pii] 10.1038/ijo.2008.212.
139. Juan CC, Chang LW, Huang SW, Chang CL, Lee CY, Chien Y, et al. Effect of endothelin-1 on lipolysis in rat adipocytes. Obesity (Silver Spring). 2006;14(3):398–404.
140. Teuscher AU, Lerch M, Shaw S, Pacini G, Ferrari P, Weidmann P. Endothelin-1 infusion inhibits plasma insulin responsiveness in normal men. J Hypertens. 1998;16(9):1279–84.
141. Ahlborg G, Shemyakin A, Bohm F, Gonon A, Pernow J. Dual endothelin receptor blockade acutely improves insulin sensitivity in obese patients with insulin resistance and coronary artery disease. Diabetes Care. 2007;30(3):591–6.
142. Anfossi G, Cavalot F, Massucco P, Mattiello L, Mularoni E, Hahn A, et al. Insulin influences immunoreactive endothelin release by human vascular smooth muscle cells. Metabolism. 1993;42(9):1081–3.
143. Gregersen S, Thomsen JL, Hermansen K. Endothelin-1 (et-1)-potentiated insulin secretion: involvement of protein kinase C and the ET(A) receptor subtype. Metabolism. 2000;49(2): 264–9.
144. Rachdaoui N, Nagy LE. Endothelin-1-stimulated glucose uptake is desensitized by tumor necrosis factor-alpha in 3t3-l1 adipocytes. Am J Physiol Endocrinol Metab. 2003;285(3):E545–51.
145. Said SA, Ammar el SM, Suddek GM. Effect of bosentan (ET_A/ET_B receptor antagonist) on metabolic changes during stress and diabetes. Pharmacol Res. 2005;51(2):107–15.
146. Wu-Wong JR, Berg CE, Wang J, Chiou WJ, Fissel B. Endothelin stimulates glucose uptake and glut4 translocation via activation of endothelin ETA receptor in 3t3-l1 adipocytes. J Biol Chem. 1999;274(12):8103–10.
147. Kirchengast M, Luz M. Endothelin receptor antagonists: clinical realities and future directions. J Cardiovasc Pharmacol. 2005;45(2):182–91.
148. McMurray JJ. Heart failure in 10 years time: focus on pharmacological treatment. Heart. 2002;88 Suppl 2:ii40–6.
149. Sutsch G, Barton M. Endothelin in heart failure. Curr Hypertens Rep. 1999;1(1):62–8.
150. Omland T, Lie RT, Aakvaag A, Aarsland T, Dickstein K. Plasma endothelin determination as a prognostic indicator of 1-year mortality after acute myocardial infarction. Circulation. 1994;89(4):1573–9.
151. Pacher R, Stanek B, Hulsmann M, Koller-Strametz J, Berger R, Schuller M, et al. Prognostic impact of big endothelin-1 plasma concentrations compared with invasive hemodynamic evaluation in severe heart failure. J Am Coll Cardiol. 1996;27(3):633–41.

152. Staniloae C, Dupuis J, White M, Gosselin G, Dyrda I, Bois M, et al. Reduced pulmonary clearance of endothelin in congestive heart failure: a marker of secondary pulmonary hypertension. J Card Fail. 2004;10(5):427–32.
153. Cernacek P, Stewart DJ, Monge JC, Rouleau JL. The endothelin system and its role in acute myocardial infarction. Can J Physiol Pharmacol. 2003;81(6):598–606.
154. Fukunaga K, Takada Y, Taniguchi H, Mei G, Seino KI, Yuzawa K, et al. Endothelin antagonist treatment for successful liver transplantation from non-heart-beating donors. Transplantation. 1999;67(2):328–32.
155. Miyauchi T, Goto K. Heart failure and endothelin receptor antagonists. Trends Pharmacol Sci. 1999;20(5):210–7.
156. Kaye DM, Krum H. Drug discovery for heart failure: a new era or the end of the pipeline? Nat Rev. 2007;6(2):127–39.
157. Vetter D, Shaw SG, Brandes RP, Munter K, Vetter W, Barton M. Beneficial cardiovascular effects of endothelin ET(A) receptor blockade in established long-term heart failure after myocardial infarction. Exp Biol Med (Maywood). 2006;231(6):857–60.
158. Kiowski W, Sutsch G, Hunziker P, Muller P, Kim J, Oechslin E, et al. Evidence for endothelin-1-mediated vasoconstriction in severe chronic heart failure. Lancet. 1995;346(8977): 732–6.
159. Cowburn PJ, Cleland JG, McArthur JD, MacLean MR, McMurray JJ, Dargie HJ. Short-term haemodynamic effects of BQ-123, a selective endothelin ET(A)-receptor antagonist, in chronic heart failure. Lancet. 1998;352(9123):201–2.
160. Cowburn PJ, Cleland JG. Endothelin antagonists for chronic heart failure: do they have a role? Eur Heart J. 2001;22(19):1772–84.
161. Kelland NF, Webb DJ. Clinical trials of endothelin antagonists in heart failure: publication is good for the public health. Heart. 2007;93(1):2–4.
162. Dupuis J. Endothelin-receptor antagonists in pulmonary hypertension. Lancet. 2001;358(9288):1113–4.
163. Dupuis J. Endothelin receptor antagonists and their developing role in cardiovascular therapeutics. Can J Cardiol. 2000;16(7):903–10.
164. Channick RN, Simonneau G, Sitbon O, Robbins IM, Frost A, Tapson VF, et al. Effects of the dual endothelin-receptor antagonist bosentan in patients with pulmonary hypertension: a randomised placebo-controlled study. Lancet. 2001;358(9288):1119–23.
165. Rubin LJ, Badesch DB, Barst RJ, Galie N, Black CM, Keogh A, et al. Bosentan therapy for pulmonary arterial hypertension. N Engl J Med. 2002;346(12):896–903.
166. Davenport AP, Maguire JJ. Of mice and men: advances in endothelin research and first antagonist gains FDA approval. Trends Pharmacol Sci. 2002;23(4):155–7.
167. Puri A, McGoon MD, Kushwaha SS. Pulmonary arterial hypertension: current therapeutic strategies. Nat Clin Pract. 2007;4(6):319–29.
168. Shapiro S. Management of pulmonary hypertension resulting from interstitial lung disease. Curr Opin Pulm Med. 2003;9(5):426–30.
169. Kaisers U, Bodil P, Deja M, Bartholdy R, Donaubauer B, Laudi S, et al. Inhalation of the ET_A receptor antagonist LU-135252 selectively attenuates hypoxic pulmonary vasoconstriction. Am J Physiol. 2008;294:R601–5.
170. Kaisers U, Busch T, Wolf S, Lohbrunner H, Wilkens K, Hocher B, et al. Inhaled endothelin A antagonist improves arterial oxygenation in experimental acute lung injury. Intensive Care Med. 2000;26(9):1334–42.
171. Kalk P, Senf P, Deja M, Petersen B, Busch T, Bauer C, Boemke W, Kaisers U, Hocher B. Inhalation of an endothelin receptor A antagonist attenuates pulmonary inflammation in experimental acute lung injury. Can J Physiol Pharmacol 2008;86(8):511–5.
172. Leuchte HH, Meis T, El-Nounou M, Michalek J, Behr J. Inhalation of endothelin receptor blockers in pulmonary hypertension. Am J Physiol Lung Cell Mol Physiol. 2008;294: L772–7.

173. Ravalli S, Szabolcs M, Albala A, Michler RE, Cannon PJ. Increased immunoreactive endothelin-1 in human transplant coronary artery disease. Circulation. 1996;94(9):2096–102.
174. Lattmann T, Ortmann J, Horber S, Shaw SG, Hein M, Barton M. Upregulation of endothelin converting enzyme-1 in host liver during chronic cardiac allograft rejection. Exp Biol Med (Maywood). 2006;231(6):899–901.
175. Lattmann T, Hein M, Horber S, Ortmann J, Teixeira MM, Souza DG, et al. Activation of pro-inflammatory and anti-inflammatory cytokines in host organs during chronic allograft rejection: role of endothelin receptor signaling. Am J Transplant. 2005;5(5):1042–9.
176. Braun C, Conzelmann T, Vetter S, Schaub M, Back WE, Yard B, et al. Prevention of chronic renal allograft rejection in rats with an oral endothelin A receptor antagonist. Transplantation. 1999;68(6):739–46.
177. Orth SR, Odoni G, Amann K, Strzelczyk P, Raschack M, Ritz M. The ET_A receptor blocker LU 135252 prevents chronic transplant nephropathy in the "Fisher to lewis" model. J Am Soc Nephrol. 1999;10:387–91.
178. Orth SR, Odoni G, Karkoszka H, Ogata H, Viedt C, Amann K, et al. Combination treatment with an ET(A)-receptor blocker and an ace inhibitor is not superior to the respective monotherapies in attenuating chronic transplant vasculopathy in different aorta allotransplantation rat models. Nephrol Dial Transplant. 2003;18(1):62–9.
179. Shennib H, Lee AG, Kuang JQ, Yanagisawa M, Ohlstein EH, Giaid A. Efficacy of administering an endothelin-receptor antagonist (SB209670) in ameliorating ischemia-reperfusion injury in lung allografts. Am J Respir Crit Care Med. 1998;157(6 Pt 1):1975–81.
180. Tang JL, Aitouche A, Subbotin V, Salam A, Sun H, Gandhi C, et al. Endothelin-1 receptor blockade and its effect on chronic rejection. Transplant Proc. 1999;31(1–2):1249.
181. Fukunaga K, Takada Y, Taniguchi H, Otsuka M, Goto K, Fukao K. Successful liver transplantation from non-heart-beating donors by blockade of endothelin and PAE. Transplant Proc. 1998;30(7):3797.
182. Burke GW, Ciancio G, Cirocco R, Markou M, Roth D, Esquenazi V, et al. Tacrolimus-related microangiopathy in kidney and simultaneous pancreas-kidney recipients: evidence of endothelin and cytokine involvement. Transplant Proc. 1998;30(2):661–2.
183. Gijtenbeek JM, van den Bent MJ, Vecht CJ. Cyclosporine neurotoxicity: a review. J Neurol. 1999;246(5):339–46.
184. Carrier M, Tronc F, Stewart D, Pelletier LC. Dose-dependent effect of cyclosporin on renal arterial resistance in dogs. Am J Physiol. 1991;261(6 Pt 2):H1791–6.
185. Cauduro RL, Costa C, Lhulier F, Garcia RG, Cabral RD, Goncalves LF, et al. Endothelin-1 plasma levels and hypertension in cyclosporine-treated renal transplant patients. Clin Transplant. 2005;19(4):470–4.
186. Benigni A. Endothelin antagonists in renal disease. Kidney Int. 2000;57(4):1778–94.
187. Benigni A, Zoja C, Corna D, Orisio S, Longaretti L, Bertani T, et al. A specific endothelin subtype A receptor antagonist protects against injury in renal disease progression. Kidney Int. 1993;44(2):440–4.
188. Bruzzi I, Remuzzi G, Benigni A. Endothelin: a mediator of renal disease progression. J Nephrol. 1997;10(4):179–83.
189. Bruno S, Cattaneo D, Perico N, Remuzzi G. Emerging drugs for diabetic nephropathy. Expert Opin Emerg Drugs. 2005;10(4):747–71.
190. Benigni A, Perico N, Remuzzi G. Research on renal endothelin in proteinuric nephropathies dictates novel strategies to prevent progression. Curr Opin Nephrol Hypertens. 2001;10(1):1–6.
191. Benigni A, Remuzzi G. Endothelin antagonists. Lancet. 1999;353(9147):133–8.
192. Gagliardini E, Corna D, Zoja C, Sangalli F, Carrara F, Rossi M, Conti S, Rottoli D, Longaretti L, Remuzzi A, Remuzzi G, Benigni A. Unlike each drug alone, lisinopril if combined with avosentan promotes regression of renal lesions in experimental diabetes. Am J Physiol Renal Physiol. 2009;297(5):1448–56.

193. Opocensky M, Kramer HJ, Backer A, Vernerova Z, Eis V, Cervenka L, et al. Late-onset endothelin-a receptor blockade reduces podocyte injury in homozygous Ren-2 rats despite severe hypertension. Hypertension. 2006;48(5):965–71.
194. Ortmann J, Amann K, Brandes RP, Kretzler M, Munter K, Parekh N, et al. Role of podocytes for reversal of glomerulosclerosis and proteinuria in the aging kidney after endothelin inhibition. Hypertension. 2004;44(6):974–81.
195. Morigi M, Buelli S, Angioletti S, Zanchi C, Longaretti L, Zoja C, et al. In response to protein load podocytes reorganize cytoskeleton and modulate endothelin-1 gene: implication for permselective dysfunction of chronic nephropathies. Am J Pathol. 2005;166(5):1309–20.
196. Morigi M, Buelli S, Zanchi C, Longaretti L, Macconi D, Benigni A, et al. Shigatoxin-induced endothelin-1 expression in cultured podocytes autocrinally mediates actin remodeling. Am J Pathol. 2006;169(6):1965–75.
197. Wesson DE. Regulation of kidney acid excretion by endothelins. Kidney Int. 2006;70(12):2066–73.
198. Collino F, Bussolati B, Gerbaudo E, Marozio L, Pelissetto S, Benedetto C, et al. Pre-eclamptic sera induce nephrin shedding from podocytes through endothelin-1 release by endothelial glomerular cells. Am J Physiol. 2008;294:F1185.
199. Pisoni R, Ruggenenti P, Remuzzi G. Renoprotective therapy in patients with nondiabetic nephropathies. Drugs. 2001;61(6):733–45.
200. Opocensky M, Dvorak P, Maly J, Kramer HJ, Backer A, Kopkan L, et al. Chronic endothelin receptor blockade reduces end-organ damage independently of blood pressure effects in salt-loaded heterozygous Ren-2 transgenic rats. Physiol Res. 2004;53(6):581–93.
201. Placier S, Boffa JJ, Dussaule JC, Chatziantoniou C. Reversal of renal lesions following interruption of nitric oxide synthesis inhibition in transgenic mice. Nephrol Dial Transplant. 2006;21(4):881–8.
202. Boffa JJ, Tharaux PL, Dussaule JC, Chatziantoniou C. Regression of renal vascular fibrosis by endothelin receptor antagonism. Hypertension. 2001;37(2 Part 2):490–6.
203. Essalihi R, Dao HH, Gilbert LA, Bouvet C, Semerjian Y, McKee MD, et al. Regression of medial elastocalcinosis in rat aorta: a new vascular function for carbonic anhydrase. Circulation. 2005;112(11):1628–35.
204. Arai H, Hori S, Aramori I, Ohkubo H, Nakanishi S. Cloning and expression of a cDNA encoding an endothelin receptor. Nature. 1990;348(6303):730–2.
205. Eguchi S, Hirata Y, Ihara M, Yano M, Marumo F. A novel eta antagonist (BQ-123) inhibits endothelin-1-induced phosphoinositide breakdown and DNA synthesis in rat vascular smooth muscle cells. FEBS Lett. 1992;302(3):243–6.
206. Davenport AP, Maguire JJ. Pharmacology of renal endothelin receptors. Contrib Nephrol. 2011;172:1–17.
207. Kirkby NS, Hadoke PW, Bagnall AJ, Webb DJ. The endothelin system as a therapeutic target in cardiovascular disease: great expectations or bleak house? Br J Pharmacol. 2007.
208. Barton M. Endothelin antagonism and reversal of proteinuric renal disease in humans. Contrib Nephrol. 2011a;172:210–222.
209. Bagnato A, Loizidou M, Pflug B, Curwen J, Growcott J. Role of the endothelin axis and its antagonists in the treatment of cancer. Br J Pharmacol. 2011. doi:10.1111/j.1476-5381.2011.01217.x.
210. Bagnato A, Rosano L. The endothelin axis in cancer. Int J Biochem Cell Biol. 2008;40(8):144351. doi:S1357-2725(08)00044-7 [pii] 10.1016/j.biocel.2008.01.022.
211. Nelson JB, Love W, Chin JL, Saad F, Schulman CC, Sleep DJ, et al. Phase 3, randomized, controlled trial of atrasentan in patients with nonmetastatic, hormone-refractory prostate cancer. Cancer. 2008;113(9):2478–87. doi:10.1002/cncr.23864.
212. Abraham DJ, Vancheeswaran R, Dashwood MR, Rajkumar VS, Pantelides P, Xu SW, et al. Increased levels of endothelin-1 and differential endothelin type A and B receptor expression in scleroderma-associated fibrotic lung disease. Am J Pathol. 1997;151(3):831–41.
213. Denton CP. Therapeutic targets in systemic sclerosis. Arthritis Res Ther. 2007;9 Suppl 2:S6.

214. Denton CP, Black CM. Pulmonary hypertension in systemic sclerosis. Rheum Dis Clin North Am. 2003;29(2):335–49. vii.
215. Denton CP, Black CM, Abraham DJ. Mechanisms and consequences of fibrosis in systemic sclerosis. Nat Clin Pract Rheumatol. 2006;2(3):134–44. doi:ncprheum0115 [pii] 10.1038/ncprheum0115.
216. Dhaun N, MacIntyre IM, Bellamy CO, Kluth DC. Endothelin receptor antagonism and renin inhibition as treatment options for scleroderma kidney. Am J Kidney Dis. 2009;54(4):726–31. doi:S0272-6386(09)00515-0 [pii] 10.1053/j.ajkd.2009.02.015.
217. Matucci-Cerinic M, Denton CP, Furst DE, Mayes MD, Hsu VM, Carpentier P, et al. Bosentan treatment of digital ulcers related to systemic sclerosis: results from the rapids-2 randomised, double-blind, placebo-controlled trial. Ann Rheum Dis. 2010;70(1):32–8. doi:ard.2010.130658 [pii] 10.1136/ard.2010.130658.
218. Sfikakis PP, Papamichael C, Stamatelopoulos KS, Tousoulis D, Fragiadaki KG, Katsichti P, et al. Improvement of vascular endothelial function using the oral endothelin receptor antagonist bosentan in patients with systemic sclerosis. Arthritis Rheum. 2007;56(6):1985–93.
219. Honing MLH, Bouter PK, Ballard DE, Padley R, Morrison P, Rabelink T. ABT-627, a selective ET_A-receptor anatagonist, reduces proteinuria in patients with diabetes mellitus. In: Regulation of vascular tone in humans by endothelium-derived mediators [thesis]. Utrecht, The Netherlands: Elinkwijk BV; 2000. p. 89–102.
220. Dhaun N, Macintyre IM, Kerr D, Melville V, Johnston NR, Haughie S, et al. Selective endothelin-a receptor antagonism reduces proteinuria, blood pressure, and arterial stiffness in chronic proteinuric kidney disease. Hypertension. 2011;57(4):772–9. doi:HYPERTENSIONAHA.110.167486 [pii] 10.1161/HYPERTENSIONAHA.110.167486.
221. Dhaun N, Macintyre IM, Melville V, Lilitkarntakul P, Johnston NR, Goddard J, et al. Blood pressure-independent reduction in proteinuria and arterial stiffness after acute endothelin-A receptor antagonism in chronic kidney disease. Hypertension. 2009;54(1):113–9. doi:HYPERTENSIONAHA.109.132670 [pii] 10.1161/HYPERTENSIONAHA.109.132670.
222. Dhaun N, Macintyre IM, Melville V, Lilitkarntakul P, Johnston NR, Goddard J, et al. Effects of endothelin receptor antagonism relate to the degree of renin-angiotensin system blockade in chronic proteinuric kidney disease. Hypertension. 2009;54(3):e19–20. doi:HYPERTENSIONAHA. 109.138263 [pii] 10.1161/HYPERTENSIONAHA.109.138263.
223. MacIntyre IM, Dhaun N, Lilitkarntakul P, Melville V, Goddard J, Webb DJ. Greater functional ET_B receptor antagonism with bosentan than sitaxsentan in healthy men. Hypertension. 2010;55(6):1406–11. doi:HYPERTENSIONAHA.109.148569 [pii] 10.1161/HYPERTENSIONAHA. 109.148569.
224. Hall JE. Louis K. Dahl memorial lecture. Renal and cardiovascular mechanisms of hypertension in obesity. Hypertension. 1994;23(3):381–94.
225. Said N, Smith S, Sanchez-Carbayo M, Theodorescu D. Tumor endothelin-1 enhances metastatic colonization of the lung in mouse xenograft models of bladder cancer. J Clin Invest. 2010;121(1):132–47. doi:42912 [pii] 10.1172/JCI42912.
226. Bagnato A, Spinella F, Rosano L. The endothelin axis in cancer: the promise and the challenges of molecularly targeted therapy. Can J Physiol Pharmacol. 2008;86(8):473–84. doi:y08-058 [pii] 10.1139/y08-058.
227. http://www.medicalquery.com/MedicalPR/Speedel-Phase-III-Study-of-SPP301.htm.

Index

D. Abraham et al. (eds.), *Translational Vascular Medicine*,
DOI 10.1007/978-0-85729-920-8, © Springer-Verlag London Limited 2012